SHEEP AND GOAT HANDBOOK
VOLUME 3

International Stockmen's School Handbooks

Sheep and Goat Handbook
Volume 3

edited by Frank H. Baker

The 1983 *International Stockmen's School Handbooks* include more than 200 technical papers presented at this year's Stockmen's School—sponsored by Winrock International—by outstanding animal scientists, agribusiness leaders, and livestock producers expert in animal technology, animal management, and general fields relevant to animal agriculture.

The *Handbooks* represent advanced technology in a problem-oriented form readily accessible to livestock producers, operators of family farms, managers of agribusinesses, scholars, and students of animal agriculture. The *Beef Cattle Science Handbook*, the *Dairy Science Handbook*, the *Sheep and Goat Handbook*, and the *Stud Managers' Handbook* each include papers on such general topics as genetics and selection; general anatomy and physiology; reproduction; behavior and animal welfare; feeds and nutrition; pastures, ranges, and forests; health, diseases, and parasites; buildings, equipment, and environment; animal management; marketing and economics (including product processing, when relevant); farm and ranch business management and economics; computer use in animal enterprises; and production systems. The four *Handbooks* also contain papers specifically related to the type of animal considered.

Frank H. Baker is director of the International Stockmen's School at Winrock International, where he is also program officer of the National Program. An animal production and nutrition specialist, Dr. Baker has served as dean of the School of Agriculture at Oklahoma State University, president of the American Society of Animal Science, president of the Council on Agricultural Science and Technology, and executive secretary of the National Beef Improvement Federation.

A Winrock International Project

Serving People Through Animal Agriculture

This handbook is composed of papers presented at the
International Stockmen's School
January 2–6, 1983, San Antonio, Texas
sponsored by Winrock International

A worldwide need exists to more productively exploit animal agriculture in the efficient utilization of natural and human resources. It is in filling this need and carrying out the public service aspirations of the late Winthrop Rockefeller, Governor of Arkansas, that Winrock International bases its mission to advance agriculture for the benefit of people. Winrock's focus is to help generate income, supply employment, and provide food through the use of animals.

INTERNATIONAL STOCKMEN'S SCHOOL HANDBOOKS

SHEEP AND GOAT HANDBOOK VOLUME 3

edited by Frank H. Baker

Routledge
Taylor & Francis Group
NEW YORK AND LONDON

First published 1983 by Westview Press

Published 2019 by Routledge
605 Third Avenue, New York, NY 10017
2 Park Square, Milton Park, Abingdon, Oxon OX14 4RN

Routledge is an imprint of the Taylor & Francis Group, an informa business

Library of Congress Catalog Card Number 81-644228

ISBN 13: 978-0-367-28722-1(hbk)

CONTENTS

Preface.. xiii

Part 1. A GLOBAL VIEW OF ANIMAL AGRICULTURE

1. A Global Food-Animal Protein System: Pipedream or Possibility? Charles G. Scruggs................ 3

2. Foundation of Civilization: Food Allen D. Tillman.............................. 9

3. World Livestock Feed Relationships: Their Meaning to U.S. Agriculture, Richard O. Wheeler and Kenneth B. Young.................................. 21

4. World Agriculture in Hostile and Benign Climatic Settings, Wayne L. Decker.......................... 29

5. The Impacts of Climatic Variabilities on Livestock Production, Wayne L. Decker....................... 39

6. Opportunities for U.S. Sheep Producers: Management and Marketing for Profit, Rodger L. Wasson......... 45

7. Use of Statistical Analysis to Improve Production in Sheep and Goats, J. Hummel and Miguel A. Galina. 49

Part 2. GENERAL CONCEPTS AFFECTING AGRICULTURE AND THE INDUSTRY

8. Political Challenges for Today's Animal Agriculture George Stone...................................... 57

9. An Agri-Woman's View of the Politics of Agriculture: National and International Perspectives Ruby Ringsdorf...................................... 65

10. Oregon Women for Agriculture Talk About the Structure of Agriculture, Ruby Ringsdorf........... 71

11. The Current Status of the Family Farm in American Agriculture, George Stone................. 79

12. Livestock Production on New England Family Farms, Donald M. Kinsman 87

13. Regulation of Agricultural Chemicals, Growth Promotants, and Feed Additives, O. D. Butler 93

Part 3. GENETICS AND SELECTION OF SHEEP

14. Breeding for Improvement of Reproduction in Sheep Maurice Shelton and Gary Snowder 101

15. The New Zealand National Flock Recording Service--"Sheeplan": Providing Flexibility in Breeding Objectives and Recording Options, R. L. Baker 111

16. Economical Performances of Suffolk Sheep for Mutton, Wool, and Breeding in Mexico Miguel A. Galina, M. Guerrero, J. Gutierrez, and J. J. Salas 129

17. Development of the Polypay Breed of Sheep Clarence V. Hulet, S. K. Ercanbrack, and A. D. Knight 137

18. Individual Records: Their Effect on Productivity Arthur Christensen 153

Part 4. PHYSIOLOGY AND REPRODUCTION OF SHEEP

19. Hormonal Regulation of the Estrous Cycle Roy L. Ax 163

20. Accelerated Lambing: Its Role in the Sheep Industry, Clarence V. Hulet and John N. Stellflug.. 171

21. Management of Sheep to Maximize Lamb Production Maurice Shelton 179

22. How Environment Affects Reproduction in Sheep Richard O. Parker 187

23. Producing Vigorous, Healthy Newborn Lambs Richard O. Parker 195

24. Managing for Increased Reproduction Efficiency in Intensive Crop-Sheep Production Systems Hudson A. Glimp 203

Part 5. ENVIRONMENT, BUILDINGS, AND EQUIPMENT

25. Measuring an Animal's Environment
Stanley E. Curtis.................................. 211

26. Measuring Environmental Stress in Farm Animals
Stanley E. Curtis.................................. 233

27. Livestock Psychology and Handling-Facility Design
Temple Grandin..................................... 245

28. Design of Corrals, Squeeze Chutes, and Dip Vats
Temple Grandin..................................... 261

29. Reducing Transportation Stresses, Temple Grandin... 279

Part 6. PASTURE, FORAGE, AND RANGE

30. Understanding Range Condition for Profitable Ranching
Martin H. Gonzalez................................. 293

31. Range Improvement Practices and Comparative Economics
Martin H. Gonzalez................................. 301

32. What Type of Grazing System for My Ranch...?
Martin H. Gonzalez................................. 313

Part 7. MEAT AND MEAT PROCESSING

33. Current Status of Meat in the Diet and Its
Relationship to Human Health, B. C. Breidenstein... 323

34. Principles of Meat Science Affecting Livestock
Producers: Microstructure, Donald M. Kinsman...... 333

35. Principles of Meat Science Affecting Livestock
Producers: Growth and Development
Donald M. Kinsman.................................. 341

36. New Technology in Meat Processing and Marketing
B. C. Breidenstein................................. 347

Part 8. COMPUTER TECHNOLOGY

37. AGNET: A National Computer System for Cattlemen
Harlan G. Hughes................................... 359

38. Rancher-Owned Microcomputer Systems: What's
Available, Harlan G. Hughes........................ 369

39. Six Steps for a Cattleman to Take in Buying a Computer, Harlan G. Hughes........................ 377

40. Direct Delivery of Market Information Through Rancher-Owned Microcomputers: A Research Report Harlan G. Hughes, Robert Price, and Doug Jose...... 381

Part 9. FARM AND RANCH MANAGEMENT AND PRODUCTION SYSTEMS

41. Opportunities for Increasing Production Efficiency in Intensive Crop-Sheep Production Systems Hudson A. Glimp................................ 395

42. An Economic Analysis of Land Use in an Intensive Crop-Sheep Production System in Kentucky Hudson A. Glimp................................ 401

43. Increasing Market Lamb Production Efficiency Clarence V. Hulet and S. K. Ercanbrack............. 407

44. Triplets Under Range Conditions? It Can Be Done Arthur Christensen............................. 417

45. Status and Use of Livestock-Guarding Dogs in North America, J. S. Green, R. A. Woodruff, and Clarence V. Hulet................................ 423

Part 10. HEALTH, DISEASE, AND PARASITES OF SHEEP AND GOATS

46. Infectious Livestock Diseases: Their Worldwide Toll, Harry C. Mussman............................ 433

47. Impact of Animal Diseases in World Trade Harry C. Mussman................................ 441

48. Biology and Control of Insect Pests of Sheep and Goats, R. O. Drummond............................ 447

49. The Internal Parasites of Sheep and Goats Thomas R. Thedford.............................. 453

50. Identification, Treatment, and Prevention of Internal Parasites of Sheep and Goats Thomas R. Thedford.............................. 461

51. Ram Epididymitis, Thomas R. Thedford............... 473

52. Bluetongue: A Disease of Ruminants Thomas R. Thedford.............................. 479

53. A Health Program for Dairy Goats Samuel B. Guss.................................. 483

Part 11. GENETICS, SELECTION, AND REPRODUCTION OF GOATS

54. DHI and AI as Aids to Dairy Goat Management
Jack D. Stout 501

55. Developments in Goat Reproductive Biology
Maurice Shelton and Janet Lawson 513

56. Reproduction Management for Dairy Goats
Samuel B. Guss 525

57. Factors Affecting Kid Production of Angora Goats
Maurice Shelton 535

Part 12. NUTRITION AND MANAGEMENT OF GOATS

58. Commercial Production of Dairy Goat Milk in the
U.S., James A. Yazman 545

59. Nutritional Management of the Dairy Goat
James A. Yazman 557

60. Dairy Goat Rearing Management with Milk and Milk
Substitutes, Miguel A. Galina and J. Ruiz 571

61. Cost Benefit of Intensive Management of a Dairy Goat
Herd Under Zero Grazing, Miguel A. Galina,
M. Guerrero, M. Gutierrez, and N. Celis 577

List of Names and Addresses
of the Lecturers and Staff 585

List of Other Books of Interest Published
by Westview Press and Winrock International 591

PREFACE

The Sheep and Goat Handbook includes presentations made at the International Stockmen's School, January 2-6, 1983. The faculty members of the School who authored this third volume of the Handbook, along with books on Beef Cattle, Dairy Cattle, and Horses, are scholars, stockmen, and agribusiness leaders with national and international reputations. The papers are a mixture of tried and true technology and practices with new concepts from the latest research results of experiments in all parts of the world. Relevant information and concepts from many related disciplines are included.

The School has been held annually since 1963 under Agriservices Foundation sponsorship; before that it was held for 20 years at Washington State University. Dr. M. E. Ensminger, the School's founder, is now Chairman Emeritus. Transfer of the School to sponsorship by Winrock International with Dr. Frank H. Baker as Director occurred late in 1981. The 1983 School is the first under Winrock International's sponsorship after a one-year hiatus to transfer sponsorship from one organization to the other.

The five basic aims of the School are to:

1. address needs identified by commercial livestock producers and industries of the United States and other countries,
2. serve as an educational bridge between the livestock industry and its technical base in the universities,
3. mobilize and interact with the livestock industry's best minds and most experienced workers,
4. incorporate new livestock industry audiences into the technology transfer process on a continuing basis, and
5. improve the teaching of animal science technology.

Wide dissemination of the technology to livestock producers throughout the world is an important purpose of the Handbooks and the School. Improvement of animal production and management is vital to the ultimate solution of hunger problems of many nations. The subject matter, the style of presentation, and opinions expressed in the papers are those of the authors and do not necessarily reflect the opinions of Winrock International.

ACKNOWLEDGMENTS

Winrock International expresses special appreciation to the individual authors, staff members, and all others who contributed to the preparation of the Sheep and Goat Handbook. Each of the papers (lectures) was prepared by the individual authors. The following editorial, word processing, and secretarial staff of Winrock International assisted the School Director in reading and editing the papers for delivery to the publishers.

Editorial Assistance

Jim Bemis, Production Editor
Essie Raun
Betty Stonaker

Word Processing and Secretarial Assistance

Patty Allison, General Coordinator
Shirley Zimmerman, Coordinator of Word Processing
Darlene Galloway
Tammy Chism
Jamie Whittington
Venetta Vaughn
Kerri Alexander, Computing Specialist Assistant
Ramona Jolly, Assistant to the School Director
Natalie Young, Secretary for the School

SHEEP AND GOAT HANDBOOK
VOLUME 3

Part 1

A GLOBAL VIEW OF ANIMAL AGRICULTURE

1

A GLOBAL FOOD-ANIMAL PROTEIN SYSTEM: PIPEDREAM OR POSSIBILITY?

Charles G. Scruggs

Throughout the long ages of man, meat has been his primary food. Further, a surface study of archaeology seems to indicate that those civilizations that primarily depended on meat have been the most advanced. The telltale signs of man dating back as far as 10,000 years ago indicate that man was primarily a hunter--and thereby a meat consumer. Today, we marvel at the drawings of muscular bulls and heavy bison found on cave walls, drawings indicating the high regard early man had for meat animals.

Approximately 9,000 years ago, man began to domesticate animals. "At Zauri Chemi, dated 9000 B.C., many wild sheep had been killed when immature, as if the inhabitants had either fenced in the grazing grounds of wild sheep, penned herds, or even tamed sheep to the extent that they could control the age at which they were killed."

The earliest evidence of tame cattle is dated approximately 7000 B.C. By 2500 B.C. to 2300 B.C., the Egyptians were milking cows.

Again, a quick observation seems to indicate that civilizations that depended most heavily on cereals were often the least advanced. When animals were not present and man was forced to depend on cereals alone, the civilizations failed to flourish and often disappeared.

As man began to concentrate in cities, he still desired--indeed, seemed to prefer--meat. And meat animals were a lot easier to transport to the cities than were cereals. Meat animals had four legs. They could be driven to people. And so they have been for thousands of years.

As man moved into new territories, he often took his meat animals with him. Cattle and sheep have traveled with the armies of invaders and in the trains of the missionaries. These animals helped civilize the world. Some of the cargo on the earliest ships was live animals--incredible, when you think about the tiny vessels and harsh long voyages made as long as 2,000 years ago. And even 200 years ago.

The earliest hunters and travelers discovered that by drying or salting they could preserve and transport meat more easily. In fact, it comes as something of a shock to us in this day and time to find that more pounds of salted or pickled meat were exported from the U.S. to Europe and the West Indies over a hundred years ago than are exported fresh today.

Then we made great progress--refrigeration was developed. This was an amazing advance--fresh meat was available year round! But refrigeration also became a shackle and a chain. In recent years, especially in the U.S., we have become chained to our refrigerators. We can go only so far as we can take our refrigerators!

On the other hand, refrigeration has also brought to the U.S. and Europe unparalleled adventures in good taste, variety, and quality of meat. Refrigeration has allowed us in the U.S. to enter the roast, steak, and hamburger era in a way never thought possible. But it has also allowed U.S. beef producers to go wandering off into the insignificant. We began looking inward and forgot the rest of the world, thinking, "Those dumb people overseas wouldn't know good beef if they saw it."

Instead of seeking wider markets, we have spent our time over the last 70 to 80 years arguing over such insignificant things as horns or lack of horns; or whether a few hairs of a different color are wrongly placed; or trying to make bulls and cows look like steers. Now we are rushing about and exchanging thousands of dollars on the difference of a centimeter or two in testicle size, trying to put legs on cattle that might look better on thoroughbred horses. And, oh, the hours we spend washing, puffing, combing tail hair so as to make a nice round-looking ball at the end of a show animal's tail! Insignificant fadism in the extreme!

We in the U.S.--the greatest food-animal producers in the world--have perhaps been so busy arguing over minute details of grade standards, which no one but a few experts understand, and over other inward-looking insignificant details, that we have missed the greatest world market for meat that has ever existed since the earliest Homo sapiens killed their first Bos primigenius (auroch or wild ox). In short, "we have been so busy fighting gnats that we let the herd get away."

There is a big world all around us--some four billion people--all hungry for meat! There are in this populous world of ours today approximately one billion people who exist at a malnourished level. Most of the people in the world today do not get the health benefit that we have proved comes from a diet in which food animal protein is the major ingredient. If, indeed, it is true that a good diet containing major elements of food animal protein makes for more intelligent, more ambitious, harder working people, doesn't it follow that U.S. livestock producers should seek to contribute that food animal protein to mankind--at a profit? Would not this contribution be greater than wars or multibillion dollar giveaway programs?

Think about it! What if the approximately 800 million Chinese could be induced to consume only one (1) pound of food animal protein per year above their present one pound? What would demand for 800 million pounds of meat do for beef, pork, sheep, and goat--food-animal--producers in the U.S.? Add to the Chinese the Russians, the Indians, and the Southeast Asians, and you have a need for food-animal protein that boggles the mind!

Thus, U.S. livestock producers must move from their present "cowboy" mentality to one in which they seek to become world marketeers of food-animal protein. We must move from being little more than herders to thinking of ourselves as food-animal producers whose goal is to sell meat protein to a hungry world--at a profit!

And this shift to world food-animal-protein suppliers can be made while we continue to supply the American consumer with desired beef, milk, pork, and mutton.

How?

Think big.

Think systems.

Let's be more specific. Here are what I believe to be the essential elements of developing a Global Food Animal Protein System.

Phase I. Search out a small or medium-sized country that has unfulfilled nutrition and food problems, heavy population density, and a reasonable amount of foreign exchange.

Study the food habits of the population in detail. Find out exactly how the people prefer to eat meat. Learn all there is to know about their overall food habits, preferences. Study their tariff laws, their customs, their religion--everything that could make an impact on their meat consumption. Learn to think as the natives think and react.

All the while your goal is a modest one: Increase per capita meat consumption, on the average, by one (1) pound.

The key to this phase is to drop the American habit of turning up our noses at the way other people like their meat. Just because we like roasts, there's no reason that all the rest of the world must eat roast the way we do. If the natives want barbecued tail bone, let's not call them stupid and say there is no market, and retreat to the nearest McDonald's. Let's sell them tail bones! The goal is a good old American custom: Study a market, decide what can be sold at a profit, then produce it well enough to earn a profit.

Phase II. Devise a production and processing system to deliver the product without waste of time or effort from conception to consumption. We in U.S. agriculture now do just exactly the opposite: We produce something and then try to peddle it in the form we want--or "to hell with 'em."

Phase III. Take the basic lessons learned in one country with modest goals, and apply them to other countries one by one--at a profit. Soon, we will have made a world contribution.

Let's explore some other dimensions of the idea of supplying food-animal protein to a malnourished world.

Let's look at Russia. Russia desperately needs meat. At present, they get their meager supplies from their own limited meat production system. They import some meat from Australia and some from their satellite European countries, but not nearly as much as the Russian population needs and wants. Instead of importing meat, the Russians are trying to do it the long, hard way: Import grains from all over the world, then process and feed them through their herds and flocks.

Has the U.S. ever seriously studied with the Russians (or the Hungarians, or the Saudi Arabians) a policy of importing ready-to-consume meat products (mutton and chicken) instead of raw grains and the resulting timelag to consumers? Wouldn't it be possible, technologically, for us to ship lean grass beef to Russia at a profit? Russians prefer beef that tastes much different from our own grain-fed beef. Mostly they get cull cow or bull beef--if they get any beef at all. Let's sell them what they are used to eating.

We can load forage-produced boneless beef on planes at Atlanta or Dallas, go up to 35,000 feet and quick-freeze it at no cost, and land it in Moscow 16 hours later. Too expensive? We don't know. So far as I know, we have never even tried it. Flying too much moisture? We can dehydrate beef fibers, reconstitute them in Istanbul or Pakistan. Remember our hunting, frontier-busting forebears? How did they transport beef supplies when traveling alone? Jerky beef. It needed no refrigeration. Wouldn't a malnourished native of Gambia be glad to have some jerky beef to mix with his root foods and maize?

American technological expertise is envied the world over. But we haven't used our skill to try to supply food-animal protein to a food-deficient world. Can we not preserve meat and milk through irradiation and/or treatment so that it can be held on a pantry shelf? Good U.S. agricultural policy, it seems to me, should be one that uses U.S. technical advantages to the benefit of the U.S.--and then for other citizens of the world. If so, we must begin to shift from a policy of shipping raw, unprocessed grains to a policy of exporting value-added, more nearly ready-to-eat food-animal products. The U.S. has the livestock. We have more feed and feedgrains than anyone. We have the nutrition knowledge. We have financing. Shouldn't we ship food-animal protein products and short-circuit the long food production chain that most countries now are trying to establish by buying our grain and soybeans?

What about the strategic military considerations of such a policy? Ship food protein ready to eat instead of supplying grain? That should be the U.S. goal. We must do this instead of supplying grains so the countries can build their own livestock infrastructure for future self-sufficiency.

The U.S. food-animal production industry has seen some quantum jumps in the last few years in development of all-tender, flavorful meat products. Meat protein cubes, dehydrated meat fibers, ready-cooked meals are just a few of the new forms and products pouring out of labs. More are about to emerge. However, we cannot hope to supply all the world with edible animal protein. Perhaps we should adopt this policy: Sell edible animal food protein to those countries whose needs are greatest and most immediate. To those countries with developed livestock industries, sell genetic and germ plasm materials to use in upgrading their production. This whole field of genetic engineering now in its infancy in the U.S. may shatter much of our previous livestock thinking. Livestock producer leaders must be equally bold in their thinking!

I have no doubt that if U.S. food animal producers--at one time called cowmen, sheepmen, dairymen--would set out to develop a Global Food Animal Protein System, they could do so rather easily. But to do so they must throw off the shackles of the past, think aggressively American, think profit, think systems, and above all think BIG.

* * *

> "Populations of the lower- and middle-income countries still are increasing rapidly, and in many countries there is a growing number of affluent people whose diets are being upgraded to include more red meat, dairy products, and eggs. As more countries are unable to meet their national requirements for staple foods, their governments are looking for outside sources of supply, often on an urgent basis. It is dangerous politically for national leaders to let food shortages occur, driving prices up; they run the risk of unrest, violence, and even overthrow of governments."
>
> <u>Beyond the Bottom Line</u>
> The Rockefeller Foundation

REFERENCES

Fagan, B. M. Men of the Earth: An Introduction to World Prehistory. Little, Brown and Co.

2
FOUNDATION OF CIVILIZATION: FOOD

Allen D. Tillman

"And he gave it for his opinion that whoever could make two ears of corn or two blades of grass grow on a spot of ground where only one grew before would deserve better of mankind and do more service to his country than the whole race of politicians put together."
--Johnathan Swift
The Voyage to Brobdingnag
in Gulliver's Travels

I chose this quotation because it is apparent to me that man now has the knowledge and power to make two ears of corn or two blades of grass grow on a spot of ground that formerly would grow one or less.

Civilization is defined as "an advanced state of human society in which there is a high level of culture, science, industry, and government." The high level of civilization that we enjoy today has resulted from many technological developments in agriculture that increased the amount of food produced and the efficiency of human labor in producing it. Each innovation freed more people for the development of human society and of the culture, science, industry, and government found in it.

In discussing some of the developments, this paper is divided into major sections as follows: (1) a brief history of agricultural development worldwide; (2) the close relationship of agricultural and industrial developments in modern societies; (3) some characteristics of a successful agriculture at national levels, and (4) reasons for optimism about the world food problem.

HISTORY OF AGRICULTURAL DEVELOPMENTS IN THE WORLD

"History celebrates the battlefield whereon we meet our death, but scorns to speak of the plowed fields whereby we thrive; it knows the names of the king's bastards, but cannot tell us the origin of wheat. This is the way of human folly."

--Jean Henri Fabre

For convenience, I have divided this section into three parts, as follows: the gathering and hunting stage (2,000,000 - 7000 B.C.), the low-technology stage (7000 B.C. - 1750 A.D.); and the scientific stage (1750 A.D. - the present).

The Gathering/Hunting Stage

"Cultural man has been on earth for some 2 million years and for 99% of this period he has lived as a hunter/gatherer. Only in the last 10,000 years has man begun to domesticate plants and animals, to use metals, and to harness energy sources other than the human body. Of the estimated 90 billion people who have lived out a life span on earth, over 90% have lived as hunters/gatherers, about 6% by agriculture, and 4% have lived in industrial societies. To date, the hunting/gathering way of life has been the most successful and persistent adaptation man has ever achieved."

--Lee and Devore (1968)

The first ancestor of man, *Australopithecus*, appeared on earth about two million years ago. His main invention was the knife, which was made by putting an edge on a pebble, an invention that permitted him to kill animals, skin, and cut meat, thereby changing him from an herbivora to an omnivora. This change was dramatic, because the addition of dietary meat reduced the bulkiness of his diet by about two--thirds, permitting him to leave the trees and to become more mobile to better utilize the rapidly developing savannas. Also, meat required less time to gather, thus he had more time for social activities - improvement in communication skills and in his tools. And so man for the first time released the brake that environment imposes on his fellow creatures. It is significant that this basic tool was not changed very much for the next million years, attesting to the strength of the invention.

Homo erectus came onto the scene about one million years ago. His ability to walk upright freed his arms, which improved his ability to hunt, and his greater ability to adapt to many ecosystems permitted him to spread out from his place of origin, Africa. In fact, the classical discovery of *Homo erectus* was the Peking man, who lived in China about 400,000 years ago. He was the discoverer of fire, which was used for warmth and cooking. The Neanderthal man, who was discovered in Europe, appears to have led directly to us, *Homo sapiens*.

The test of the ability of *Homo erectus* to adapt came about 500,000 years ago when the *Pleistocene* Ice Age covered much of the earth. Clans of 40 or more moved to caves for protection and work, a move that required a new organization - the young and stronger men (usually 10 per clan - Willham, 1980) were the hunters, while the remainder were assigned duties in keeping with their abilities. Some of the dwellers even had time to paint on the cave walls. Bronowski (1973) felt that these are saying to all - "This is my mark, this is man." Man is now saying that he has the ability to shape the world and is not a mere creature to be shaped by the environment.

The great glaciers began to retreat about 30,000 years ago. Left in their wakes were great savannas that were soon filled with grasses of all kinds and with cloven-hoofed ruminants to consume these. In response, the clans came out of the caves and spread out over the plains following the animals, going north in the summer and returning south in the winter. This was the beginning of a transhumant way of life, which later became dominant in many areas of Eurasia. In fact, there are cases of this way of life even today--East Africa, North Africa, Finland.

The dog was domesticated in about 20,000 B.C., and this greatly increased man's ability to hunt. Also , there developed a symbiotic relationship between man and animals: animals furnished meat, skins, and other products for man, while man furnished some protection and salt to the animals. Urine of meat-eating man contains salt, a valuable commodity to ruminants in many salt-deficient areas. Man developed oars in about 20,000 B.C. and the bow and arrow in about 15,000 B.C.; both inventions increased his efficiency as a hunter. Man domesticated the gregarious sheep and goats during the latter part of this stage, and benefited greatly by the increased food supply from these animals.

At the end of this stage, about 10,000 to 7,000 B.C., and for a long time after man had already established village agriculture, animals continued to be an important source of food for man. The live animal represents, until it is sacrificed, a reserve food supply. This fact is often forgotten or overlooked by planning economists, who plan for the aid programs that are given to the developing countries. Early man recognized the importance of animals: so much so that the root word used for money in many languages reflects the importance of animals to man (Leeds and Vayda, 1965).

Low Technology Agriculture (7000 B.C. - 1750 A.D.)

> "The greatest single step in the ascent of man is the change from nomad to village agriculture."
>
> --Bronowski (1973)

With the receding of the Pleistocene ice, there came great environmental changes in the Old World. The hot and dry winds off the Eurasian Steppes eliminated all but the hardier grasses in much of North Africa and some of Asia, thereby turning some of the lands into semideserts or deserts. As a result, wild animals migrated to the river valleys of the Euphrates, Tigris, Indus, Nile, and Yellow rivers. Agriculture began when the nomads decided to stay put to exploit plants. Whether the nomads planned to develop agriculture or were the benefactors of two genetic accidents is not clear. What is clear is the fact that the early ones came to hunt and to gather wild wheat. By a genetic accident (Harlan, 1975), wild wheat containing 14 chromosomes, crossed with wild oat grass, also containing 14 chromosomes, to produce a fertile hybrid, called Emmer. It contains 28 chromosomes, thus the grain was much larger than wild wheat. Therefore, man began to cultivate it. Emmer's grain is so tightly bound to the husk and chaff that it is easily dispersed by the wind. As a result, it spread over wide areas. It appears that by another genetic acident, bread wheat came to the settled agriculturalists: Emmer crossed with another wild oat grass to produce still another fertile hybrid, bread wheat, containing 42 chromosomes, which has a large grain. In contrast to Emmer, when bread wheat is broken, the chaff flies off, leaving the grain in place, thus it is not spread by the wind. With the advent of bread wheat, which man has to plant and cultivate, man and wheat developed a symbiotic relationship that remains up to this day (Heiser, 1978).

Farming and husbandry in a settled agriculture creates an atmosphere from which technology and science take off (Bronowski, 1973). The first tools used by village agriculturalists were the digging stick, which was invented about 7000 B.C. This later evolved to a crude plow, the footplow that used human labor, which appeared in about 6000 B.C. The cow was domesticated in about 6000 B.C., and when man yoked the ox to the plow he for the first time began to utilize a power source greater than the human muscle. This was undoubtedly the most powerful invention of this early period, making it possible for man to wrest a great surplus of agricultural products from nature. The surplus food released more men to create, invent, innovate, and to build great civilizations--something for which the nomads had never had time. Civilizations, with their specific cultures, developed on the flood plains on the Nile, the Tigris-Euphrates, the Indus, and the Yellow Rivers. Many of their activities, such as irrigation, required cooperation by many men; therefore there developed administrative systems that led to the building of empires. Law and government developed. The great cities of that period thrived on a cereal-based agriculture. Trade between cities developed in order for them to acquire necessities - salt, spice, metals, etc. (Thomas, 1979).

The animals found in the settled villages, up to about 3000 B.C., were goats, sheep, the oxen and the onager, a kind of wild ass. As long as the animals were the servants of agriculture, all went well. But some time after 4000

B.C., the horse was domesticated and the nomads learned to ride in about 2000 B.C. Thus, the nomad was transformed from a poor wanderer to a threat to the settled villages. Warfare of that period was intensified by the discovery of how to ride the horse, and warfare became a nomad activity. The nomads battered on doors of the settled villages from about 2000 B.C. until the early part of the 14th century A.D. Sometimes the nomads were successful and took over villages, but in all instances, the nomads were absorbed into the villages. Historians have made all of us aware of the great wars waged by the nomads, recording the names of the famous nomads--the Huns, the Phrygians, and the Mongols. The Mongols were defeated in about 1300 A.D., thereby ending the threat of their making nomad life supreme throughout sections of the Old World.

When the horse collar was discovered, the horse became an important draft animal, especially in northern Europe were great teams of horses turned the heavy sod. Without these teams of horses, which permitted the Vikings to produce great surpluses of grains, they could not have been such a military power and threat to much of northern Europe.

The low-technology stage continues right on through the European Renaissance (Thomas, 1979), during which time there were many contributions to agricultural innovations, each with its consequent improvements in food production and the efficiency by which man produced it. As there were many such improvements over time, for sake of time and space, let us summarize the advances:

- Man developed methods for the systematic exploitation of plants.
- Man developed methods for the cultivation of plants for the production of grain--wheat, barley, millet, and rice (Heiser, 1978).
- Man domesticated animals--dog, cow, sheep, goat, and the horse.
- Man developed systems of irrigation.
- Man developed some degrees of mechanization--the digging stick, the plow, the ox-drawn plow, the wheel, and others.

The developments in limited technology were not continuous but came in ebbs and flows throughout the period up to about 1750. There appeared to be a _de facto_ technological ceiling upon agricultural production throughout the entire stage. At the core was the simple Malthusian element--population expansion ultimately pressed against the land, thereby producing malnutrition, famine, disease and, finally, a decrease in population. In China, Hung Liang Chi (Rostow, 1978) the predecessor of Malthus, wrote "during a long reign of peace--the government could not prevent the people from multiplying themselves." Rostow said it well--"During this period of limited technology, if war did not get you, peace did." And so ended the second era.

<u>The Scientific State (1750 - present)</u>

The scientific state began in about 1750 and continues until the present. In this period, the western nations for the first time broke the ceilings on agricultural and industrial technology so that invention and innovation came at a regular flow. The key to these, I feel, was the advent of the "scientific revolution which brought with it experimental science." Experimental science, for the first time, permitted and motivated the formulation of scientific laws to describe general and natural scientific phenomena. This led scientists to design experiments that would lead to the manipulation of nature to man's advantage. This exciting time saw the advent of scientific agricultural societies in which agriculturalists met together for the discussion of scientific discoveries and how these could be put to use by farmers. Innovations came at a faster pace, and as in the past, each invention or innovation increased the level of food production and its efficiency of production. The rates increased rapidly and we now have the development of high-technology agriculture. Some of the basic advancements that are characteristic of this stage are as follows:

- Classification of soils, along with estimates of their fertilities.
- Improved plants by gene manipulations (genetic engineering).
- Improved animals by gene manipulations.
- Scientific utilization of fertilizers.
- Proper use of irrigation.
- Proper use of fermentation and other means of food preservation (Tannahill, 1973).
- Use of insecticides, fungicides, herbicides, vaccines, etc.
- Use of modern techniques in farm mangement.

Many names stand out during this third era; however, it is significant to mention that King George III (better known to Americans for other reasons) gave much support to the newly developing agricultural research in England. Some feel that the innovations resulting from this agricultural research greatly increased agricultural production in Great Britain. In fact, some suggest that it was the English agricultural revolution that permitted that nation to defeat Napoleon at a time when agricultural imports were essentially cut off by France.

Agricultural research is so important that every modern nation has now developed a national agricultural production plan in which agricultural research is a powerful component. Those countries that are lagging in agricultural production are the ones that have been slow in developing good agricultural research, teaching, and extension programs.

How about the United States? Our country had its beginning in 1776, or sixteen years after the third stage began. When our Declaration of Independence was signed, the Revolutionary War was fought, and our people set out to build a nation, fully 90% of our people were directly engaged in agricultural production. If time and space permitted, it would be useful to point out the significant inventions, such as the first cotton gin, the first wheat thresher, the first steam or gasoline-propelled tractor and many others, all American inventions or innovations, and to note the effect of each upon food production and efficiency. However, I will end this section by saying that during the 200 years since Independence, our farm population has decreased from 90% of the total population to only 5%. However, these fewer people are producing food of improved variety and quality, providing nourishment for a vigorous population. Our national agricultural program has been one of the modern success stories.

THE CLOSE RELATIONSHIP OF AGRICULTURAL AND INDUSTRIAL DEVELOPMENT IN MODERN SOCIETIES

In reviewing agricultural developments worldwide, Rostow (1978) noted that the modernization of agriculture and industry have gone hand-in-hand in successful national development programs in the past. In fact, one finds that the modernization of agriculture must precede industrialization in most countries. There are many reasons for the close interrelationship; successful agriculture (Foster, 1978) provides:

- An ever-expanding supply of high quality foods to nourish its people, increasing their vigor.
- An adequate supply of high quality food, available at a reasonable price, to combat inflation which, if left uncontrolled, hampers national development.
- Capital for the expansion of the nonagricultural sector of the economy. This is very critical in the early stages of industrial development because at least 90% of the population in every country studied were farmers at this critical period.
- More food for the nonagricultural population. Therefore, all through the modernization process, more and more people are freed for work in the nonfarm sector.
- Educated and motivated people for work in the nonagricultural sector in the developed countries.
- Land for the nonfarm sector--for highways, railroad, airports, shopping areas, etc.
- A market for the nonfarm sector--tools, machines, medicines, insecticides, clothing, gasoline, etc. Again, this market is most critical in the early stages of industrialization.

SOME CHARACTERISTICS OF A MODERN AND PRODUCTIVE AGRICULTURE AT THE NATIONAL LEVEL

In general, a successful national agriculture results from a good agricultural plan or programs that provide farmers with relevant production information and assures an adequate infrastructure that is needed for both production and marketing of agricultural products. In addition, farmers must receive a fair return from their investment of land, labor, and capital. Otherwise, production will be sporadic rather than continuous. Some specific requirements are:

- Adequate government financial support for research, teaching, and extension (public service) for agriculture.
- Infrastructure: The national government also has to provide certain components of the infrastructure needed by farmers--such as roads, harbors, railroads (in some cases). (Many of the infrastructures in the U.S. are now furnished by industry.)

Some inputs needed by modern farmers are as follows:

- Farm machinery and spare parts
- Farm tools and spare parts
- Power source--gasoline, electricity, diesel fuel
- Fertilizers
- Insecticides, fungicides, herbicides, vaccines, etc.
- Irrigation tools--pumps, pipe, valves, etc.
- Credit
- Others

Those who are familiar with the ready availability of inputs on the American scene might well question why private industry hasn't made these available in many developing countries and in developing agricultural industries. In many cases, private industry will not take the risk of production and distribution of many necessary inputs unless the volume and price justifies the risk. In such situations, the government has to subsidize these inputs until the industry is well along toward development. The writer has spent over 10 years in four developing countries and found that the unavailability of certain inputs, such as vaccines, insecticides, and fertilizers, has served as a severe constraint on production. In the same vein, lack of marketing services results in severe losses of the harvested produce.

Marketing services needed to assist production increases are:

- Farm produce collection and storage
- Processing and subsequent storage of the processed produce
- Wholesale distribution of produce
- Retail distribution of produce
- Marketing--farm markets, small stores, supermarkets, and others.

Even with all of the above components in place, farmers will not produce continuously unless the price for agricultural products pays for the costs of inputs and allows them a fair return on their investments.

REASONS FOR OPTIMISM ABOUT THE WORLD FOOD PROBLEM

The world food problem has two components--the demand side (population) and the supply side (production). Many reports today emphasize that the demand side is growing faster than the supply side. Let us analyze the population problem first.

Population

The demographic facts at first glance are frightening! If we plot population against time, we see that the population growth from about 8000 B.C. until about 1650 A.D. was almost a straight line function, a very slow growth rate. In 1650 A.D., there were about 0.5 billion people, and the rate of increase was only 0.3% per year. At that rate, the population would require 250 years to double. Instead we have had an exploding population since about 1700 A.D., and now some authorities estimate that we will have 6 to 7 billion people in the year 2000.

It is my belief that the population estimates have all been wrong, from Malthus down to the United Nations' recent estimates. For example, the U.N. estimated in 1976 that the world population by the year 2000 would be 7 billion. However, in 1979 that estimate was reduced to 6 billion. Therefore, in 3 years, we "lost" 1 billion people. There are many other examples of inaccuracies. Why is this so? Population projections require assumptions about the choices people have and make in regard to the size of their family. Therefore, it is easy to see why all past assumptions have been wrong.

Simon (1981) maintains that there is a built-in, self-reinforcing logic that forces the rate of population growth to respond to resource conditions. It does so by reducing population growth and size when food resources are limited, and expands it when resources are plentiful. Others have studied population changes in Europe for 1400 A.D. until 1800 A.D., and found that the population did not grow at a constant rate, and that it did not always grow. Instead, there were advances and reverses, provoked by many forces--famine and disease, however, were not the major forces. We know that increased incomes, associated with economic developments, reduce birthrate and population growth. For example, the populations in Singapore, Hongkong, Japan, and other places have tended to stabilize within the last 20 years.

In summary, these facts lead me to suggest that population size tends to adjust to the production conditions.

Following a technological advance in agriculture, there is an increase in production followed by an increase in population size. However, the rate of population increase levels off as the new technology is "used up".

Food Production

If I am optimistic about the demand side of population problem, I have even more reasons to be optimistic about the supply side--food production. As pointed out by Schultz (1964), the world food problem exists because of low productivity in the poor countries. It is estimated that the amount of crops produced per ha of land in the poor countries was only about one-fourth of that in the industrial countries, and the production of animal products/animal unit was even lower. Therefore, the potential for greatly increasing production lies in the poor countries where it is needed.

Starting with the success of the Green Revolution in India during the 1965-70 period, there has been an awareness worldwide that a nation can increase its food supply if it has the will to do so.

Since the World Food Congress in Rome in the early 1970s, the FAO, the World Bank, the bilateral aid programs and other organizations have given priority in their aid programs to agricultural development in the poor countries. Up to 1975, the proportion of foreign aid dedicated to increasing food production was less than 10% of the total program of aid and the level has now more than doubled.

Some of the reasons for my optimism about future world food production are:

- The nature of the food/population problem as a whole is now becoming understood.
- The complexity of technology transfer from the industrial countries to the developing countries is also becoming understood; the international research and development centers have made excellent progress in these endeavors.
- The potential for increasing yields per land unit in the poor countries, most of which are located in the tropics, is enormous. We call this "payoff on research." I have made some estimates of the annual rates of return on monetary investments in agricultural research, both by commodities and by countries. If we consider rice research in the tropics, the figures run from 46% to 71%; however, if only Asia is considered the figures vary from 74% to 102%. the International Rice Research Institute, in the Philippines, has done an excellent job.
- Fertilizers and the knowledge of how to use them are now available to farmers in the poor countries. Many national governments are now subsidizing the price of this input.

- New and better adapted plant varieties are now available to most poor countries; many of these plants are resistant to certain diseases.
- Many national governments in the poor countries are now using aid and loans from the FAO, World Bank, bilateral sources and philanthropic organizations to support agricultural production by improving some or all of the following: (a) The infrastructure (such as roads, communication, and harbors), (b) The teaching, research, and extension structure facilities and activities. The research is now directed at solving their own production problems. (c) The credit structure. (d) Water resource utilization (e) Active intervention programs to increase the production of certain critical commodities. For example, in 1976-75, Indonesia initiated an active intervention program in rice production using subsidization of inputs, price stability, increased research, increased extension inputs, etc.

In 1969-71, the average production of padi was 2346 kg/ha in Indonesia. By 1978, this figure increased to 2921 kg/ha, an increase of 27.5% in 8 years, or about an increase of 3.5% per year. Preliminary figures for 1980 show a dramatic increase and give hope that Indonesia is about to gain self-sufficiency in rice production.

An interesting aspect of Indonesia's successful efforts on a single commodity, rice, is the fact that production improvements in other commodities have not improved, and some have decreased. To me, this is a god sign--meaning that with increased effort, Indonesia could increase production in any agricultural commodity chosen. Their leaders now know this and are putting forth successful efforts to increase production of other selected plants and animals.

REFERENCES

Bronowski, J. 1973. The Ascent of Man. Little, Brown and Co., Boston, Mass.

Burke, J. 1978. Connections. Little, Brown and Co., Boston, Mass.

Foster, P. 1978. Food as foundations of civilization. In Food and Social Policy. G.H. Koerseiman and K.E. Dole (Ed.) Iowa State University Press, Ames, Iowa.

Harlan, J.R. 1975. Crops and Man. Crop Science Society of America, Madison, Wisconsin.

Heiser, C.B. Jr. 1978. Seed to Civilization. W.H. Freeman and Co., San Francisco, CA.

Lee, R.B., and I. DeVore. 1968. Man, the Hunter, Aldine, Chicago, Ill.

Leeds, A., and A.P. Vayda. 1965. Man, Culture and Animals. AAAS, Washington, D.C.

Rostow, W.W. 1978. Food as foundation of civilization. In: Food and Social Policy. G.H. Koerseiman and K.E. Dole (Ed.). Iowa State University Press, Ames, Iowa.

Simon, J.L. 1981. World population growth. The Atlantic Monthly 248 (2):70-76.

Tannahill, Rey. 1973. Food in History. Harper and Row, New York, N.Y.

Thomas, Hugh. 1979. A History of the World. Harper and Row, New York, N.Y.

Willham, R.L. 1980. Historic development of use of animal products in human nutrition. Mimeo. Rpt., Iowa State University, Ames, Iowa.

3

WORLD LIVESTOCK FEED RELATIONSHIPS: THEIR MEANING TO U.S. AGRICULTURE

Richard O. Wheeler,
Kenneth B. Young

INTRODUCTION

Livestock producers both in the U.S. and worldwide have an important stake in the operation of the world grain economy. The continued availability of relatively low-cost grain in the world economy would tend to foster further livestock development in grain-deficit countries and provide competitive advantages in all countries for livestock and livestock production practices more dependent on intensive grain feeding. On the other hand, if world grain supplies become more restricted, grain-deficit countries would be more likely to import livestock products rather than grain, and livestock production practices less dependent on grain feeding would gain a competitive advantage.

Prior to the early 1970s, the world trend was toward increased grain supply and continued buildup of large stocks in the industrialized exporting countries, primarily the U.S. and Canada. These large stock levels helped to maintain relatively low prices and assured a stable supply for importing countries. For example, U.S. wheat export prices deviated very little from $170 per ton from the mid-1950s to the early 1970s. This long period of stable grain supply, sold at very attractive prices, encouraged the widespread use of grain feeding to increase livestock production. In addition, some grain exporters offered other inducements, including liberal credit arrangements and Public Law 480 assistance for countries unable to compete in the international grain market.

During this era, many developing countries adopted intensive grain-feeding practices for poultry and swine production, and cattle feedlots became a prominent feature of the U.S. agricultural system. On a worldwide basis, use of all cereals for animal feed increased from 37% in 1961-65 to 41% in 1975-77 (Harrison, 1981). Eastern Europe and the Soviet Union registered the largest increase in grain feeding of all countries--a change from 48% in 1961-65 to 69% in 1975-77. Use in Latin America increased from 32% to 41%.

At the present time, the world grain market has recovered from the 1972/73 shortfall and world stocks have been restored to the level of the 1960s. Nevertheless, the shortfall did mark a major shift in the supply-demand balance of the world grain market, a situation that had not occurred previously.

The structure of the world grain market has changed dramatically since the 1950s. World trade increased over 200% from 1960 to 1980. Currently, over 100 countries are dependent on grain imports from a few exporters. The U.S. dominates the world grain trade, accounting for about 60% of global coarse-grain exports and 44% of world wheat exports. About 40% of the total U.S. grain production is now exported and the annual rate of increase in exports reached 7% per year during the 1970s. Projections of future U.S. crop exports available from the Economic Research Service of USDA (1981) indicate that the growth in foreign demand will continue through the 1980s, although not as rapidly as in the 1970s (table 1). Annual average export demand for corn and rice is projected to increase about 4½% compared with 2% for wheat and soybeans.

TABLE 1. PROJECTED INDICES FOR U.S. CROP EXPORTS, 1981-1989

Commodity	1981	1982	1983	1984	1985	1986	1987	1988	1989
					(1981=100)				
Corn	100	106	111	115	123	127	131	135	139
Wheat	100	96	99	101	103	105	107	110	115
Rice	100	109	113	117	120	124	127	131	135
Cotton	100	107	103	103	103	104	106	106	107
Soybean	100	100	101	104	107	111	113	116	119
Peanuts	100	123	140	147	150	153	157	160	163

Source: These projections have been calculated from Problems and Prospects for U.S. Agriculture, ERS-USDA (1981). They are not official USDA projections.

There is some question now about the U.S. ability to keep up the recent pace of expanding exports. Most of the available cultivated land is currently in production and we are losing about a million acres of cropland per year to nonagricultural uses. The rate of soil erosion has increased substantially with more intensive cultivation and use of marginal cropland formerly not used for crop production. The same problem is occurring in other countries and average crop yields are leveling off over much of the world.

PROJECTIONS ON WORLD GRAIN SUPPLY

A Winrock International study was completed in 1981 on world use of grain and other feedstuffs (Winrock International, 1981). The study was designed to evaluate the interaction between the world livestock system and the feed- and food-grain system.

Estimates of current world use indicate that poultry consume 27% of all grain fed; swine--32%; draft animals--4%; sheep and goats--2%; and cattle and buffalo, including dairy animals--35% (table 2). Feed use in table 2 is expressed in terms of megacalories of metabolizable energy

TABLE 2. ESTIMATED ANNUAL WORLD FEED USE FOR DIFFERENT TYPES OF LIVESTOCK, 1977

Livestock category	Livestock Output		Feed use				
	Meat	Other	Grain	Protein meal	By-products	Forage & other	Total feed
	(million metric tons)		(billion mcal ME)				
Poultry	22.8	23.3[1]	387.9	91.1	73.4	51.7	604.1
Sheep & goats	7.3	--	23.6	5.3	35.8	993.9	1,058.6
Cattle & buffalo	46.8	415.0[2]	507.3	42.8	204.2	4,101.0	4,855.5
Swine	41.0	--	460.9	56.0	213.1	157.2	887.2
Draft animals	13.4	--	57.9	5.9	23.5	1,214.9	1,302.2
All livestock	131.3	--	1,437.6	201.1	550.0	6,518.7	8,707.4

[1] Eggs.
[2] Milk.
Source: Winrock International (1981).

rather than metric tons due to variation in the quality and variety of feed used in different countries. For example, grain-feed use in the Soviet Union is reported on a "bunker weight basis" generally containing excess moisture and extraneous matter. The percentage of grain use in poultry rations is estimated to be similar for developed, centrally planned, and developing countries since the technology of modern poultry production has been readily adopted all over the world. Developing countries feed less grain and more forage and by-products to swine. The ruminants in developing countries subsist almost entirely on forages. However, nearly half of the world grain feeding occurs in developing and centrally planned countries dependent on grain imports.

Grain feeding was projected to continue increasing according to recent trends evaluated in the Winrock study. There will be occasional setbacks for countries with severe foreign exchange problems and domestic recession. Most of the centrally planned countries have set target levels of increased livestock production requiring additional grain feeding. The Winrock study projected that total world feed use of wheat and coarse grains would surpass direct human and industrial consumption by 1985. Recent trends also indicate that total world grain use is increasing at a faster rate than world production. World grain demand has been increasing steadily due to continued growth in world population and rising per capita consumption of livestock products dependent on grain feeding while the growth in supply is slowing due to limitations on development of new cropland and reduced productivity gains on existing cropland. This increased tightening in the world grain market implies that the grain export price should increase substantially by 1985.

GRAIN SUPPLY OUTLOOK FOR U.S.

Winrock International is currently initiating a study of both the potential for and implications of additional crop production in the U.S. The 1977 Natural Resource Inventory compiled by Soil Conservation Service of USDA shows a total of roughly 460 million acres of cropland available in 1977 containing about 70% prime land in the Class 1 and 2 categories. Heady and Short (1981) of Iowa State University have projected that the 1977 cropland base will dwindle to 353 million acres by the year 2000, but that there are 37.6 million acres of high-potential land and 90.1 million acres of moderate-potential land that could be converted to cropland. This land area for potential development is located primarily in the South Atlantic, South Central, Great Plains, and North Central regions of the U.S. However, other economists in the U.S. have serious doubts whether it would be feasible to convert this much additional land to crop production. Some limitations to development of additional cropland and current use of this land are shown in table 3; the data indicate that much of this potential cropland is currently used for pasture and timber production and that there are definite erosion hazards and probable high conversion costs to develop this land area for crop production. Such limitations imply that there will be a major increase in production cost to bring these new lands into crop production after we reach full capacity on existing cropland.

TABLE 3. ESTIMATED POTENTIAL CROPLAND AND LIMITATIONS TO DEVELOPMENT IN THE CONTINENTAL UNITED STATES

Type of limitation	Percent of potential cropland	Present use	Percent of potential new cropland
Erosion	59	Pasture-range	79
Drainage	23	Forest	17
Soil	7	Other rural	4
Climate	4		
No limitation	7		
Total	100	Total	100

Source: 1977 Natural Resource Inventory, Soil Conservation Service, USDA.

IMPLICATIONS FOR U.S. LIVESTOCK PRODUCTION

World population is projected to increase 50% between 1975 and 2000. There is increasing emphasis on livestock production to improve the quality of human diets, particularly in centrally planned and developing countries, and increasing pressure on cropland worldwide. In the short term, there may be temporary swings between shortages and surpluses in the world grain market. This is expected due to greater year-to-year variation in world crop production as a result of expansion of cultivation on marginal lands with increased drouth stress and other climatic variability. Stability of supply may also be reduced due to mounting pressure on exporting countries to reduce the carryover of grain stocks from year to year. Grain prices are projected to increase with gradual tightening of world grain supplies. Increased export volume will require eventual conversion of at least some pasture and timber land in most of the key exporting countries, with an associated rise in grain-production cost.

Some implications for U.S. livestock producers include increased grain-feeding costs eventually rising above the general inflation rate and the loss of some pasture and rangeland area converted to cropland as indicated in table 3. Higher grain prices will be translated into somewhat higher meat prices, particularly for poultry, swine, and fed cattle because these enterprises are highly dependent on grain feeding. However, it will become more profitable to utilize additional crop residues and by-products in livestock feeding, particularly for cow maintenance, to replace present grain use. The biggest deterrent to using these low-quality feeds is cost--primarily for labor, equipment, and interest. To date, the availability of a stable supply price for grain is analogous to the situation we had 10 years ago for oil and natural gas. It has not been cost effective to utilize many alternative sources of feed energy, although there is an abundant physical supply available in the U.S. The amount of corn crop residue physically available was estimated to be 231 million tons in 1977 (Ensminger and Olentine, 1978). This would support 117 million cows for a 4-month grazing period on a purely physical supply basis. The nutritive value of crop residues can be enhanced with special processing techniques, and some very promising research work has been done on ammonia treatment of straw.

Cattle, sheep, and goat producers could potentially utilize these alternative sources of feed to substitute, at least partially, for grain or to enlarge breeding herds even on a drylot basis if it became more economical to do so. If meat prices increase along with grain prices, some livestock producers may regain a competitive advantage over poultry and swine producers who are more vulnerable to rising grain prices. Under the current regime of depressed grain prices, poultry producers, in particular, have been gaining a significant competitive advantage over beef producers. Poultry meat prices have now declined to 30% of average beef prices compared with 80% a few years ago (National Cattlemen's Assocation, 1982).

A reduction in grain feeding of cattle in the U.S. would mean increased competition for use of existing pasture and range lands. With increased grain exports, there would be an associated reduction in the grazing land area, particularly in the Southeast and Great Plains regions of the U.S. Increased dependence on crop by-products and residues implies that more livestock will be produced in traditional cropland areas to utilize these waste products, as was the practice in the U.S. before the feedlot era began. This is the situation now in most developing countries where the bulk of livestock production is found in mixed crop/livestock systems (Winrock International, 1981).

Increased use of crop residues to reduce the amount of grain feeding would have a significant impact on the management system for livestock, especially cattle. Levels of annual offtake would decline as cattle would have to be nearly a year longer to reach market weight on a less intensive feeding program. The cattle operator would be forced to move cattle to crop-production areas and to lease crop residues from crop farm owners as is now done for wheat pastures. Additional use of feed supplements would be necessary as crop residues are generally lacking in total nutrient requirements. The cattle operator would also have to invest in additional fencing and equipment to utilize crop residues. Thus the overall implications are that major adjustments would be required in the U.S. cattle industry, including shifts in the location of production, the composition of herds, and feeding programs.

CONCLUSIONS

The general outlook for the international grain market points toward continued price variability for feedgrains, a gradually rising price level for grains as the world market continues to tighten, and eventual loss of some grazing land in the U.S. when additional cropland is needed to meet expanding export requirements. It is possible that the problem of price variability may be alleviated by additional government intervention such as paid acreage reduction or other methods of supply control on the market, but this does not appear likely in view of the current emphasis on curbing spending for most agricultural support programs.

Expected consequences of the grain-market outlook for cattle producers include continued fluctuations in feeder cattle prices and returns from cattle feeding during the next few years and a general trend toward higher feeding costs. Although price variability is nothing new to livestock producers, the sharpness and range of price movement will likely be increased as long as we continue to be the shock absorber for the world grain market. The U.S. is one of the few nations that exposes domestic producers to price fluctuations of the international market.

Short-term effects of the expected swings in prices will be of more immediate concern to most livestock producers than the longer term upward trend in feeding cost, particularly for those in a weak financial position. To some extent, producers may be able to reduce the financial risk through greater participation in the futures market or by direct contracting. However, their most urgent need to survive in the livestock business will probably be to secure alternative methods of financing to provide more flexibility on repayment of loans. Other possible options for reducing or spreading the risk of price movement include the development of programs for outside investors to assume partial ownership of livestock and other creative financing schemes to shift at least part of the risk from producers to other outside parties. There may also be an opportunity for further revision of the tax laws to encourage more outside investment in the livestock business.

Long-term implications of changes in the world grain market, as well as expected increases in transportation cost, are that the structure of the U.S. cattle production system will change. Projected world food-system trends suggest increasing prices for all livestock products due to rapidly rising consumption in most countries and upward pressure on grain prices. However, there may not be much increase in livestock-product consumption in the U.S. market because per capita consumption rates have stabilized. A continuing problem for beef will be competition from pork and poultry in the U.S. meat market. To recapture its former market share, beef will require more efficient production and marketing throughout the system.

Increasing grain prices may provide some opportunity for beef producers to improve their production-cost relationship relative to pork and poultry by changing to less intensive grain feeding. Additional research and development is needed on the utilization of crop residues and by-products to reduce cost in cattle production.

REFERENCES

Economic Research Service, USDA. (1981). Problems and Prospects for U.S. Agriculture, Washington, D. C.

Harrison, P. 1981. The inequities that curb potential. FAO Review on Agriculture and Development. Food and Agricultural Organization of the United Nations, Rome, Italy.

Heady, E. O. and C. Short. 1981. Interrelationship among export markets, resource conservation, and agricultural productivity," Agr. J. Agr. Econ. 63:840.

National Cattlemen's Association. 1982. The future for beef. Special Advisory Committee Report, Englewood, CO.

Soil Conservation Service, USDA. 1977. 1977 Natural Resource Inventory. Washington, D. C.

Wheeler, R. O., G. L. Cramer, K. B. Young and E. Ospina. 1981. The World Livestock Product, Feedstuff, and Foodgrain System. Winrock International, Morrilton, Ark.

Winrock International. 1981. Report on Livestock Program Priorities and Strategy. Winrock International, Morrilton, Ark.

4

WORLD AGRICULTURE IN HOSTILE AND BENIGN CLIMATIC SETTINGS

Wayne L. Decker

CHARACTERIZATIONS OF CLIMATE

Health, nutrition, and suffering of the human population are determined, in part, by weather and climate. Regional wealth and the levels of economic development are impacted by the natural resources, including the climate resource. But agriculture and the associated food production industries are more directly affected by weather and climate than any other sector of the economy. Climate determines production potential of both grain and livestock producers, identifies strategies available to the producer for resource allocations and marketing, and determines the feasibility of plans for exports and imports of commodities. An improved understanding by agriculturalists of the nature of the climatic resource is essential if the impacts of climatic risks are to be minimized.

Climate is defined by the space and time distribution of weather events: temperature, precipitation, wind, humidity, and sunshine. In spite of the unpredictability of weather events, climate occurs systematically in both the space and time scale. As a result of these consistencies, climatic zones are easily recognized. For example, in the tropics and subtropics some regions are consistently rain-free in summer, others are smaller areas with an even seasonal distribution and abundant rainfall. In the temperate latitudes, the continental regions also demonstrate regional consistencies in climate. The west coasts of continents are mild with abundant winter rainfall, while the continental interiors tend to exhibit summer maximum of precipitation and marked seasonal temperature variations (Mather, 1974).

In most climatic regions, there are periods during the year with hostile climates for agriculture. These climatic hostilities are associated with temperature stresses (both high and low) and with deficiencies of rainfall. In many of these regions, there are periods of the year during which the weather is consistently dry, thus producing a hostility. This climatic hostility can be removed by irrigation, or avoided by adopting an enterprise with a low water need. For the hostile climates produced by temperature stress, shelters may be constructed to protect animals from the critical temperatures, and crop production can be scheduled to avoid the consistent occurrence of high or low temperatures.

Many regions have climates that are consistently favorable for agricultural production. These climates are usually characterized by dependable water supply (rain or irrigation supply). Benign climates are also characterized by moderate temperatures without a high probability of extreme temperatures during critical times for sensitive plants.

CLIMATIC CHANGE

Climatic change and the impact of climatic change on man have become controversial issues in recent years. Articles on the subjects appear regularly in technical and popular magazines, and both paperback and hardback books have been published dealing with climate change. The written opinion concerning climatic change and its impact are as different as day and night. Even scholars of climatology are confused by the diversity of opinion.

The evidence to support the existence of major climatic changes through geologic time is well documented. Long periods of geologic history are characterized by mild climates, i.e., benign climates. These periods were interrupted by relatively short intervals when glaciers extended into the middle latitudes (the ultimate in hostile climates). It is generally accepted that the current climate of the world is more like that of the glacial period than the warmer "climatic optimum." Climatic researchers do not agree about the mechanisms causing these major climatic changes. Current thinking focuses on long-term oscillations in the slope of the terrestrial axis, but continental uplift and the associate volcanic activity appear to be necessary conditions for the glacial climates.

Variations in the climate of the earth also have been documented from historical records. The rise and decline of civilizations during the past 4,000 years appear to be related to changes in climate. Plagues, famine, and migrations have been linked to shifts in climate. In modern history, the period corresponding to the North American settlement and the establishment of the United States was a period of climatic stress, frequently called "the little ice age." Again, meteorologists do not agree on the physical processes that caused the climatic variations in historical time. Volcanic activity, variability in the solar output, and combinations of both these factors are mechanisms receiving prominent attention.

It was not until the late 1800s that a worldwide network of weather observing stations was established. Although records of weather observations can be traced into the 18th century at selected points, networks of observational stations did not generally exist until the early and mid-nineteenth century. In the United States, for example, it is difficult to find documented weather records prior to the establishment of the Weather Bureau in the Department of Agriculture in 1890. For this reason studies of climatic change based on meteorological observations are confined to the most recent 90 years.

Attempts have been made to establish the trends in climate from the meteorological observations. The best documented estimate of the trend in climate is shown in figure 1 from Waite (1968). During the first 40 years of this century, the average air temperature near the earth's surface increased, but about 1940 this trend was reversed. Figure 1 verifies that these trends in temperature apply to regions of different size and are the most pronounced in the polar and subpolar regions of the northern hemisphere.

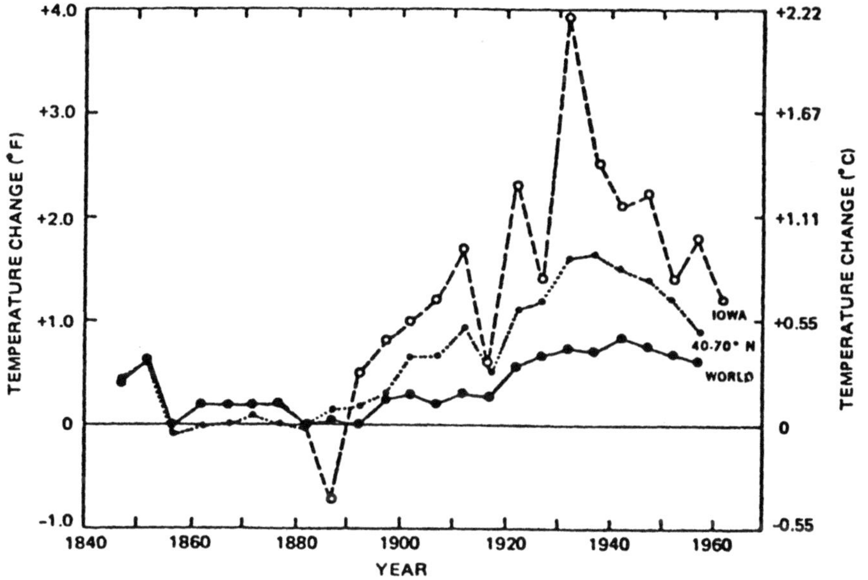

Figure 1. Worldwide trend in mean annual temperature (Waite, 1968)

For agriculture it is more instructive to look at the climatic variabilities associated with precipitation. In figure 2 and figure 3 the historical records of summer rainfall are summarized for the prairie provinces of Canada and the winter wheat region of Russia. It is difficult to identify trends from these data, although many interesting pulsations in the annual rainfall statistics, lasting for a decade or so, are apparent.

The atmospheric mechanisms producing the climatic trends and fluctuations that extend for a decade or a few decades have probably all been identified. These mechanisms include ocean-atmosphere interactions, volcanic activity, man's interference (CO_2 and particulate matter), and solar activity. The analytical contribution of each individual mechanism and the interaction between the mechanisms have not been defined; a major objective of the meteorological community to mathematically describe these processes based on known physical relationship efforts.

Efforts to research the physical causes for climate change led Dr. B. J. Mason, Director General of the British Meteorological Office, to observe in a recent article in the New Republic (1977):

> The atmosphere is a robust system with a built-in capacity to counteract any perturbation. This is why the global climate, despite frequent fluctuations, is fairly stable over periods of 10,000 years or so. Sensational warnings of imminent catastrophe, unsupported by firm facts or figures, not only are irresponsible but are likely to prove counterproductive. The atmosphere is want to make fools of those who do not show proper respect for its complexity and resilience.

FLUCTUATIONS IN CLIMATE

Climate variability adds an additional stress to the agriculture system and adds a component to climatic hostility. When the weather of one or more years departs markedly from the expected, a farm management strategy that has been successfully used becomes inappropriate for the agricultural enterprise for a particular year or growing season. Several years of drought (such as the 1930s on the U.S. Great Plains) is hostile to the farm enterprise adopted to nondrought enterprises. On the other hand, periods of years with benign climates often lure farmers into strategies not adapted to the hostile and stressed condition that follows. The type of fluctuation most often used by climatologists deals with the variation of climatic events between years. This variability refers to the variation in annual or seasonal temperatures and precipitation totals. McQuigg (1973), for example, demonstrated the low variability in climate between 1955 and 1970 for the major agricultural production regions of the U.S. McQuigg simulated

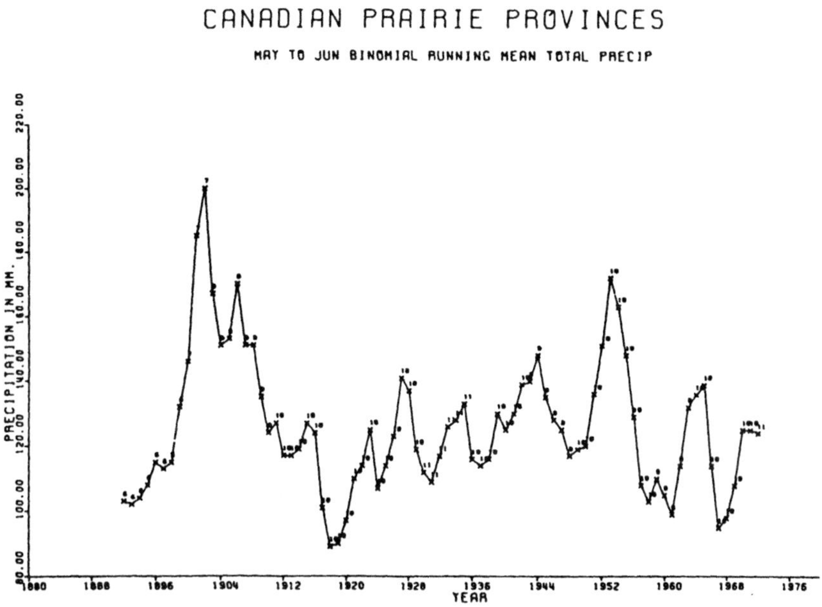

Figure 2. The year to year variability in the total precipitation for May and June (smoothed by a binomial technique) in the spring wheat producing area of Central Canada.

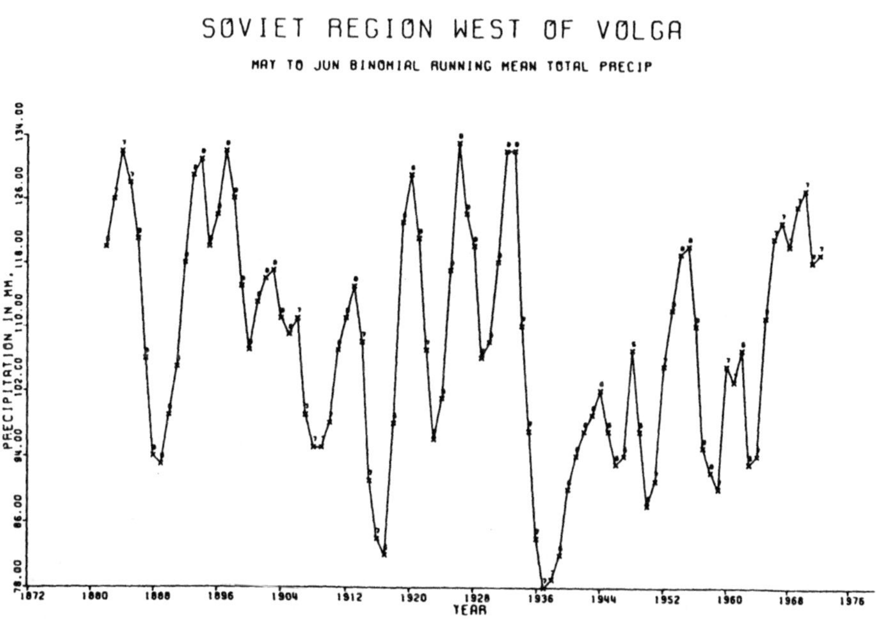

Figure 3. The year to year variability in the total precipitation for May and June (smoothed by a binomial technique) in the winter wheat producing area of the Soviet Union (west of the Volga River).

yields of grain throughout this century from climatic data at a constant technology. He showed (figure 4) that the yields simulated from climate during the period extending from the late 50s through the 60s were remarkably constant and relatively high, i.e., the climate of the U.S. Corn Belt was benign. The year-to-year variability in climate in the U.S. has been greater in the 1970s and the early 1980s than the preceding decade and a half.

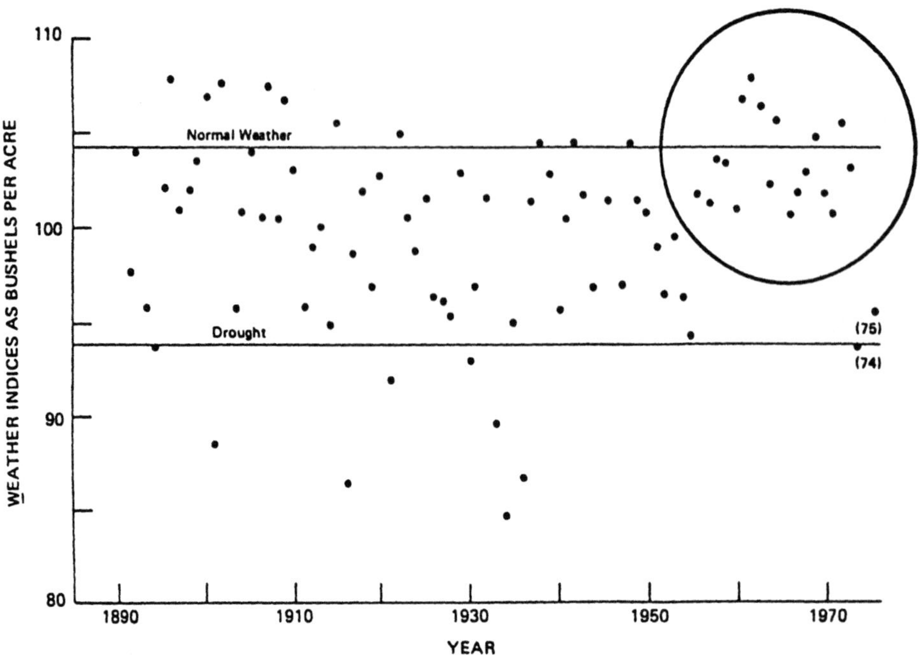

Figure 4. Year-to-year variability in climate expressed in simulated corn yields for the U.S. (NAS, 1976)

To examine how climatic fluctuation varies through time, the variances of climatic elements by decades have been computed for major agricultural production areas in the world. Figure 5 shows the 10 year variances in the May plus June precipitation in the Canadian prairie region, while figure 6 presents the same values for the Soviet winter wheat region. The May and June rainfall totals are vital to wheat production in these two regions. In both cases, the variability in May and June precipitation during the most recent decades were below average; however, the tendency for a lower variance does not appear to depart from that expected from the normal variability. The high year-to-year variability in the May-June rainfall in Canada just after the turn of the century is quite striking. There was also a period of high variability for the May and June precipitation in the Soviet Union between 1925 and 1950.

FIGURE 5

Climatic Variability
Canadian Wheat Region
May-June Precipitation
Deviation from mean variance by ten year periods

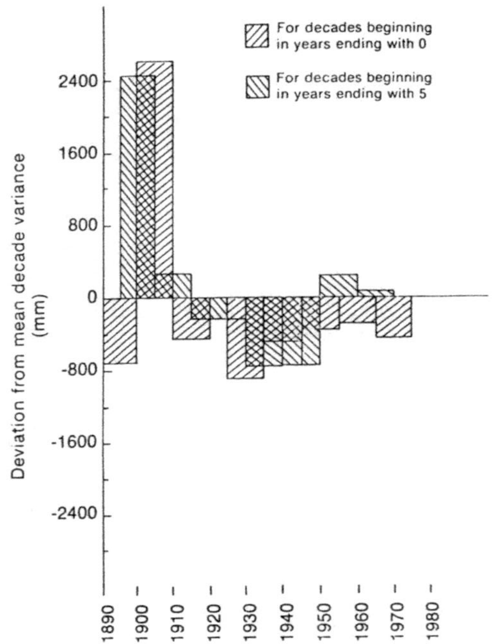

FIGURE 6

Climatic Variability
Soviet Union, Wheat Region West of Volga
May-June Precipitation
Deviation from mean variance by ten year periods

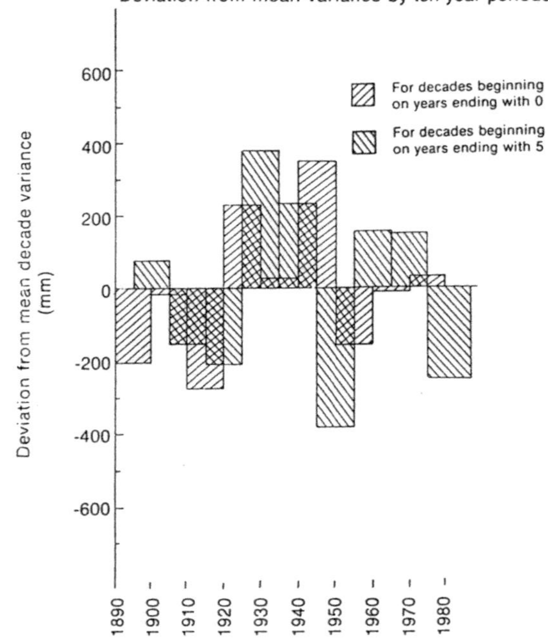

Figures 5 and 6 show that climate did not experience unusually low variability everywhere in the world during the period 1955 to 1970. A second, and more important, lesson concerns the recognition that periods of 1 or more decades in length with high (and low) year-to-year variability in climate do occur.

A report by the National Research Council (1976) identifies sudden and unexpected climatic fluctuations as the greatest climatic hazard to agricultural production. The managers of agricultural systems are forced to use practices that are not adapted to the regional climate. The farm manager has difficulty in finding the "best" strategy during periods with fluctuating climates. A large year-to-year variability forces the manager to choose management options that are "out of step" with the season's weather.

RESPONSES TO CLIMATE CHANGES AND FLUCTUATIONS

The policy makers and agricultural managers could easily respond to weather and climate fluctuations, if events could be anticipated before they occurred. Although several climatologists, with impressive credentials, regularly make seasonal forecasts, most meteorologists agree that the science of meteorology has not advanced sufficiently for these forecasts to be considered valid. The thirty-day outlook, which projects the general trends in mean temperature and precipitation, is correct only about 55% to 60% of the time. Seasonal forecasts (cold winter, dry summer, etc.) have an even lower accuracy. Complex computer models, which simulate the atmospheric circulation, offer the best promise for the development of rational forecasting schemes. As these atmospheric models are improved, weather and climate forecasts will have improved accuracy and be extended for longer periods. The improvement in forecasting skills will be slow. In the next two decades there appears little hope for major and sudden breakthroughs in our understanding and interpretation of atmospheric circulation.

Weather modification provides an additional strategy for removing weather risks. Modification of the surface energy budget through changes in the surface color, drainage of the land, or shaping of the soil surface through tillage, offers many important options for improved technologies for agriculture. Increased rain through cloud seeding is more often considered as a weather modification option. There are several difficulties that reduce the potential for cloud seeding to respond to fluctuations in climate:

- Proof of small increases in amounts of rain are almost impossible to obtain because variations in area and time affect the amount of rain falling over a region.
- Rain-making only augments the natural rainfall, so it is not a "drought stopper."

- The possibility exists that the manipulation in the clouds will reduce the rainfall from some clouds within a given weather system.

Cloud seeding appears to be most favorable for use in the mountainous regions--to increase the winter snow pack. Increased snow in the mountains improves the water supply for the agriculture of the adjacent semiarid and arid regions.

Disaster insurance spreads the risks of "bad" weather. An international program for grain and food storage should stabilize supply between the "lean" and "bountiful" years. A marketing cooperative, or even the individual farmer, may establish an "ever-normal granary" by withholding grain from the market. For the individual farmer, the best opportunity for spreading the risk is through disaster insurance. A national food policy must include an insurance program using both governmental and private agencies as underwriters. Insurance programs may stabilize farm incomes, but, in the long run, will not provide increased productivity of food for the expected increase in world population.

Technologies developed through private and public research provide an additional response to provide for a stable food supply and farm income under a fluctuating climate. Meteorological science cannot be expected to deliver completely reliable warnings of pending shifts and fluctuations in climate. The meteorologists will not save us from the adversity of a variable and often hostile climate. Agricultural strategies and technologies must be developed to respond to the expected climatic variabilities. These developments will emerge from integrated, interdisciplinary agricultural research.

REFERENCES

Mason, B. J. 1973. Bumper crops or droughts. Mimeo. NOAA, U.S. Dept. Commerce, Washington.

Mather, J. R. 1974. Climatology, Fundamentals and Applications. McGraw Hill, pp 112-131.

National Research Council. 1976. Climate and Food. Report on Climate and Weather Fluctuations and Agricultural Production. National Academy of Science, Washington.

Waite, P. J. 1968. Our weather is cooling off. Iowa Farm Science 23:13.

5

THE IMPACTS OF CLIMATIC VARIABILITIES ON LIVESTOCK PRODUCTION

Wayne L. Decker

CLIMATE AND LIVESTOCK

Regional climates contain factors that need consideration in determining the kind of profitable livestock production for a region. The variability of weather and climate provides a component in determining the profitability of the livestock enterprise. Climate imposes both direct and indirect effects on commercial animal agriculture. The direct effects include weather events producing physical injury (lightning, wind, flood, temperature extremes), occurrences associated with physiological stress (such as heat and humidity), and weather events promoting insect or disease episodes. Indirect climate events are those that impact on availability of forages and the supply of feed grains. These indirect impacts of the weather and climate are generally imposed by chronic deficit in water and occasional droughts.

LOSSES IN ANIMAL PRODUCTION DUE TO CLIMATE EVENTS

Animals, grown commercially on farms and ranches, are normally subjected to ambient environmental conditions. On many occasions, the atmospheric conditions are less than ideal and the animal is subjected to stress. This stress, which is usually related to the heat and energy balance of the animal, reduces the production. The stress causes declines in egg or milk production and reduced weight gains for swine, beef, or broilers.

Over the years there have been repeated attempts to mathematically define the impact of stress imposed by atmospheric conditions (temperature, humidity, wind, etc.) on animal production. The experimental basis of these efforts comes from two sources: (1) barns or chambers with controlled environmental conditions (Brody, 1948) and (2) field experiments measuring animal performance as related to observed ambient environmental conditions. From the observations obtained through the experimentation, functional relationships between the weather event (or events) are derived. The resulting mathematical formulas are usually obtained through standard statistical procedures. Strickly speaking, the expressions are only applicable to the experimental conditions from which the relationship is derived, so each relationship must be tested against independently collected data.

Literature has many examples of relationships between the performance of domestic animals and weather and climate events. Two mathematical expressions for relating cattle performance to environmental conditions are discussed below.

Milk Production

Using data obtained from controlled experiments, Berry et al. (1964) related the decline in milk production to an index involving both atmospheric temperature and humidity. This index, which is called the temperature-humidity-index (THI), is shown in equation (1).

$$THI = T + .36\ T_d + 41.2 \qquad (1)$$

where

T is the temperature in °C,

T_d is the dew point temperature in °C.

The relationship between THI and the decline in milk production (MD) is shown in equation (2).

$$MD = -2.37 - 1.74\ NL + .0247\ (NL)(THI) \qquad (2)$$

where

NL is the normal production of a cow under thermoneutral conditions.

THI is the temperature-humidity-index.

Since negative values for MD do not make sense under the definition in this equation, decline in milk production is assigned the value of zero for all negative values in equation (2). This means that a zero production decline is expected until a critical value of the THI is reached. For higher values of THI, the decline in production decreases linearly. This critical value of THI is between 70 and 74 for normal production levels of between 25 and 30 pounds of milk per day.

Meat Production in Cattle

Bolling and Hahn (1981) and Bolling (1982) report the results of a regression analysis relating climatic variables to rates of gain for beef animals. The functional relationship, which best explains the reduction in weight gains, contained terms related to cold stress, heat stress (THI), precipitation, and wind. The results of the regression analysis are shown in figure 1, as taken from the work of Bolling (1982). The author concluded:

> "Analyses consistently suggested that stress resulting from the direct effects of cold, combined with the effects of precipitation, have a greater impact on feedlot cattle in Nebraska than does any

other type of atmospheric stress studied. Interestingly, heat stress rarely appeared to be a significant factor affecting cattle performance."

Of course, this conclusion concerning heat stress is counter to the one for milk production. This difference may be due to the difference in climate of the regions where the milk production and rate of gain experiments were conducted.

Hahn et al. (1974) noted that beef cattle were able to overcome heat stress. In the Missouri Climatic Laboratory, beef cattle were stressed by being subjected to 5 weeks of temperatures of 30°C. A marked decrease in rate of gain as compared to a control group resulted; when the animals were returned to optimal conditions, the stressed animals outgained the control group. This result, which Hahn calls "compensatory growth," is demonstrated in figure 2. No such compensation occurred after animals were subjected to a greater stress (35°C). Hahn (1976) indicates that similar compensatory growth occurs with swine and broiler chickens. It is significant that (1) the research indicates that compensatory growth does not occur after exposure to a high degree of heat stress, (2) the laboratory experiments on compensatory growth were done at constant temperature and may not apply to temperatures experiencing a diurnal range, and (3) no experiments have been made to discover whether "compensatory growth" occurs after cold stress.

MORTALITY OF ANIMALS DUE TO CLIMATIC STRESS

Hostile climates do cause mortality to domestic livestock. This hostility occurs as a result of both heat and cold stress. In summer, high temperature and humidities (THI values of 80 or higher) produce stress that can lead to death. This condition is, of course, aggravated by other stress factors associated with handling and/or shipping. In winter, the cold stress can cause tissue to freeze and the animal to die. For U.S. cattlemen of the open ranges in the High Plains and eastern slopes of the Rocky Mountains, the cold stress may be further aggravated by high winds with snow. These blizzards cover the winter food supply, make access to the herds difficult (if not impossible), and bury herds under the drifting snow.

Bolling (1982) presents analyses that document the weather impacts on cattle mortality under feedlot confinement. Strong winds and cold stress were the "best" predictors of mortality under Nebraska conditions. The author was apparently unable to document the mortality due to stress imposed by hot and humid weather. Of course, one would not want to use these results to estimate mortality under range or pasture exposures.

The National Weather Service has established policies for issuing weather advisories to stockmen. These advisories are issued by Weather Service Forecast Offices located

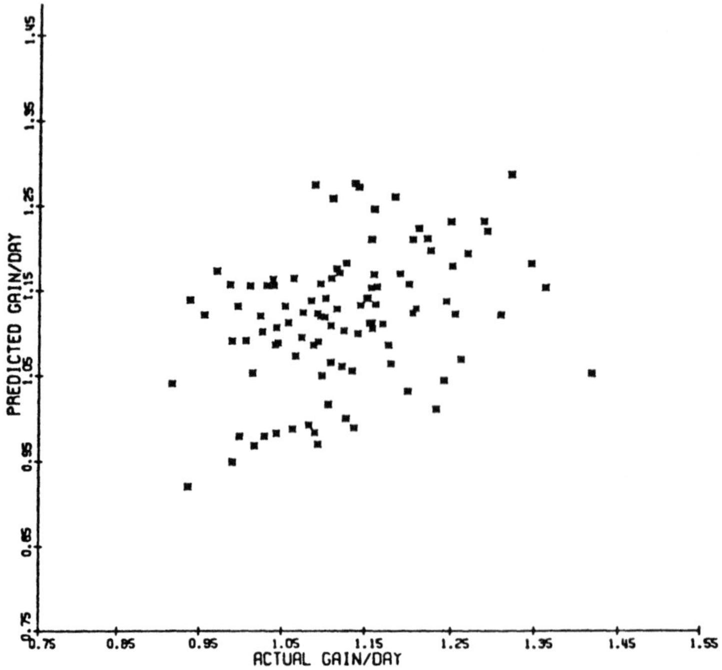

Figure 1. Relationship between measured gain of beef cattle and that predicted from weather data using a statistical relationship (Bolling, 1982)

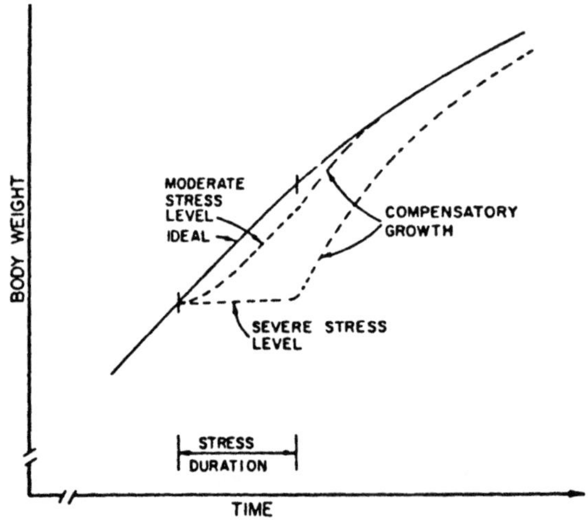

Fig. 2. Schematic to Illustrate the Principle of Equifinality in Animal Growth (Hahn, 1982)

in each state. These advisories are:

- Heat stress advisories are issued whenever high temperatures combine with high humidities to present danger to livestock. The THI is used as the index for danger to livestock. Two categories are recognized: (1) livestock danger (THI 79 to 83) and (2) livestock emergency (THI higher than 83). When air temperatures are below the body temperature, wind provides cooling. Because of this, wind is mentioned in the stockman's advisories when velocities higher than 20 mph are expected.
- Winter storm watches are issued when there are strong indications of a blizzard, heavy snow, freezing rain, or sleet. A blizzard is defined as a condition with high winds (in excess of 35 mph), with falling or blowing snow (visibilities less than 3 miles). A winter storm warning or blizzard warning is issued when the storm's development is a virtual certainty.
- Chill indices are released routinely. This index combines wind and termperature in an effort to approximate the equivalent temperature under light wind (4 mph) conditions. At Columbia, Missouri, in the winter of 1981-82, there were 398 hours (11% of the hours) with chill factors below 0°F, 181 hours (5%) with chill factors below -10°F and 72 hours (2%) with chill factors below -20°F. The number of days with chill factors below 0°, -10° and -20° sometime during the day was 30, 18, and 7 days, respectively.

REFERENCES

Berry, I. L., M. D. Shanklin and H. D. Johnson. 1964. Dairy shelter design based on milk production decline as affected by temperature and humidity. Trans. Amer. Soc. Agr. Engr. 7:329.

Bolling, R. C. 1982. Weight gain and mortality in feedlot cattle as influenced by weather conditions: Refinement and verification of statistical models. Progress Report 82-1. Center for Agricultural Meteorology and Climatology, Univ. of Nebraska, Lincoln.

Bolling, R. C. and G. L. Hahn. 1981. Climate effects on feedlot cattle: Growth and death losses. Proc. 15th Conference on Agr. and Forest Meteorology: 86-89. Amer. Meteorological Soc., Boston.

Brody, S. 1948. Environmental physiology with special reference to domestic animals, I: Physiological background. Res. Bull. 423. Mo. Agr. Exp. Sta., Columbia.

Hahn, G. L. 1982. Compensatory performance in livestock: Influences on environmental criteria. Livestock Environment, Proc. 2nd Internatl. Livestock Environment Symposium. Amer. Soc. Agr. Engr., St. Joseph, Michigan. (In Press.)

Hahn, G. L. 1976. Rational environmental planning for efficient livestock production. Proc. 7th Intern. Biometeorological Cong., College Park, MD. pp 106-114.

Hahn, G. L., N. F. Meador, G. B. Thompson and M. D. Shanklin. 1974. Compensatory growth of beef cattle in hot weather and its role in management decisions. Livestock Environment, Proc. Internatl. Livestock Environment Symposium, pp 288-295. Amer. Soc. Agr. Engr., St. Joseph, Michigan.

6

OPPORTUNITIES FOR U.S. SHEEP PRODUCERS: MANAGEMENT AND MARKETING FOR PROFIT

Rodger L. Wasson

Today the sheep industry is challenged to grow in new directions. Economic factors, global competition, high operating costs, and a shifting consumer market have caused the industry to reevaluate its objectives and place increasing emphasis on one goal--assuring a livable and profitable return from sheep operations.

The more fully the various segments of the sheep industry pull together in support of this aim, the more fully it will be realized. Producers, feeders, breeders, allied industry, research, and extension all have valuable roles to play.

Within the context of that goal, let's take a look at the role of producers and the American Sheep Producers Council (ASPC). Together, through interrelated efforts, they can transform this goal into a reality.

One major opportunity that we have as a sheep industry is to manage our business like the billion-dollar-a-year agribusiness that it is. This involves taking in the total perspective - not just parts of the whole, but the whole picture. When we do this an enormous range of resources and industry talent comes into view. A greater strength immediately becomes apparent - the strength that comes of a unified effort focused on collective goals. There's nothing we can't achieve when we think and work this way.

Part of the producer's role within this is the management of his private operation at a level of optimum profit. Thousands of well managed operations--in both production and business--build a strengthened industry.

Then there's the network through which all well-managed operations relate to the larger industry - procurers, processors, distributors, retailers: how do producers maximize their profitability within this context? And further, how does each of these various components of the American Sheep Industry realize a beneficial and equitable return?

We need to think like a business at the collective national level. A business thinks marketing when it wants to achieve its greatest return and yield its greatest contribution. In the words of Peter Drucker, the highly regarded management authority, "Marketing is so basic it cannot be considered a separate function of the business, on a par with manufacturing or personnel. Marketing is the whole business, seen from the point of view of its final results."

ASPC can provide this overall perspective from the viewpoint of final results through the marketing function that keeps its eye on the collective bottom line. When ASPC was first chartered through the Wool Act in 1954, its purpose was to increase the consumption of lamb and wool through programs of advertising and sales promotion. This was well and good, but today we know that the foundation of any advertising/promotion program lies in marketing--that intelligence by which a firm selects the most appropriate blending of product and markets, distribution, pricing, timing and competitive positioning. After this fundamental evaluation is completed, then strategies are identified and budgets allocated.

After the allocation of budgets comes advertising, sales promotion, packaging, merchandising, and field sales to implement the program within the guidelines established by the marketing plan. This system sets advertising and sales promotion within the framework of a marketing plan and enables them to be more effective since they become means to a pre-established end rather than just an end in themselves.

But none of this structure - marketing planning, advertising or sales promotion - is of any value if it doesn't satisfy the customers. All marketing programs begin with the customers and their reasons for buying as they do - their attitudes, opinions, perceptions. A good marketing plan turns a consumer-perceived benefit into a benefit for the organization; by providing that consumer-desired product or service the organization establishes its basis for industry profit.

Now we can restate ASPCs original objective as "creating a product engineered to consumer requirements that, when done effectively, maximizes industry profitability." We can achieve this through a comprehensive marketing plan that will offer optimum profit for all industry segments.

To do this we need to rethink our approach to selling lamb and wool. As an industry we can approach this at two general levels--the farm level and the industry level. At the farm level, we can tighten our standards, assuring a consistently high quality, uniform product in both lamb and wool. Producers receive a direct incentive from the marketing system when they receive a price that returns to them the benefits of their higher-quality products. At the level of an integrated lamb industry itself, we can help each processor receive an equitable return for his value added to the product by working hand-in-glove with each stage of product finishing.

As an industry, then, we are seeking to directly reward quality and excellence throughout each line of the marketing chain and to maximize our mutual benefits through cooperative interrelatedness. This is what a large corporation would be doing with each of its profit centers and through each of its interrelated marketing stages.

As an example of how this could work, let's focus for a moment on lamb. The total marketing approach would suggest that we reposition our product so that we are selling what we are, not what we aren't. Currently, we are a seasonal product with production quantity varying by location. We are not beef, pork, or chicken, and so we fill somewhat different consumer needs. The more closely we style our product to consumer needs, the more we, as well as the consumer, benefit.

We have high-consumption, general market areas on the East and West Coasts, in the Midwest and emerging and localized market areas throughout the nation. This suggests that we need a national-level promotion and overall-product image--specifically we need promotion to handle specialized demand. Looking beyond the national level, we see an international market that must be considered in our planning. Other opportunities are increasing at the local level with a growing number of producers direct-merchandising most of their production through stores or locker plants within 20 miles of their operation.

We need to market our product in "levels" to meet our tiered markets. These levels are:

- Local promotions involving local producers.
- Regionalized promotions in which producer efforts are coordinated with national promotion in small areas, states, and regions.
- National promotion through ASPC.
- International promotion.

At the same time, we need to be thinking about specific marketing strategies to gear our seasonally fluctuating supply to these markets. A diversity of product approaches could answer our need for effective product movement and demand with good producer returns. This diversity would take into account several different product strategies that answer the needs of our tiered market and fluctuating supply. We might, for instance, position lamb nationally as the most succulent meat - the one to buy for a savored change of pace. Within that stance we might take an additional position for intermountain lambs - the most succulent meat, grazed in nature for prime quality - available only in the fall. People would buy lamb year-round and look for a special kind of lamb in the fall like some anxiously awaited Bock beer every spring. This strategy has effectively increased the market share for branded products by filling two consumer needs, not one. Why couldn't it work for lamb by increasing perceived value and demand at peak supply times while maintaining price. When a promotion like this is used - in concert with a marketing program that distributes lamb to preidentified areas - we gain much more price influence. It benefits everyone in the marketing chain.

These promotions need to be brought back to the producer through producer-supported local and regional promotions. This plan leaves plenty of opportunity for grassroots participation through events like supermarket taste sampling which pique consumer demand.

This brief survey of marketing possibilities can help us formulate a response to the challenge of better profits. Whether these or other marketing strategies prove best is not as important as the fact that the sheep industry must come to grips with an integrated marketing approach. When we can do that, we will have taken more of our destiny into our own hands and have created a better livelihood and better opportunities for our industry.

7
USE OF STATISTICAL ANALYSIS TO IMPROVE PRODUCTION IN SHEEP AND GOATS

J. Hummel,
Miguel A. Galina

INTRODUCTION

By understanding the relationship between the variables affecting a farm environment and by setting up a mathematical model of this relationship, it is possible to control management methods better and, therefore, maximize production (Clow et al., 1974).

A production deficiency was discovered in the area of small ruminants in Mexico so that it became necessary to set up a complete farm manageent program (Hummel, 1982). A report was done on on the status of meat, milk, and by-products of sheep and goats, and it was found that there was inconsistency between the agencies that reported the numbers (Galina et al., 1981a; 1981b). It was apparent that there would be no progress in investigation until there was a suitable system of statistical analysis; therefore our program was initiated.

The importance of small ruminants has taken on a new dimension within the last few years because of various factors:

- Their emergence as an important source of meat and milk protein within the third world (Galina et al., 1982).
- The change of status from a "backyard" animal (as in the case of the goat) to an economically important enterprise (Galina et al., 1982).
- The increase in demand for mutton and kid (as a substitute for beef), fresh goat milk and cheese, and the return to natural fibers for clothing (Galina, 1980).

Most of the research done on production has been applied to dairy cattle because of their economic importance to the world community (Finley et al., 1978). The genetic evaluation program for dairy sires and cows at the University of California at Davies has shown the American farmer that there can be dramatic gneetic gains for milk production in the dairy cow (Finley et al., 1978). In fact a Norwegian program reports expected genetic improvement in milk yields of 1% to 5.2% per year as a result of their genetic evaluation program (Steine, 1980).

In Mexico it has been demonstrated that a program of milk registration and mastitis control (INL, 1982) can increase milk production in a country in desperate need of fresh milk.

MATERIAL AND METHODS

The program developed through three separate stages with each one interdependent and overlapping the other two. They were 1) the preliminary investigation of the productive performances of sheep and goats under various conditions to determine the present level and needs of the Mexican farmer, 2) the programming of a computer to accept and analyze data from all the participating locations, 3) and the application of the results on model farms/ranches.

Our model ranches/farms were from various areas and represented small ruminant producers with various economic successes. Results from some of the ranches/farms have been published elsewhere, while some of our results will be published in a paper accompanying this presentation (Galina et al., 1983a; 1983b; 1983c).

The farms in this program were both in the State of Mexico and specialized in Suffolk sheep. One was government owned and located at Chapa de Mota, the center for low-cost sheep management methods and traditional systems (Guerrero, 1982). The other farm was privately owned with moderate-cost management methods and a tendency toward experimental systems (Gutierrez, 1982). Details on the management procedures have been published elsewhere (Guerrero, 1982; Gutierrez, 1982).

The computer system used in the second stage was a Digital 11/34 medium-sized computer located at the National Autonomus University of Mexico in Acatlan in the State of Mexico. We used a "sort and merge" system of programming.

RESULTS AND DISCUSSION

Experimental data was obtained from these two farms and used to correct and program our computer's analytical method. Adjusted factors were developed to evaluate the nutritional supply as related to the productive performance and the selective methods used for genetic improvement of the herd.

Also computed into the overall evaluation were sanitation and selective management techniques (such as grazing). We demonstrated that program data from sheep and goat farms with different management and conditions can be computer analyzed. The program applied to these farms showed that imported sheep performed better than local sheep when nutritional and sanitary environments were similar to those in the U.S. A direct relationship was demonstrated between productivity, feed energy provided, and sanitation. This evaluation was eased by the use of accumulative data into the computer. The computer made the assimilation analysis and gave the print out of results more rapidly than was possible with any other method.

Our preliminary investigations led us to set up five separate computer programs to study all areas in this program:

- Reproductive behavior of the herd, i.e., fertility, prolificacy, lamb survival at 15 days and lamb survival at weaning, age of puberty, daily weight gains, and overall reproductivity of the herd.
- A genetic adjusted weight evaluation of each animal at birth, weaning, and first breeding, with adjustments for type of birth, age of dam and sire.
- A genetic evaluation of each animal, male and female, as potential breeders.
- Evaluation of the production of by-products, i.e., milk, meat, and hair of goats, and meat, wool, and skins of sheep.
- Analysis of diseases and causes of mortality or low-productive performances within sheep and goat herds.

It was felt that these five programs covered the needs of the rural communities and would supply a mathematical model of relationships between the variables facing the Mexican, small-ruminant farmer.

Application of the results will begin as soon as the computer has begun to implement the programs. Each owner will be given a monthly report on his herd along with any evaluation of other farms that he requests. By the owner working closely with the technical advisers, the results can be applied to the conditions of the individual farms.

Already the interest among farmers is good, and we expect participation from various areas of Mexico. The producers understand the need to have reliable statistics and control on management methods to improve their production and increase their profit.

REFERENCES

Clow, D. and N. Urguhart. 1974. Mathematics in Biology. W. W. Norton & Company Inc., New York.

Finley, C., F. Murrill, B. Kennedy and E. Bradford. 1978. University of California Buck Summary. University of California, Davis.

Galina, M. 1980. Proyecto para la creacion de la maestria en produccion animal (ovionos y caprinos). Facultad de Estudios Superiores, Cuautitlan, National Autonomous University of Mexico (UNAM), Mexico.

Galina, M., O. Rojas and J. Hummel. 1981. Diagnostico y perspectivas de la produccion ovina en Mexico. Memorias Ovinas, i Encuentro Nacional sobre produccion de ovinos y caprinos. Facultad de Estudios Superiores, Cuautitlan, National Autonomous University of Mexico (UNAM), Mexico.

Galina, M., O. Rojas and J. Hummel. 1981. Diagnostico y perspectivas de la produccion caprina en Mexico. Memorias Caprinas, i Encuentro Nacional sobre produccion de ovinos y caprinos. Facultad de Estudios Superiores, Cuautitlan, National Autonomous University of Mexico (UNAM), Mexico.

Galina, M., M. Guerrero, V. Rojas, M. Ruiz and V. Vasquez. 1982. Social status of the goat industry in Mexico. Proc. III Int. Conf. on Goat Production and Disease. Dairy Goat J.:420.

Galina, M., M. Guerrero, M. Gutierrez and N. Celis. 1983. Cost benefit of intensive management of a dairy goat herd under zero-grazing., (In print.)

Galina, M., M. Guerrero, M. Gutierrez and J. Salas. 1983. Economical performances of Suffolk sheep for mutton, wool and breeding in Mexico. (In print.)

Galina, M. and J. Ruiz. 1983. Dairy goat rearing management with milk and milk substitutes. (In print.)

Guerrero, M. 1982. Evaluacion de la eficiencia productiva del Centro Nacional de Fomento Ovino en Chapa de Mota. Estado de Mexico (SARH) de 1979-1981. Thesis. Facultad de Estudios Superiores, Cuautitlan, National Autonomous University of Mexico (UNAM), Mexico.

Gutierrez, J. 1982. Evaluacion de la eficiencia productiva de un rebano Suffolk en Huchuetoca, Estado de Mexico, Mexico de 1980-1981. Thesis. Facultad de Estudios Superiores, Cuautitlan, National Autonomous University of Mexico (UNAM), Mexico.

Hummel, J. 1982. Unpublished data.

INL Instituto Nacional de la Leche. 1982. Evaluacion de perdidas economicas por mastitis subclinico detectada por la prueba de California (CMT) durante 20 purebas realizadas una cada mes. MVZ. Noticias IV (14).

Steine, T. 1980. Norwegian goat improvement program emphasizing selection based on performance and progeny testing, International Goat and Sheep Research 1(2)108.

Part 2

GENERAL CONCEPTS AFFECTING AGRICULTURE AND THE INDUSTRY

8

POLITICAL CHALLENGES FOR TODAY'S ANIMAL AGRICULTURE

George Stone

Not all of the challenges to today's animal agriculture are political. Some of the challenges are internal. They relate to the nature of our industry and to the nature of the people in the livestock industry.

We seem to have an allergy to change. And we have a disposition to go our own way as individuals, even if we could improve things for ourselves by working together.

ROLE OF FEDERAL GOVERNMENT

Part of the time, we want the government to leave us alone. Part of the time, we get very impatient when the government is too slow with helping us. This is not the first, nor the last, speech to be made in this nation on the role of the federal government in food and agriculture. We talk about that subject as if we were about to make an original choice. But, as a matter of fact, the choice was made long ago. As early as 1796 and as recently as 1977, and many times in between, the federal government has decided that the family farm should be fostered and encouraged.

Our society has decided that the federal government should take measures to help assure that land remains in the hands of family farmers.

Our society has decided that the federal government should be involved in the conservation and protection of land and water resources.

Our society insists on a federal involvement in environmental protection.

Our society has dictated a federal role in assuring that food supplies are safe and wholesome.

Our society requires federal supervision to see that pesticides and chemicals are safe for farmers and consumers.

Our society provides for federal supervision of the marketing system to try to keep it fair and competitive.

There are more kinds of government intervention in agriculture today than ten, twenty, or thirty years ago. There will probably be more federal involvement in farm and food policy in 1990 or the year 2000 than there is today.

The real question should be in regard to the nature, extent, and purpose of government involvement, and the degree to which farmers have a voice in decisions that will affect them.

Government intervention has often been justified when there was no other effective way to cope with problems. The government's role in agricultural research and education is widely accepted and advocated. Various federal farm credit programs had to be initiated because the private sector could not take the entire risk in financing agriculture. Commodity exchanges and boards of trade had to get federal supervision because they could not be left to police themselves. The federal rural electrification system had to be created because the private sector could not, or would not, do the job.

In many agricultural sectors, government involvement can be good or bad:

- Rules on farmers' handling of pesticides or chemicals can be reasonable or ridiculous, depending on how much farmer input there has been in the process.
- Feedlot pollution abatement rules can be workable and effective, or unrealistic and oppressive.
- Dredge and fill regulations can be a protection or a harassment for farmers.
- OSHA regulations can be a godsend or an aggravation.

Everything depends on how well farmers have involved themselves in the process and how much voice they have had in shaping these laws and regulations. That is not an easy task for us as farmers and livestock producers. The leading codification of laws affecting agriculture now runs to 14 volumes of about 500 pages each. That is 7,000 pages of laws.

MAJOR POLITICAL CHALLENGES

Having said this much in the way of background, let me now turn to what I perceive as some major political challenges for livestock agriculture. One thing that is quite obvious, but not generally appreciated, is that we do not function in a vacuum as livestock people. We depend upon consumer purchasing power and demand for our products, and we cannot expect to be stable and prosperous if there is high unemployment and weak buying power.

Recession and Unemployment

Much of the difficulty that has faced the livestock producer in the past three years has been attributable to recession and unemployment. High interest rates the past three years have diverted about 30 billion dollars a year of consumer purchasing power from food and other necessities. High interest rates have an impact on the cattle producer as well. Some calculations done at Oklahoma State University reveal that the interest cost per head of livestock sold would be $133 per head at 9 percent interest; $237 per head at 16 percent; and $297 per head at 20 percent interest. The difference between the two extremes of 9 percent and 20 percent is $164 a head, more than enough to wipe out any potential profit.

When national economic conditions are difficult and there is a pinch on consumer buying power, the tendency is for a reduction in the higher-priced meat purchases and, to some extent, for the consumer to buy other products. In a time of recession and unemployment, meat purchases are the first thing affected.

Specifically, because of the tendency for the American diet to be hurt by recession and unemployment, the Congress in its wisdom developed and implemented the food stamp program. Of course, one can expect that when times are tough, the food stamp program becomes costly. Each additional one percent of unemployment adds one million people to the food stamp rolls, not because it is a bad program, but because it is doing what it is supposed to do--maintain a healthful diet for the lowest-income people in our society.

About 27 percent of the food stamp benefits are spent by recipients to buy meat and meat products. This means that if the food stamp program makes available 12 billion dollars in food subsidies to low-income families, over 3 billion dollars will be used to buy meat. If the food stamp program is cut by 3.7 billion dollars, as it has been in fiscal budgets for 1982 and 1983, that means a 1-billion-dollar reduction in the demand for meat. That is a political decision and we live with it as livestock producers.

Competition Between Humans and Farm Animals

One of the political challenges that will become more serious as time goes on will be the competition for living space between humans and farm animals. Twenty years ago, a livestock economist at a midwestern land grant university was making the prediction that there would be little room in the world for livestock 100 years in the future. Human population was increasing so rapidly that the land would be needed for living space. There would only be room for enough livestock to provide meat flavoring for synthetic meat substitutes, he predicted. Although there are still 80 years to go to see if the good professor was right, we doubt that he was on the mark. Still, his kind of thinking has surfaced in some other forms.

One noted futurist has looked in his crystal ball and concluded that if we fed the grain to humans instead of to livestock, we could perhaps support a population more than three times greater than at present. Some world hunger activists were suggesting a few years back that if each of us would give up one hamburger a week, it would help feed the starving of the world. Still others have suggested that Americans should quit fertilizing their lawns and golf courses and send the fertilizer to the developing nations to help them grow more of their own food supply. These ideas are simplistic and, even if carried out, would have little measurable effect on hunger in the world. Basically, outside of famine caused by natural disasters, there is no actual shortage of food in the world, nor of land or fertilizer. There are more than adequate supplies of land, fertilizer, and food. What is lacking is the purchasing power to pay for the food. What is lacking is effective consumer demand. Wherever the cash is available to pay for the food, the food becomes available.

Politically Prescribed Diets

I rather expect that we will have increasing frustrations ahead with those who wish to try to influence the diets of the American people by political prescription.

For forty years, the Food and Nutrition Board of the National Academy of Sciences has been the widely recognized and respected source of dietary guidance. The Food and Nutrition Board has been the agency that has issued the "Recommended Dietary Allowances" or RDA that have been the basis for most nutritional education aimed at the consuming public. Just two years ago, the Food and Nutrition Board issued a report, entitled "Towards Healthful Diets," in which it advised that the average adult American whose body weight is under reasonable control should feel free to "select a nutritionally adequate diet from the foods available, by consuming each day appropriate servings of dairy products, meats or legumes, vegetables and fruits, and cereal and breads."

In the same report, the Food and Nutrition Board recommended that dietary change or therapy should be undertaken under a physician's guidance. Aware of some of the political headline hunting being done by self-appointed guardians of American diet, the Food and Nutrition Board plainly warned that it is "scientifically unsound to make single, all-inclusive recommendations to the public regarding intakes of energy, protein, fat, cholesterol carbohydrates, fiber, and sodium."

You are aware, of course, of the rash of studies and reports on diet and heart disease, diet and cancer, and other topics that have singled out animal fats and meat as the causes of human health difficulties. There was a surgeon general's report in 1979 and a study by the Senate Select Committee on Nutrition and Human Needs. Last summer, a special panel of the National Academy of Sciences on "Diet, Nutrition, and Cancer," issued a report which differed in important respects from the position of the Food and Nutrition Board and that had not, in fact, been submitted to the Food and Nutrition Board for review and evaluation. Like many of the other political diet studies, the cancer study issued sweeping recommendations that included a 25 percent reduction in the consumption of fats, fatty meats, and dairy products. The report recommended avoidance of smoked sausages and fish, ham, bacon, frankfurters, and bologna.

The cancer and diet panel admitted it did not know what percentage of cancer risks are attributable to diet or how much the risk could be reduced by modifying one's diet. Still, while admitting that there was considerable uncertainty about the scientific basis for its findings, the diet and cancer panel issued its recommendations anyway and, as you might expect, the press treated it in a sensational manner. It was entirely in order for the livestock industry to ask that a review be held to reconcile the contradictory advice that was originating at the same time from the National Academy of Sciences. In our belief, the practice of medicine should neither be carried out by politicians or advertising agencies. The practice of medicine should be left to the medical profession.

Livestock Marketing Revitalization

At a time when all other industry seems to be centralizing, livestock marketing is disintegrating--breaking up into bits and pieces--with no central system for determining price.

As a result, live cattle prices are being based on data reported in the Yellow Sheet or the USDA meat news--sources that may represent as little as 2 percent of the market volume. Farmers have been looking at other options, such as electronic auction markets, to restore some competition into the system. But, it will take time, considerable capital, and major organizational efforts to establish an effective producer-controlled system of that sort. We may need some federal help and encouragement to get the job done.

International Trade in Meat and Meat Products

Another political challenge we may face will be in regard to international trade in meat and meat products. We appreciate the desire of some in the livestock industry to expand foreign markets for U.S. meat and related products. Foreign market development ought to be pushed in any constructive way. There may well be some potential for gains, particularly as some of the developing countries increase their purchasing power and seek to upgrade their diets. However, care must be taken that nothing we do in seeking to expand world trade reacts to undermine our own meat-import control laws. We ought to recognize that if we attack the quotas, the nontariff barriers and protectionist devices of other countries, this will certainly expose our own Meat

Import Act of 1964, as amended by the Meat Import Act of 1979, to attack from abroad. You may recall that when the 1979 Act was adopted, it was criticized by some who look to the U.S. market to dump their oversupplies. While we now export almost one billion dollars' worth of meat and meat products, we import 2.2 billion dollars in meat and meat products, plus another 1.5 billion dollars in animals and animal products.

The expansion of U.S. meat exports will be a gradual, long-term proposition. It will take a long time before exports offset imports, and this will be particularly true if we go to a free market in meat trade. At any time that world meat supplies are excessive in relation to effective demand, the U.S. will tend to be the magnet for oversupplies. That situation will tend to prevail most of the time. So, in a free market situation, U.S. meat imports will tend to expand more rapidly than meat exports.

We have a good law in the 1979 Meat Import Act. It is a responsible measure that helps us retain a domestic livestock and meat industry. The countercyclical factor that determines the allowable level of imports is particularly important. When the U.S. cattle industry is in the liquidation phase and beef production is relatively high, the countercyclical factor will tend to reduce the allowable level of imports. When the cattle cycle is in the rebuilding phase and domestic production is low, the allowable imports will be increased. If we can hold foreign imports to the minimum figure of 1,250 million pounds or near to the figure, the situation will remain in control. But, in a free market situation, it is easy to imagine that without any controls, meat imports, now subject to the law, would quickly advance to new all-time record levels.

Animal Welfare Concerns

Another political challenge livestock producers may have to face would be from the animal welfare lobby. Up to a couple of years ago, few farmers were concerned about the animal welfare lobby--many had not heard of it. Thus far, there has been no serious effort in the Congress to delete the several words from the Animal Welfare Act that would end the exemption of farm animals and birds from that statute. However, there have been bills in the 96th and 97th Congresses that would regulate confinement feeding of animals. Some activist groups have emerged and have gotten some coverage from farm magazines and the media but, so far, it seems they have been open to dialogue with farm and livestock groups. We should be realistic enough to expect that concerns of nonfarmers about this area of farm production will increase but, if we can keep some lines of communication open with responsible citizen groups, perhaps the discussions can be kept on a reasonable basis.

National Farmers Union has had some concerns over the harmful effects of excessive concentration of poultry and animals. To be frank about it, our concerns have been more in terms of environmental, health, and economic effects of such concentration, rather than with humane treatment of livestock and poultry. Up to this point, here in the U.S., we have been able to avoid the confrontations between consumers and farmers that have been common in Western European countries over animal rights. Most of the difficulty arises with people who are almost totally lacking in knowledge about livestock production methods.

By attempting to carry on a dialogue, we may be able to avoid misunderstandings that tend to polarize viewpoints on their side or ours. If we can win some understanding, we may be able to keep the issue out of the political arena.

Sensible and Effective Regulations

In the time that we have had on the program today, it has not been possible to touch upon all the political decisions that will affect our business and livelihood. There is a whole array of potentially serious problems, if the hysteria for deregulation is carried to the extreme. Reducing needless paperwork burden is desirable, of course, but to eliminate needed regulatory measures is something else. The Packers and Stockyards Administration covers a whole range of competitive issues. What we need is sensible and effect regulation, not the elimination of regulation altogether.

Most of my adult life was spent in Oklahoma as a farmer and farm leader. We found at times that there were governmental decisions so bad that we had to fight them with all our might. But, we also found that we could reason with people and that if we got involved early enough and got a voice in shaping the decisions, we could avoid situations that otherwise might have become desperate. The key is to get involved.

Rest assured, the government is involved and will continue to be involved in our agriculture and our society. If we don't take part in the process, the results will be worse. If we curse the government, the government will not go away. We will just be destined to live under scarcely bearable regulations devised by someone who doesn't really understand livestock farming. Fortunately, we do have a choice and a voice, if we want it.

9

AN AGRI-WOMAN'S VIEW OF THE POLITICS OF AGRICULTURE: NATIONAL AND INTERNATIONAL PERSPECTIVES

Ruby Ringsdorf

As a city girl turned farm wife some 27 years ago, learning all about the various aspects of agriculture has been very interesting and educational as well as very rewarding for me. Because so many have left either the farm or rural area and relocated in the city, farm community or farm bloc no longer carries the clout it once did.

When we look at the statistics we see that in 1949 25% of the populations were farm folks--today we constitute about 2.7%. From a significant segment of society that was once courted, soothed, and cared for, we have become a minority of the voting public--our "clout" practically nil. Our collective ability to feed and clothe our nation and significantly affect this nation's economic status of favorable contributions to the balance of payments still exists, but now one producer feeds 78 people--not just a simple household. And therein lies the rub. Production capability and output efficiency do not vote. Contributions to the common good carry little political weight. This has created our current predicament. We, as producers, find ourselves in the unenviable position of providing a product with public utility overtones without even the restrictive protection that public utility status would bestow.

From a public relations standpoint we are viewed as either land-hungry, dollar-greedy, unpatriotic, rich farmers (witness the writing coming from some of our church groups--writings such as "Strangers and Guests in the Heartland"), or as not-so-smart farm folk in overalls or print dresses (note the celery farmer and his wife on the Baggies commercial), or as sheltered, dull, farm children, apprehensive of the outside world (M & Ms make friends and the hayseeds are readily won over with a hand full). We are none of these or all of these. Farm folk run the gamut--as do shop keepers, manufacturers, and other American entrepreneurs. We are business people. We love agriculture and perhaps are willing to sacrifice some "return on investment" in monetary measures for the "return on investment" in satisfaction and the personal pleasure that come from involvement in production agriculture. <u>But</u> agriculture is too expensive to keep as a hobby. We are commercial farmers. "Commercial" is not a dirty word--nor is "profit." The politics of agriculture has economic overtones and very real economic repercussions for the producer.

I would like to touch on a few of the political influences with which we have had to deal over the past few years. I apologize for the negative tone of this presentation, but other than our own attitude and the placement of a free marketer in the office of Secretary of Agriculture, there has been little to be positive about for commercial production agriculture.

Past grain embargoes and trade restrictions have worked to the detriment of the free-market system and have, as a by product, encouraged grain and oilseed production in other countries, in direct competition with the U.S. producer, thus diverting sales arrangements, agreements, and traditions to these countries and away from the U.S. We have become, in many cases, the residual supplier--the supplier of last resort. We have a reputation as an unreliable source of grain, a nation whose past political philosophy has allowed food to be used as a weapon and the farmer as a political pawn in the game of foreign relations. The 1973 embargo sent Japan to South America with the technology and economic necessity to develop a source of soybeans not dependent upon U.S. philosophical and political attitudes. Now, rather than competing with the protein supplied by the anchovies harvest of the coast of Peru, our U.S. soybean farmers must contend with a full-blown, mirror-image harvest in the southern hemisphere. We welcome healthy competition, but resent our own government policies encouraging such, while at the same time inadvertently and directly threatening the economic livelihood of our American producers.

The most recent embargo reinforced this scenario, left us a residual supplier to the U.S.S.R., and brought agripolitical America to its economic knees. The U.S. farmer cannot continue to produce in abundance without the reasonable expectation of an adequate return for the labor and capital involved; prices commensurate with this return are probable only if the world market is available to the producer. Ours is a global society. It makes no sense to subject the producer to the unpredictable risks of the world while confining his free marketplace to the narrow national confines of the U.S. and placing constraints on his participation in the world economy.

The political intrigue surrounding the long-term trade agreement with the U.S.S.R. and the threat to employ trade sanctions against our enemies have kept the farmer in the dark. Any long range capital plans, any cash flow projections are subject not only to the economic ebb and flow of supply and demand but to the implementation of foreign policy philosophies either to further our political ideologies in another country or to achieve some internal goal. our soybean farmers held their breath for possible repercussions when our government talked of imposing import taxes on Japanese automakers.

In 1979 we felt the political influences nipping at our heels as the Secretary of Agriculture Bob Bergland called for hearings on the "Structure of Agriculture." (Our official response as American Agri-Women is included as an Appendix.) We had women testifying at hearings held all over the country. Our basic premise was that agriculture does have some problems but we're basically sound. We were profoundly disturbed as we heard witness after witness talk of agriculture's problems and propose solutions bordering on political and economic systems inconsistent with our free-enterprise democratic society.

Other players in the political game are the food lobbyists. We, as producers, are part of the picture. With representatives of both general farm organizations and commodity groups in our state capitals and in Washington, we pay individuals to serve as our eyes and ears and also to pass along our concerns to the powers in political and economic decision-making.

Often working at cross-purposes with us, and just as often uninformed as to the probable repercussions of actions they propose, are various church groups, hunger organizations, social activists, and do-gooders. A recent example is the animal-rights lobby and their campaign against proven and protective commercial agricultural practices. In recent years we've dealt with land groups who favor land redistribution. There is also a strong school of thought that land belongs to everyone and should not be privately held nor used to make a profit.

We've had farm price-support programs since 1933. Although the concept may have initially been a necessary emergency measure, the fantastic productive capability of the agriculture sector today has been accomplished in spite of government interference via the farm programs, not because of them.

We would rather see the emphasis of government involvement redirected toward programs that will encourage a healthy, contributory agriculture. We see government commodity programs as an alternate source of credit rather than as a guaranteed price. The trend should be toward market development and a broader use of trade barriers and guarantees (supported by action) that the U.S. farmer is a reliable supplier.

Food is the primary need of all mankind. Efficiency in agricultural production has allowed portions of the labor force to enter industry, the arts, government, religion, and other human endeavor while depending on others to provide their daily requirements of food. The U.S. has always held individual and religious liberty to be a right of all its citizens and has chosen to protect this liberty by establishing a democratic republic and by fostering the private ownership of land and other resources. The ownership of this land is openly accessible to all who will work for it. American agriculture is the most efficient in the world, and its productivity has reduced the cost of food to the point that only about 17% of our average disposable income is now spent for this most basic human need. This monumental productivity has relieved 97% of our population of any need to produce food, leaving less than 3% of our people directly employed in farming. This separation of most people from everyday agricultural production has raised a great deal of public questioning of agricultural practices, farmland ownership, labor, and similar issues. These are political issues and affect every individual in this country. They deeply affect farmers and their methods of production and could ultimately affect both the structure of American society and the total supply of food available to the people of the world.

APPENDIX A

AMERICAN AGRI-WOMEN
POSITION ON STRUCTURE OF AGRICULTURE HEARINGS

Whereas, R. Bergland, Secretary of Agriculture has requested input on the Structure of Agriculture in twelve areas, we, American Agri-Women, submit positions on the following subjects:

1. Land ownership, tenancy and control.
   - Land ownership, control and tenancy belong in the private sector.
   - The majority of ownership will be limited to U.S. citizens.
   - Since unlimited leasing is not restricted in any other business, ownership and productive use of agricultural land need not be with the same party.
   - Agricultural landowners need the flexibility to manage their land as their expertise guides them. A residency requirement for agriculture is archaic - just as it would be for any other business.
2. Barriers to entering and leaving farming.
   - Barriers to entering are basically determined by the lending institution.
   - Leasing can be a vital avenue for entry into farming and therefore must be unrestricted.
   - We object to government support of "life-style farmers" (non-commercial) because productive incentive will be lacking.
   - Capital gains taxes and environmental zoning restrictions are barriers to farming.
3. Production efficiency, size of farms, role of technology.
   - Each farmer should be able to determine his farm size. Technology should be uti-

lized to the fullest. Agriculture must not be singled out in use of size technology, and methods of production.

4. Government programs.
   - Much of the success of American agriculture is due to agricultural research. This represents only a small portion of the USDA budget. To feed a hungry world, we vitally need accelerated research programs.
   - Food assistance programs represent 56% of the total USDA budget. A cut in agricultural research, extension, soil and water conservation programs would not be in the national interest.
   - We oppose USDA funding of persons to testify at regulatory hearings.
5. Tax and credit policies.
   - Inheritance tax laws need changing, particularly the carry-over provision.
   - More tax credit incentives for conservation measures are necessary to justify the required capital outlay.
   - No taxes on transfers between spouses.
6. Farm input supply system.
   - Leave in the private sector. The current system works well. Leave it alone.
7. Farm product marketing system.
   - Leave in the private sector. The current American marketing system is the best in the world.
8. Present and future energy supplies.
   Develop multi-faceted energy supplies and aggressively work to decrease our foreign dependence.
   - Agriculture uses only 2.9% of U.S. energy supplies and we needn't apologize to anyone.
9. Environmental concerns, including conservation and the use of soil and water.
   - Farmers were the first environmentalists and conservationists as their income and livelihood depend on it. Let us continue with less regulatory interferences.
10. Returns to farmers.
    - Farmers need to develop collective bargaining. But farmers, like any other independent businessmen, accept the responsibilities and assume the risks.
11. Cost to consumers.
    - The American farmer produces the least expensive, most bountiful, and highest quality food and fiber for the consumer anywhere in the world.

- Farmers must be allowed to pass on their costs as in any other business.

12. Quality of rural life.
    - There is a difference between rural America and farming America. The economic quality of life of agripolitician America is dependent upon agriculture - just as the economic quality of life of metropolitan America is dependent upon the central business district.
    - Quality of life in rural America is as good as the people who live there and make up the rural community. As people from rural areas, we can attest to the fact that it is a good life if you work at it, and are allowed to work at it without government interference.

Whereas, we American Agri-Women, representing 24,000 people, present these proposals with a diversified history of farming success and failures in each of our 50 states; that the aforementioned proposals are based in fact and are the positions American Agri-women takes on the structure of agriculture.

Adopted November 3, 1979 by the governing body of A.A.W. at annual convention, San Diego, California.

10

OREGON WOMEN FOR AGRICULTURE TALK ABOUT THE STRUCTURE OF AGRICULTURE

Ruby Ringsdorf

Let it be known that Oregon Women for Agriculture do believe that it is the inherent right of every child born today to have adequate nourishment; and that the American farmer will continue to feed the hungry if not strangled with bureaucratic rules and regulations.

We furthermore feel that it is neither our duty, nor even our right, to enter into the internal policies of a foreign country whose political system, or local corruption, are preventing food from reaching their hungry masses. Neither are we prepared to let those countries' Marxist-oriented political ideologies creep into and destroy our free enterprise system. The free enterprise system is the propelling factor that has made American agriculture the envy of the world!

LAND OWNERSHIP, CONTROL, AND TENANCY

Fifty-five percent of the land in the state of Oregon is already publicly owned (L.C.D.C.). The number of all commercial farms (farms with sales of $2,500.00 or more) in Oregon increased rather than decreased from 1969 to 1974.

The number of commercial farms in Oregon with sales greater than $40,000.00 increased 67% from 1969 to 1974 and comprised 30% of all of Oregon's commercial farms. At the rate of inflation over the past 10 years it is surprising that this percentage is not greater. It doesn't take much of a farm to produce $40,000.00 in sales today, but the net probably isn't enough to keep the family dog in dog food for the year.

According to an Oregon State study (EM N:23), family farms, nonincorporated, comprise 96.4% of all commercial farms with 3.6% being corporate farms. Of the corporate farms, 87.3% are family farms (94% have 10 or fewer shareholders, 44% are controlled by one stockholder).

Many family farmers own some land and lease more from retired farm relatives or neighbors in order to make their units more economical.

BARRIERS TO ENTERING AND LEAVING FARMING

Inflation, high interest rates, inheritance tax laws, FHA regulations, EPA regulations, and our national cheap food policy are all barriers to entering and leaving farming. Inflation is not only driving up the cost of land to an impossible price for a beginning farmer (the interest alone for each acre of ground is much more than what he would have to pay to lease the ground), but also the cost of the equipment needed to start farming. Inheritance tax laws, especially the carry-over provision, make it almost impossible for children to inherit a farm without selling a portion of it or splitting it up to meet the tax obligation so that they no longer have an economical unit.

EPA regulations are becoming increasingly more difficult to cope with and discourage many young people from even thinking of farming. FHA regulations restrict the amount of money available to a young farmer for a small acreage because it is not an efficient, economical unit that could produce enough net income to service debt and provide a decent living for his family without off-the-farm income. And yet, if the unit is large enough to do both, the cost is far more than FHA is allowed to cover for one farm.

It is difficult for a farmer to retire and leave farming. The land cannot be sold because of strict zoning laws, because of capital gains tax on the appreciated land value (often the only net savings realized from the farmer's investment in time and labor), and because his acreage is no longer large enough to be an economic unit for a family farmer. He stays on the land, rents to others, and does the best he can, too often becoming another rural-poor statistic.

PRODUCTION EFFICIENCY AND SIZE OF FARMS

Production efficiency and size of farm are tied together. Our Oregon State study shows that the average size of all commercial farms in Oregon increased very little from 1969 to 1974. It would seem the trend to larger farms has already peaked because of production efficiency.

Size and number of farms in Oregon vary greatly from one geographical area to another. In the Lake Labish area near Salem, where land sells for $10,000.00 per acre and up, a 20 A farm is considered large. In the southern end of the Willamette Valley, in the grass seed capitol of the world, a thousand acre farm is not considered large, and in the cattle grazing lands of eastern Oregon a 5,000 A ranch is not large. A dairy farmer can have only 100 A but milk 500 cows and be considered a large farmer, but another dairy farmer can have 50 cows on 500 A and be considered a small farmer.

Because Oregon produces over 170 different marketable commodities, it is unfair to use a gross dollar amount figure for sales to define large and small. Different crops show different net results. It is impossible for a 20 A berry farm to net more than a 500 A wheat ranch. Most often a farmer cannot convert his acreage to higher value crops because of soil types, marketing limitations, increased risks, and increased operating and capital requirements.

The efficiency of the American farmer is the envy of the world. After American farmers feed the U.S. they export 60% of their wheat and rice, 50% of the soybeans, one-fourth of their grain sorghum and one-fifth of their corn. The U.S. provides half of the world's wheat (Oregon Grange Bulletin 9-4-78).

Agricultural products are the second largest category of U.S. exports. Agricultural exports returned $23 billion to our country in 1976. In 1975 agricultural exports provided the foreign exchange to cover 83% of our petroleum imports.

American farmers provide all this despite the fact that the number of U.S. farms and farm workers has decreased by two-thirds since 1940.

One American farmer can now produce enough to feed 60 people. From 1950 to 1978, farm productivity increased at an annual rate of 5.3%--more than twice as much productivity as compared to any other nonagricultural business (Oregon Grange Bulletin 9-4-78).

GOVERNMENT PROGRAMS

Much of the success of American agriculture is due to agricultural research. This represents only a small portion of the USDA budget. Food assistance programs, including Food Stamps and Child Nutrition (programs benefiting from past research programs), represent 56% of the total USDA budget. A cut in agricultural research, extension, and soil and water conservation programs would not be in the national interest.

As to having a national or world grain reserve, why not establish a worldwide monetary food fund (a required UN fund, a contribution fund with participation by churches and other interested groups, internationally funded and administered) to be used for international food crises. Reserves have a history of depressing prices to producers and stabilizing prices at the lower levels. Reserves have also acted as a disincentive to production so that farmers change (if possible) to producing crops that will hopefully yield more net return. Any grains in reserve should be isolated from world markets and used as aid rather than trade. The cost of this grain reserve should be shared by all people internationally.

Another program that is under consideration for USDA funding is the plan to pay the expenses of low-income and nonprofit groups that testify at regulatory hearings. We are very much opposed to such a plan; even though Women for Agriculture would qualify for funding.

TAX AND CREDIT POLICIES

Inheritance tax laws need changes, particularly the carry-over provision. Prior to enactment of the carry-over provision, beneficiaries inheriting appreciated property received a stepped-up tax basis on property at the time of inheritance, and each generation of a farm family was subject to capital gains tax only on the appreciation that occurred while they owned the property. This procedure was radically changed by the carry-over provision that bases capital gains on inherited property on the descedent's acquisition price and not the market value at the time of transfer.

There should be more tax credit incentives given for conservation practices, since these can be very costly for one individual.

FARM INPUT SUPPLY SYSTEM

Government regulations and inflation have had a strong influence on the farm-input-supply system. Labor costs have spiraled, inventory taxes have prevented smaller manufacturers and suppliers from keeping a full inventory; land costs have spiraled and it is becoming more difficult to obtain necessary capital.

FARM PRODUCT MARKETING SYSTEM

There is a lot of talk about direct marketing from farmer to consumer. This works only in agricultural areas. Most crops sold in this manner are perishable, thus limiting choice, variety, and quality. It takes much more time and energy to drive all over to pick up vegetables here, eggs there, milk at that place, and fruit at still another stop. We already have the most energy-efficient distribution system.

Farmers are being told, "You can get a better price for your produce than the processor gives and the consumer can get it for less. Let's cut out the middle-man." Just who is the middle man? If we cut out all the middle men in the food-processing chain, our unemployment rate would probably be closer to 30% or 40% than the 8% or 9% it is now. Also, at the same time we are hearing rumblings about vertical integration (selling your own produce as a finished product and cutting out all middle men). This is what multinational corporations are accused of doing. Direct marketing is the same.

Probably close to 100% of Oregon's fruits (not including tree fruits) and vegetables are sold by forward contracting or contractual arrangements with a processor. This method is preferred over freedom of decision-making at harvest time. There are not many farmers who would care to wait until harvest time to search for a market for their perishable products. They would be at the mercy of the processor who would then know the farmers have no choice but to take whatever price is offered. If an equitable price is not offered at planting time, the farmer has the freedom of decision to plant or not to plant.

Contractual arrangements and forward contracting on seed and grain crops isn't bad either. It can certainly provide some freedom from fear. The farmer who stays in business all his life usually is the one who contracts ahead whenever he feels the price is such that he can make a fair return. It takes some of the gamble and risk out of farming.

PRESENT AND FUTURE ENERGY SUPPLIES

In the early 1920s we had 25 million horses to pull the plow, the wagon, and the carriage. We fed about one-fifth of our grain and roughage to those horses. (Today it would take one-third of our crop land plus 20 years to breed enough horses and mules for today's needs.)

It is time we look into using biomass or agricultural products as a future energy source. The liquid energy that we import is priced at $1.50 to $2.50 per gallon in most major industrial nations. We are nearing these world prices now, which will make the production of biomass fuel profitable. There are tons and tons of grass straw in Oregon alone that can be used for fuel pellets or biomass conversion.

We feel it is unfair for American farmers to be told to conserve fuel and energy when farming uses only 3% of the total energy consumed in this country. It takes more energy in the home for food preparation than it does for agricultural production (including fertilizer and other energy-intensive inputs).

Productivity per man-hour in agriculture has been increasing about twice as fast as the rate of productivity per man-hour in manufacturing.

ENVIRONMENTAL CONCERNS, INCLUDING CONSERVATION AND THE USE OF SOIL AND WATER

Oregon farmers are also environmentally concerned. We are also concerned because we who are engaged in agriculture are such a minority. Even though Department of Environmental Quality tests gave proof to the fact that smoke from field burning, a practice used in the Willamette Valley to sanitize our seed crop grass fields, really had little or no effect on the quality of the air in the Valley, the EPA would not allow easing of regulations that are now strangling the Grass Seed Industry in Oregon. Why? Because all other sources of pollution are increasing yearly!

Bureaucratic rules and regulations often conflict with each other. In the Silverton Hills area, for example, 25 to 30 years ago farmers were losing tons of top soil every winter as a result of water run-off on cultivated fields. They discovered that their soil and climate was suitable for perennial grass seed production, which held the soil on the hillsides. Now EPA has restricted field burning because of air quality. Without burning, the grass fields become diseased and seed production is no longer economically feasible. We are again faced with soil erosion and resulting probability of water pollution.

Oregon has been working on Water Quality Management Programs, or non-point-source pollution. There are six major agricultural pollutants: sediments, nutrients, salts, organics, pesticides (including herbicides) and disease-producing organisms. So far only sediments have been found in our streams and these have been coming primarily from nonagricultural sources.

America's agricultural engineers are coming up with new and better seed drills that utilize low-till and no-till methods. Oregon farmers have been using grassland drills for many years, but we cannot use a grassland drill in a grass field that has not been burned. Here again, air quality versus water quality.

RETURNS TO FARMERS

Because our government has long endorsed the cheap-food policy, returns to farmers for the last 35 years have not kept up with the rest of the economy.

The American farmer was forced to become more and more efficient and only those who were efficient, excellent managers stayed in business. (Often even the most efficient were wiped out because of extraordinary conditions, such as weather. The margin in good years was so slender that a poor year wiped them out.)

Naturally, when the returns per acre became less and less, we had to expand our acres if we were going to live off the land. Now, suddenly, we are all called "commercial farmers" and that seems to be the wrong kind of farmer to be. "Commercial farmer" seems to be a dirty word in many circles. At the Rural America Conference in Washington, D.C., in June of this year, commercial farmers were being blamed even for dope addiction because farm mechanization caused these people to be out of work.

The people in Rural America meetings defined the family farm as a unit that produces food only for the people on the farm unit and who do not sell any farm product for profit.

In 1945 our fathers sold rye grass seed for 12 cents a lb. (They also sold wheat for $100 a ton or $3.33 per bu.) A tractor at that time cost around $3,000 and the first self-propelled combines came out for around $5,000. Today a tractor costs between $30,000 and $60,000; a combine from $65,000 to $70,000. Guess what the price of rye grass seed is today? Twelve and one-half cents per lb. Last year it was only 10 cents per lb so there are big headlines in the newspapers that agricultural wholesale prices increased almost 25% over a year ago! It is usually buried in the back pages when our farm product prices decrease or simply stay at the level of 30 years ago.

COSTS TO CONSUMERS

We as farmers are always hearing from our city friends about farm subsidies and how we are being paid to keep land out of production, etc. The truth is, the American farmer has been subsidizing the consumer for the last 40 years.

The trend toward greater efficiency in farming has benefited the consumer most. People have never had a greater variety of safe and nutritious food at so low a cost. The consumer can purchase more food for his hour of labor than at any time in history or in any other country. When the housewife complains about high food prices, she is often paying for a maid-in-a-box by buying prepared and semiprepared foods that cut down on food preparation time at home. It all depends on your priorities--time versus cost.

We also feel that if it is a policy to serve the public interest, by drastic and disruptive actions (such as export controls and import floods), then the general public, not just the producer, should pay or help pay for this policy.

QUALITY OF LIFE IN RURAL AREAS

This is interesting! Rural areas, not agricultural areas! Rural America: Educational Needs of Rural Women--these programs are not referring to farmers or farm women. Rural America means the urban population that has moved to the less-populated areas of the U.S. from all socioeconomic levels and who most likely have never farmed and never will. The problems of these people should come under the jurisdiction of H.E.W., not USDA.

We farmers have noticed that this migration to the rural areas is causing problems. They tell us, "We are moving out to the country because we want peace and quiet and we want to be one with nature." And then they complain about the noise and dust from farm machinery and activities; so now we have noise pollution and dust-control laws. Now rural residents are suing because of smells coming from swine production and cattle feeders, forcing farmers to put in costly equipment to take care of the smell.

The environmentalists insist we use too much commercial fertilizer, but they complain about the smell when the local dairy farmer puts the liquid manure on his fields with an irrigation gun. There are over 100 known toxic substances occurring "naturally" in the environment, yet they complain any time they see our spray rigs come out of the yard.

Rural America says there are 131 rural towns in the U.S. without a doctor, that many people have to drive from one to two hours to a medical center. This is not always bad. We would rather drive for two hours to a good medical center with full facilities than fifteen minutes to a small facility where sometimes a local doctor tries to do only what a very specialized doctor should do.

Rural America also says one-half of the maternal deaths occur in rural areas. We believe that. Much is due to the back-to-nature trend, which is currently popular along with do-it-yourself childbirth. This is fine if everything proceeds normally. In the home we do not have the back-up facilities to aid a difficult delivery; it is too late then to rush to the hospital. Many of these same people have no prenatal care of any kind and their diet is often very inadequate because of their chosen life style and eating habits.

If we talk about the quality of life of real farm and rural people, we are talking about something entirely different. Genuine farmers seem to have fewer divorces per thousand population and they generally have a very strong family unit because everyone learns to work together.

A farmer has a lot of respect for his Creator; he is too closely involved with growing and living things to think that we are here purely by chance.

Generally speaking, the quality of life depends on the individual involved. The socialite who grew up in the large city might be quite disenchanted and bored living in a small rural community. She might complain bitterly about the lack of culture! On the other hand, the people who grew up in that same small area are quite content and feel that they have the good life. If they were to move to the large city they might then cry bitterly about the hustle and bustle, the unfriendliness, the foul air, the crime rate, etc., etc.

The kind of life we want in America can be found by any one who intends to earn it. If we expect it to be given to us, we will never find it.

<u>Oregon Women for Agriculture</u>

11

THE CURRENT STATUS OF THE FAMILY FARM IN AMERICAN AGRICULTURE

George Stone

Almost everyone knows what a family farm is, but hardly anyone is able to define it on paper to the satisfaction of other people. A particular farm may meet most of the criteria that might be suggested, but there will always be some differences of opinion on such things as size, ownership, and control. There is not time for a debate on the fine points of a family-farm definition. But if I am to talk to you today on family-farm agriculture, you are entitled at least to know what I think I am talking about.

FAMILY FARM DEFINED

I like the National Farmers Union's definition of a family farm. It says:

> "A 'family farm' is, ideally, one which is owned and operated by a farm family, with the family providing most of the labor needed for the farming operation, assuming the economic risk, making most of the management decisions, and depending primarily on farming for a living."

That is probably as well as it can be explained in less than fifty words.

PUBLIC POLICY RELATED TO FAMILY FARMS

Our national public policies have endorsed and advocated a family-farm structure of agriculture for almost 200 years, dating back to the Ordinance of 1785, the Land Act of 1796, the Pre-emption Act of 1841, the Homestead Act of 1862, the Reclamation Act of 1902, and a half dozen major statutes in this century as recently as the Food and Agriculture Acts of 1977 and 1981.

The 1977 Act includes a declaration by the Congress that it "firmly believes that the maintenance of the family-farm system of agriculture is essential to the social well-being of the nation"...and that "any significant expansion of nonfamily owned large-scale corporate farm enterprises will be detrimental to the national welfare." The 1977 Act also mandated that the Department of Agriculture should issue an annual report on the "Status of the Family Farm." This has been done, supplying some continuing data on the structural trends and changes in agriculture.

In March 1979, speaking at the national convention of the Farmers Union, Secretary of Agriculture Bergland called for a national dialogue on the structure of agriculture, declaring:

> "We are at a point in our history where a broad-based public discussion of the issues that shape national policies is needed to promote the kind of agriculture and rural living this nation wants for the future."

In that Kansas City speech, Secretary Bergland observed that "we really don't now have a workable policy on the structure of agriculture," and warned:

> "We can act now to insure the kind of American agriculture we want in the years ahead. Or we can let matters take their course, with the probable result that we will wake up some morning to find that we have forfeited our last chance to save those characteristics of the farm sector we believe are worth preserving. I, for one, do not want to see an America where a handful of giant operators own, manage, and control the entire food production system. Yet that is where we are headed, if we don't act now."

In late 1979 and early 1980, Secretary Bergland conducted this national dialogue at a series of regional hearings. Numerous economic papers and a comprehensive report were eventually published. It was a worthwhile exercise in stimulating Americans to think about what they want in an agricultural system and what they want their federal government to do to assure such a system. But, while there may be better public understanding of our agricultural system, there is little in the way of agricultural legislation or administrative decision-making that can be attributed to the Bergland study.

ASSESSING CURRENT FARMING CRISIS

We find ourselves here, early in 1983, still trying to assess the current situation of family farms. Through our 80-year history, National Farmers Union has been totally dedicated to the family-farm system. We believe that the family farm represents the best choice for the American people on every score:

- Assured abundance
- Efficient production
- Best care and use of land and water resources
- Rural employment
- Quality of life in rural communities

- Highest export earnings
- Most favorable balance of trade

During the 80 years of Farmers Union, the family farm has proven its durability and staying power. Family farms have survived wars, natural disasters, and a total of 14 recessions, panics, and depressions, including the most recent. During most of this century, it was usually assumed that the family farm would survive as the dominant form of agricultural structure. Now, although it is readily acknowledged that the family farm is the most efficient agricultural production unit, it is no longer that certain that it will survive much longer.

In a spirit of candor, one must admit that there have been other times when the survival of the family farm appeared to be in doubt. Calamity seemed at hand. Yet, while some farm operators were lost in the crises of the past, and their loss was regrettable, the system as a whole survived and continued to produce for the nation.

Having expressed the caution that at times in the past things have appeared worse than they turned out to be, there are signs that the current challenge to the survival of family farms is the most dangerous, at least since the years of the Great Depression.

Not since the early 1930s has the nation had three such bad years in agriculture in succession. Many signs indicate the magnitude of the farm crisis. Net farm income dropped from $32 billion in 1979 to $19.8 billion in 1980, $18.9 billion in 1981, and while we do not yet have final figures, appears destined to be still lower in 1982. In terms of purchasing power, the farm parity ratio in much of 1981 and into 1982 has been the most unfavorable suffered by farmers since 1933. In 1933, U.S. farmers had $3 billion in net income, but only $9.1 billion debt. That was a ratio of $3.10 in debt for each dollar of net income. Today, we have something over $11.00 in debt for each dollar of net income.

In 1981, for the first time in recorded history, U.S. farmers paid out a total of $19 billion in interest outlays on their debt, a sum that exceeded their net income for the year. At the worst of the Depression of the 1930s, the interest rates paid by farmers averaged 6.4%, while recently the rates paid by farmers on loans to commercial banks averaged over 18%, as reported in the Federal Reserve Bulletin.

In these last three years, farmers have been substituting credit for income at an alarming rate. Years ago, farmers were able to generate much of their capital needs internally. In 1970, for example, farmers depended on borrowed capital for only 5% of their cash operating funds. By 1980, the proportion was up to 21% and, in 1981, it was almost 23%.

Another important measurement is the liquidity ratio of farmers. In 1950, as an example, U.S. farmers had $13.8 billion in cash assets such as deposits, currency, and savings bonds. Against this, they had $12.4 billion in debts. That was a liquidity ratio of 111%. In 1960, there were cash assets of $13.9 billion and $24.8 billion in debts, a liquidity ratio of 56%. By 1970, there were $15.6 billion in cash assets and $53 billion in debts, a liquidity ratio of 29%. In 1981, the cash assets totalled $19.9 billion against a total debt of $194.5 billion, a liquidity ratio of about 10%. Behind all those statistics are human families trying to earn a living in a productive and useful endeavor.

FACT-FINDING HEARINGS

To document the human side of the farm crisis, the National Farmers Union held a series of nine regional fact-finding hearings in March and April of last year. We heard testimony from 230 witnesses, including farmers, farm wives, main street businessmen, cooperative officials, teachers, bankers, and community leaders. The summary report, which we published on these hearings, is entitled "Depression in Rural America." It did not deal just in generalities or endless statistics, but told the personal story of families beset with hardship and despair because of conditions over which they had no control. The report showed how the desperate economic conditions were affecting the lives and survival of working farmers, their business communities, and the fabric of life in rural America.

The purpose of the hearings and the report was, of course, to mobilize opinion and develop a sense of urgency about farm legislation that would help family-scale farmers survive.

Of course, there are some who say that the federal government should not intervene on behalf of family farmers --that we should just let nature take its course. The theory is, if we just let the decline in family farms continue, then, after a while, just the efficient farmers will remain and they will be able to prosper.

But those who have been involved in agriculture for a lifetime have yet to see such a scenario work. In 1960, for example, there were about four million farming units and they were earning farm income and purchasing power equal to 80% of parity. In the decade of the 1960s, the nation lost one million farmers, but farm income did not go up. We have been as low as 57% and 58% of parity. At this rate, it might be asked, how long will it take to get to 100% of parity?

The truth is you won't reach some sort of ideal economic situation for farmers by that route. The truth is that the economic hardship is not weeding out small, marginal, or innefficient farmers. The farmers who have been hurt most in these past three years of low farm prices and high interest rates have been the good, efficient operators in the middle of the scale in farm size.

Further, the projections are that this kind of attrition of our best farmers will continue. In 1980, USDA economists did a projection of what will happen to farm size and structure by the year 2000. The report projected that the number of farming units would drop by 30% to 1.8 million by the year 2000, with most of the decline in middle-sized farms. Small farms, with less than $20,000 annual gross, will still make up 50% of the total farming units, with large farms, with $100,000 or more in gross sales, edging out the middle-sized operators.

Along with these structural changes, the USDA officials foresee an increase in concentration in both farmland ownership and production. The USDA specialists project that it will take $2 million in capital assets to run a farm capable of grossing $100,000, and that these large capital requirements will tend to concentrate farm wealth in the hands of a relatively few.

Young beginning farmers will have increasing difficulty entering the industry. USDA projects that there will be fewer than 300,000 farmers under the age of 35 years in the year 2000, a drop of 200,000 from the current level. The number of individual ownerships and partnerships in farming will decline by the year 2000, while the number of corporate farms and multiownership units will increase, the report indicates.

In another report associated with the farm structure dialogue, entitled "Another Revolution in U.S. Farming," USDA economists predict that there will be further declines in the number of farms, but not at as sharp a rate as in the 1950s and 1960s. However, there will be increasing concentration of production among the largest producers, along with strong pressures for the separation of ownership and use of farming resources. Because of taxes and other factors, off-farm investors can get higher overall returns by investing in farmland than they can by investing in common stocks of business and industry.

In the Farmers Union, we view the separation of land ownership and farming operations with a great deal of concern. It seems to us that such a trend will have the tendency to create a new generation of sharecroppers--people who have little control over their own destiny. That is why Farmers Union in the past several years has taken the leadership in seeking to limit absentee ownership of agricultural land. The threat has come from three different sources:

- Investments in U.S. farmland by American business corporations, conglomerates, and off-farm investors. This has included efforts by individual business firms and such spectacular schemes as the Ag-Land Trust Company.
- Investments by foreign corporations and investors.
- Proposals to invest pension fund assets in U.S. farm cropland.

In regard to the threat of domestic corporations and investors to take over farmland, several midwest states in the past several years have enacted limitations on corporate ownership of farmland and corporate farming. Most of the states in the Mississippi Valley now have restrictions of some sort on corporate farm ownership. Largely because of the vigorous campaign by the Farmers Union and the opposition raised at a Congressional hearing in Washington, D.C., the Ag-Land Trust proposal was dropped.

Because American farmland had become a magnet for foreign corporations and investors, the Farmers Union successfully won adoption of the Agricultural Foreign Investment Disclosure Act of 1978, under which foreign persons or foreign-controlled firms acquiring U.S. farmland must report such holdings to the USDA. Although there may be some evasion of the disclosure law, we now have some hard data on the extent of foreign holdings. It is now clear that there has been more foreign investment in American farmland in the past five years than in the previous fifty years. The latest annual disclosure report by USDA shows that almost 5 million acres of land were acquired by foreign persons during 1981. The foreign acquisitions were equal to about 25% of the total of 18.1 million acres of farmland sold during the year.

The third proposal, that of the American Agricultural Investment Management Corporation of Chicago, proposed to facilitate the investment of nonprofit pension funds in U.S. farmland. We became concerned because pension funds represent a huge pool of capital earning very modest returns on the order of 3% to 4% a year. We thought the opportunity to invest and take advantage of the rapid appreciation of farmland values would be irresistible, even if the profit from farming were modest. As a matter of fact, pension fund assets now total about $700 billion and are expected to rise to $1.5 trillion by the year 1990. Obviously, there would be enough capital to buy all of the farmland in the nation.

We don't expect that to happen. But even if only 3% or 4% of the pension funds were invested in farmland, that would total $18 to 24 billion a year--about the total of farmland sales values in recent years.

The ability of beginning farmers--or existing farmers seeking to expand their operations--to bid for land would certainly suffer by the presence of institutional investors who would not have to pay for the land from their agricultural earnings. Young farm couples who hope to acquire a viable farming unit would be virtually fenced out of the competition for farmland by vast amounts of absentee capital. Of course, the promoters of the pension fund scheme claim they would be doing farmers a huge service by relieving them of the necessity to own farmland. Such a separation of land ownership and farming operations would enable farmers to use all of their limited capital in production.

This is a phony argument. The operating farmer pays a land cost whether he owns or rents. He pays land costs whether he is a cash renter or a share renter. We take this attitude in the Farmers Union because we believe that public policy, whether federal or state, ought to be helping families become owners of the land they farm, not separating them from that possibility.

In conclusion, we regard our efforts to keep farmland in the hands of operating farm families as very important. But, it should be pointed out, the challenge of these outside forces is most damaging because of the weak economic position of our farmers. If farm prices and income were maintained at a more satisfactory and stable level, farmers would be able to withstand more easily the competition of outside investors. Low farm prices and income, accompanied by high interest rates, compound the problems of farmers in sustaining themselves in land ownership and farming. Because this is true, we cannot simply go on as we are and let nature take its course. We must act positively on farm income and other measures to assure that we continue to have a predominantly family-owned, family-operated farming system in our nation.

12

LIVESTOCK PRODUCTION ON NEW ENGLAND FAMILY FARMS

Donald M. Kinsman

INTRODUCTION

The Northeast region of the United States encompasses the 12 states of Delaware, Maryland, New Jersey, New York, Pennsylvania, West Virginia, Connecticut, Maine, Massachusetts, New Hampshire, Rhode Island, and Vermont; with the last 6 named constituting the New England states. Table 1 cites the basic facts and figures for the total Northeast region, which contains 128 million acres of land or 5.6% of the land area of the U.S. The 6 New England states represent 40 million acres (about one third of the Northeast) or 1.7% of the U.S. land area, yet the Northeast contains 21.6% of the nation's population or about 49 million people. New England has 12.3 million people or 5.4% of the U.S. total. The average annual precipitation is 40 to 46 in., and the mean temperature variation is from 20°F to 40°F in January to 70°F to 80°F in July. Temperature and snowfall vary considerably with elevation, which is dominated by the Appalachian Highlands. The Atlantic Ocean serves as a moderating influence along the coast. The frost-free period ranges from 90 to 150 days.

Sixty-four percent (64%) of the area is forested, compared with a U.S. average of 32%. The Northeast contains 11.4% of the nation's forest lands; New England represents 4.5%. The land suitable for agricultural production is primarily gray and brown podzolic soils, and agriculture on these lands is intensive.

Being the most highly urbanized region in the U.S., over 12% of the land is city, urban, and industrial compared to 9% U.S. average. Grassland pasture represents 3.2% of this area versus 26% for the U.S., and 15% is crop land as compared to 21% for the nation.

LIVESTOCK PRODUCTION

Against such a background, one might wonder about the livestock potential for this Northeast region of which New England is a microcosm. Approximately 6.1% of the U.S.

TABLE 1. COMPARATIVE AGRICULTURAL DATA* FOR NEW ENGLAND, THE NORTHEAST, AND THE U.S.A.

	New England		Northeast		
		% of U.S.		% of U.S.	U.S.A.
States	6	-	12	-	50
Land area (million acres)	40	1.7	128	5.6	2,264
Population (million)	12.3	5.4	49	21.6	227
Number of farms (thousand)	26.4	1.1	179	7.7	2,333
Average farm size (acres)	171	-	183	-	450
Cropland (million acres)	2.2	0.5	19	4.2	456
Forestland (million acres)	32	4.5	82	11.4	718
Livestock numbers (thousand):					
Cattle (beef & dairy)	753	0.6	5,750	5.0	115,013
Sheep	42	0.3	450	3.6	12,492
Swine	106	0.2	1,687	2.6	64,520
Agricultural cash receipts ($ billion):					
All commodities	$1.4	1.1	$8.0	6.1	$131
Livestock products	$0.98	1.4	$5.4	7.8	$69
Dairy products	$0.55	3.7	$3.1	20.7	$15
Cattle & calves	$0.08	0.2	$0.7	2.0	$35

*Data selected from USDA reports.

agricultural commodity value is produced in the Northeast. The major portion of the livestock receipts is from dairy products--3.1 billion dollars. Beef cattle and calf receipts account for $700 million annually or approximately 2.0% of the U.S. total. All cattle in the Northeast represent 5.0% of the U.S. total, sheep 3.6%, and hogs 2.6%; for New England alone, those percentages are an infinitesimal 0.6%, 0.3%, and 0.2%, respectively. Therefore, it goes without saying, the Northeast is a meat deficit area, and New England in particular produces but 4% of its meat consumption, thus it imports 96% of its meat supply, chiefly from other sections of the nation. Livestock production, exclusive of dairy, in New England must be considered as consisting primarily of small-livestock farm operations. Using the USDA definition of a small farm ($20,000 or less gross sales), 42% of all farms in New England are considered small. Most producers are part-time farmers. These part-timers are on family farms and do contribute to their owners' well-being in addition to producing meat for a ready market.

CAN NEW ENGLAND LIVESTOCK FARMS CONTINUE?

Although dairy farms in New England do continue to decrease in numbers and dairy cow numbers diminish accordingly, the milk production per cow and per herd or farm continues to climb. This, in turn, maintains the milk supply, perhaps at too high a level, but also makes available land, facilities, and expertise for other agricultural pursuits. To protect the better agricultural land, several northeastern states, namely Connecticut, Massachusetts, and New Jersey, have instituted Farm Land Preservation Acts that set aside for perpetuity the best agricultural lands to remain forever available for food production. This needs to be done nationally before any more of our precious, highly productive agricultural land comes under control of the developers and their paved jungles. We in New England have felt the pressure first and gladly share our experience with all to maintain a viable agriculture. Our aim is to achieve greater self-sufficiency in producing more of our food requirements as we recognize the danger of possible isolation in the paths of energy crises, weather, transportation strikes or failures, and our climate restrictions. With the courage, fortitude, industry, and imagination of our early forefathers, we are moving toward narrowing the gap between dependency and self-sufficiency.

Multiple land use is important to livestock production in New England. In the one instance, where our forestlands account for twice the proportion of total acres compared with the U.S. average, we have a large potential to utilize forages within these forestlands for grazing. Secondly, with the decrease in dairy numbers, there is a tendency to replace them with beef cattle, or sheep, or even hog operations--thus utilizing existing land, facilities, and labor for a combined or replacement operation. Additionally, the availability of inexpensive by-product feeds and the use of unconventional feedstuffs encourages these livestock operations. Because the ruminant especially can be maintained on lower quality or by-product feeds, New England as a natural, cool-weather grass country can produce forage-fed beef and lamb with a minimum of purchased feed through wise grazing and forage harvesting management.

Furthermore, the markets are prevalent in New England, with its population density (albeit concentrated in large cities along the sea-coast), thus allowing rural production within easy access to the consumer. This proximity also permits direct marketing from producer to packer or to consumer. This proximity also permits direct marketing from producer to packer or to consumer, development of a freezer trade, and utilization of farmers' markets as well as the existing auction markets. New England does lack a major terminal market within its confines. Livestock marketing pools are becoming more prevalent. Specialty marketing, catering to the natural or organic food interests, also is practiced to some degree. Some cater to specific ethnic demands, often of a seasonal nature, such as Easter lambs.

THE TIME IS NOW

In general, New England does have much in its favor for the production and marketing of livestock for meat purposes. New England farmers historically have been excellent livestock men. The first U.S. meat packer was Captain John Pynchon (established in 1645) in Springfield, Massachusetts. The Brighton (Massachusetts) Stock Yards were developed to feed George Washington's Continental Army in 1775--the nation's first and oldest terminal market. "Uncle Sam" Wilson followed suit supplying the U.S. troops with meat during the War of 1812. The first agricultural and livestock show or fair in the U.S. was held in Pittsfield, Massachusetts, in 1810. Many of the early imports of livestock from Europe funneled into the New England states, and in time this breeding stock of all species was disseminated west. The first major U.S. importation of Merino sheep was to Weathersfield Bow, Vermont, in the early 1800s by Hon. William Jarvis, then U.S. Consul to Lisbon, Portugal. By 1865, there were 1.5 million sheep in Vermont alone. These sheep were the foundation of the great Merino flocks of Ohio and now Texas. In 1875, Herefords from the Bodwell and Burleigh herd of Vassalboro, Maine, sold to the Hon. William F. Cody of Scout's Rest Ranch, North Platte, Nebraska, and to other prominent breeders of that day.

Although there are some large-acreage livestock farms in New England, most are small, family farms. The average farm size in New England is 171 acres compared to a U.S. average of 450 acres. Some of these are registered, purebred breeders supplying breeding stock to the area and throughout the U.S. A limited few even sell breeding stock or semen internationally. New England has long been a seedstock producing area, and its livestock compete very successfully in the show and sales rings of the nation's major expositions and sales.

The New England farmer has often been faced with the quandary of how to make a living under ofttimes less-than-desirable conditions. Frequently he has survived by living on "not what he earned, but what he did not spend." Through the vagaries of climate, weather, topography, land capabilities, and pressures, the New England livestock producer has developed a unique capability in growing, managing, harvesting, and preserving forage in the form of grass or legume hay, haylage or silage, and corn silage where possible. The producer has realized maximum TDN per acre through wide pasture management and has obtained maximum livestock production on his precious land. New England has been forced to take the lead in forage production of meat and dairy animals as its ability to raise grain and protein supplement has been very limited to practically nonexistent. This has perhaps been the salvation of the New England livestock producer and especially the family farm where homegrown labor and homegrown feed have been the major resources for survival.

Some New England farmers combine livestock operations with other agricultural pursuits such as:

Major Enterprise	Supplementary Enterprise
1. Cucumbers (pickles)	Hogs
2. Dairy	Sheep, feed cattle
3. Forestry, firewood	Beef cattle, sheep
4. Hogs	Sheep
5. Landscape and bedding plants	Beef cattle
6. Maple products (syrup, sugar)	Sheep, beef cattle
7. Orchard	Sheep, beef cattle
8. Poultry	Beef cattle, sheep, hogs, veal
9. Tobacco	Beef cattle
10. Vegetable gardening	Hogs, sheep, cattle
11. Vineyard	Beef cattle

Generally, these are family farm operations that have diversified to utilize surplus feed, labor, facilities, or alternatives that best fit the existing situation and provide additional homegrown products for family use as well.

NEW ENGLAND, WHAT'S AHEAD?

With the advent or resurgency to greater self-sufficiency, we have already noted a greater number of "backyard" meat animals being produced for the home meat supply. More family farms and part-time farmers are turning to this program--not only for their own meat requirements but also for producing "a few extra to sell." Dairy farms are replacing some of their cull cows with dairy steers or other livestock to utilize homegrown roughage that can be marketed through these animals. Some operators are expanding their programs to satisfy the continuing and expanding demands for fresh and processed meat of all species, as well as from the fast-food chains. Most of these increasing needs require leaner meat that favors a forage-fed program. The challenge is to develop animal-forage management systems that will maximize the utilization of forages through grazing. Additionally, with the recognized growth efficiencies and greater muscle production of intact males, and with the great availability of dairy bull calves in this region, New England has the opportunity to utilize these surplus (to the dairy herd) bulls and feed them out for a specialized market. These bull calves provide an alternative veal production system that presently serves as a viable program for the small family farm that may feed out 100 to 500 vealers in confinement systems. Over one million dairy bull calves are produced annually in the Northeast with approximately 161,000 of these being New England-reared.

The future belongs to those who prepare for it. New England livestock producers, though small and often diversified, are facing the future with courage, adaptability, innovation, and confidence that they will continue to do a respectable job in maintaining their families and farms and contributing to the nation's meat supply.

(Statistical data presented herein has been derived from "Beef Research Program for the Northeast," [in progress, 1982], of which the author is a member of the Steering Committee.)

13

REGULATION OF AGRICULTURAL CHEMICALS, GROWTH PROMOTANTS, AND FEED ADDITIVES

O. D. Butler

Agricultural chemicals, from fertilizers to pheromones, help make U.S. agriculture the most productive in the world. Discovery, testing for efficacy and safety, manufacturing, marketing, and proper use all represent the ultimate in biological sciences, in ingenuity, and in exercise of the free enterprise system.

Some say that in this case the enterprise system is not very "free." Thalidomide, DDT, aldrin, dieldrin, arsenic, and many others did not pass safety tests. The thalidomide tragedy may have aroused the most fear in public minds, but the diethylstilbestrol use in the 1950s for sustaining pregnancy in women, which apparently resulted in increased incidence of cancer in their daughters 20 or more years later, would have to be rated a close second in the world, and first in the U.S.

Public demand expressed through members of Congress the last couple of decades caused ever-more-strict federal regulations on development and use of agricultural chemicals. During the past year, however, the Food and Drug Administration, the USDA, and the Environmental Protection Agency (the major responsible agencies) have shown good evidence of more reasonable postures concerning laws, regulations, and interactions with manufacturers and users of agricultural chemicals.

President Reagan's appointment of a cabinet-level committee chaired by Vice-President George Bush with a mission for reducing burdensome regulations, gave an unmistakable signal to the agencies. Now we see Congress considering revision of the Federal Insecticide, Fungicide, Rodenticide Act (FIFRA), and the Food Safety Laws, especially the extremely strict 20-year-old Delaney anti-cancer clause. This clause was made obsolete by almost unbelievable advances in assay procedures that now detect parts per trillion of materials in foods that were considered to have zero residue with the parts-per-million capability of assays in the 1960s. Assays are now as much as a million times more sensitive.

Strict laws that were formerly written to ban toxic substances on the basis of risk alone are being reconsidered. A couple of reasons derive from the issue of essential elements--such as selenium required by the body at a low level, but toxic at higher levels, and nitrite used for centuries in meat curing to give the characteristic color. Derivatives--for example, nitrosamines that may be developed during cooking of bacon--have been shown to cause an increased incidence of cancer in susceptible laboratory animals. More recently, the finger of suspicion has been pointed at nitrite itself, in a highly disputed experiment with laboratory animals. Nitrite produces color, but more importantly, it protects against the deadly botulism bacteria, so use of nitrites has not been banned, but has been strictly limited. Critics of the regulations point out that many natural foods contain nitrites and that human saliva does also. Avoiding cured meats would reduce nitrite consumption by a very small and negligible amount, critics say. But the "scare" stories certainly reduce demand for ham, bacon, and hot dogs.

What are producers' primary concerns about agricultural chemicals? I believe that you should have a general idea of how they are discovered, tested for efficacy and safety, and used in a safe and effective way. You should also know the direct cost of materials, as well as the indirect cost, if consumer concerns affect demand for products marketed.

Good basic biological research done primarily by public institutions, such as the Land Grant Universities, usually provides the foundation for development of an effective product. The need for products to control pests or diseases usually is expressed by producers reinforced by producer organizations, by extension specialists and research workers who interact with producers, and by supplier representatives.

Because of the similarity of all living cells, there must be a good understanding of the biology of both species affected to be able to kill a parasitic living organism without consequent toxic effect on the host. Then, for food producing plants and animals, there must be great concern about residues that might have an effect on consumers.

Animal producers are served well by a group of competing companies seeking profit by manufacturing and marketing drugs, biologicals, pesticides, and related materials. Most of the companies belong to an industry trade association, the Animal Health Institute (AHI), headquartered in the Washington area. It serves the industry the same as the many other trade associations there, trying in every way to protect the opportunity for the industry to produce products that customers will buy and use because of benefits and thereby earn a profit for investors.

Almost inevitably it seems, any position taken or change advocated by the AHI is opposed by one or more organizations that classify themselves as consumer protectionists. Lawmakers and regulators usually have to make decisions between opposing viewpoints without the benefit of absolutely conclusive evidence. In the last decade such controversy has been a major stimulant to the formation of the American Council on Science and Health (ACSH) and Council for Agricultural Science and Technology (CAST), both of which I support.

"The American Council on Science and Health (ACSH) is a national consumer education association directed and advised by a panel of scientists from a variety of disciplines. ACSH is committed to providing consumers with scientifically balanced evaluations of food, chemicals, the environment, and human health." This is quoted from their March 1982 publication, "The U.S. Food Safety Laws: Time for a Change?"

The Council for Agricultural Science and Technology (CAST) is an organization sponsored and managed by twenty-five scientific agricultural societies. Its major purpose is to assemble and report the scientific information on important issues of national scope for the benefit of lawmakers, regulators, and the general public. It is not an advocacy organization. Most of its task force reports, now numbering about a hundred, were prepared at the request of members of Congress, some by government agencies, and some because the 47 officers and directors, all representing the scientific societies, decided that there was a need to assemble and print the scientific evidence on an important issue. CAST celebrated its tenth birthday anniversary in July 1982 at a directors' meeting at its headquarters. I have the privilege of serving as president of CAST in 1981, as did Frank Baker, the Director of this International Stockmen's School, in 1979. (I want to especially recommend CAST task force reports mentioned in the references.)

Some of the scientific societies work directly with regulatory agencies. I served as chairman of the Regulatory Agencies Committee of the American Society of Animal Science for about 10 years until 1981. The Institute of Food Technologists, like the American Society of Animal Science, has been very active in identifying and nominating qualified scientists to serve on CAST task forces and has also produced independent papers on various aspects of food safety.

Drug manufacturers have been very critical of the Food and Drug Administration (FDA) for taking so long to consider new animal drug applications (NADAs) before approval. A recent report entitled "The Livestock Animal Drug Lag" by the AHI describes the problem and suggests solutions. U.S. manufacturers have been able to obtain approval to market their products in the United Kingdom and European countries in a fraction of the time required for U.S. approval. An example is albendazole, a broad spectrum anthelmintic effective against gastrointestinal roundworms, lungworms, tapeworms, and liver flukes in cattle. Approval was obtained in 5 months in England in 1978. The same application filed in the U.S. in 1977 is still pending, though strong producer pressure resulted in limited approval in 1979 under a special investigative New Animal Drug authorization in a limited number of states. After the Food and Drug Administration banned hexachloroethane for liver fluke control, cattlemen had no approved drug. Texas and Florida producers, with pastures along streams and low-lying areas that have snails (the intermediate fluke host), just had to have an effective drug. Cattle producer organizations rallied to the cause and helped obtain the limited approval.

The AHI sponsored a Forum on Regulatory Relief in Alexandria, Virginia, in June 1982. Dr. Arthur Hull Hayes, Commissioner of the FDA, announced there that "I've decided that all activities in the Review of Animal Drug Applications, including issues of Human Food Safety, will be consolidated within the Bureau of Veterinary Medicine." That is certain to allow faster decisions. The Bureau of Food review has been blamed for much of the delay in the recent past.

Dr. Hays gave a definition of safe as "a reasonable certainty of no significant risks based on adequate scientific data, under the intended conditions of use of a substance." More and more we are realizing that there is no such thing as <u>absolute</u> safety, or <u>zero</u> risk. His speech gave some reassurance concerning "sensitivity of method" regulations that have been under consideration for several years by FDA. The bureau now seems willing to accept foreign data in support of New Animal Drug Applications under certain restrictions and also to consider cross-species approvals. It is not a good investment for drug companies to spend several million dollars to obtain approval of an anthelmintic for goats, for instance, that is very important in Texas (which has about 95% of U.S. goats) because the market is so limited. Other minor species, even sheep, fall in that same category. I believe it will be necessary for publicly supported institutions like the Texas Agricultural Experiment Station to assist in developing drugs and obtaining approval for use in such minor species.

The FDA is also considering some liberalization of restrictions on feed manufacturers. Dr. Lester Crawford, recently reappointed to the position of Director of the Bureau of Veterinary medicine (BVM) of FDA spoke to the American Feed manufacturers' 74th Annual Convention at Dallas in May 1982. He reported that "The Subcabinet Working Group, chaired by USDA Assistant Secretary Bill McMillan, has proposed the total elimination of FD 1800s, the notorious application required to authorize manufacturing and sale of medicated feeds. Instead, the BVM would have authority to deny registration of feed manufacturers that lacked adequate facilities and controls to assure safety."

Even the Environmental Protection Agency (EPA) is trying to "simplify the regulatory burden on industry and reduce unnecessary costs." So said John A. Todhunter, Assistant Administrator for Pesticides and Toxic Substances, at the 1982 Beltwide Cotton Production Mechanization Conference, January 1982, at Las Vegas. He described a reassuring response to Vice-President Bush's task force, especially that FDA has instituted a plan to improve the quality of scientific assessment, including a peer review system for major scientific studies and reports. There is, therefore, hope for maintaining availability of the herbicide 2, 4, 5T and even reapproval of compound 1080 for predator control. The states also have regulatory authority and enforcement responsibility. We are all aware of Governor Brown's reluctance in California to institute effective control measures for the Mediterranean fruit fly because of the political pressure of environmentalists.

Food Chemical News, a weekly publication, keeps you up-to-date on what is happening in Washington.

For those of you mixing your own feed, and for feed distributors, I recommend the annual Feed Additive Compendium, a guide to use of drugs in medicated animal feeds with monthly, up-to-date supplements.

In conclusion, I want to make a plea to agricultural producers for closer adherence to label requirements and restrictions on use of agricultural chemicals. The Agricultural Extension Service in every state has a responsibility for assisting producers in the proper use of chemicals. More attention is being devoted to that. Very few people deliberately break the laws, but many are not aware of the precautions necessary to prevent cross contamination of products and elimination of residues in feeds and foodstuffs. The USDA state producers' effort to eliminate sulfa drug residues in pork is an example of the kind of cooperation required to maintain availability of chemicals so important to modern food production. Let us resolve to intensify the effort for safe use of agricultural chemicals in order to gain greater public confidence in the safety of our abundant food supply.

REFERENCES

American Council on Science and Health (ACSH). 1982. U.S. food safety laws: Time for a change? 1995 Broadway, New York, N.Y.

Council for Agricultural Science and Technology (CAST). 250 Memorial Union, Ames, Iowa 50011.

CAST. 1977. Hormonally active substances in foods: a safety evaluation. Report No. 66.

CAST. 1981. Antibiotics in animal feeds. Report No. 88.

CAST. 1981. Regulation of potential carcinogens in the food supply: the Delaney clause. Report No. 89.

CAST. 1982. CAST-related excerpts from U.S. House of Representatives hearing on the Federal Insecticide, Fungicide, and Rodenticide Act (FIFRA). Special Pub. No. 9.

CAST. 1982. CAST-related testimony on the food safety amendments of 1981. Special Pub. No. 11.

Feed Additive Compendium. Miller Publishing Co., 2501 Wayzata Boulevard, P.O. Box 67, Minneapolis, Minnesota 55440.

Food Chemical News. 1101 Pennsylvania Ave., S.E., Washington, D.C. 20003.

Part 3

GENETICS AND SELECTION OF SHEEP

14

BREEDING FOR IMPROVEMENT OF REPRODUCTION IN SHEEP

Maurice Shelton,
Gary Snowder

The sheep industry in the U.S. gains its greatest income from lambs or meat, and the major opportunities for increasing income and/or improving the efficiency in the sheep industry can be obtained by increasing the efficiency of lamb production. Current or potential production levels confirm these statements. The mean fleece weight of a breeding ewe in Texas is about 8 lb valued at $8 per head. An 80% lamb crop marketed as 70 lb feeder lambs ($.60 per lb) yields an estimated gross of $33.60 per ewe. On a national basis, the lamb crop approximates 100%, which would further increase income (expected from lamb production).

In many of the smaller farm flocks, producers donate the wool clip in exchange for getting their flock shorn, further suggesting that the potential for increasing income is greatest in lamb production. Whereas, the potential to increase wool production or income appears to be very limited. There is a potential for producing and marketing larger lambs (up to 110 lb) and for a lamb crop of over 200%. Combining these latter two factors, it is easy to project a potential income of $100 to $120 per ewe.

Although these examples lead to a correct conclusion that efforts should be made to maximize lamb production they tend to minimize the role of wool production, at least for range producers of finewool flocks. Misconceptions about the role of wool production occur because all sheep produce a wool clip, but not all are efficient lamb producers. The examples given assume that all lambs are sold and that none are kept for replacements. Actual live-animal sales are reduced by number of replacements kept, the extent of death loss, and the lower market value of aged ewes culled from the flock. More of the cash costs, especially feed, are associated with the production of lamb than of wool.

Shelton and Kensing (1980) contrasted the gross income to Texas producers from meat and wool for the years 1940 to 1978. Sixty percent of the income was obtained from wool in 1940 and between 20% to 30% in 1978, depending on whether incentive payments were included as wool income. On a national basis, the relative portion of the income obtained from lamb production is even greater than that in Texas.

The reader should not conclude from the above comments that wool production has no future. However the long-term picture is not clear for production in the U.S. and the economic efficiency of fiber production has not yet been determined. In cold climates, or during cold-weather stress in more temperate regions, the energy saved by the fleece more than offsets the nutrient cost of its production. In tropical climates, particularly in humid regions, the fleece cover is a burden to sheep requiring increased energy for heat dissipation. Between these extremes the desirability of a heavy fleece might be questioned. It is the writer's opinion that under conditions of good nutrition (positive nutrient balance), fleece production is almost a bonus to the producer. However, under conditions of nutritional stress, such as the lactating ewe, the nutrient requirements for fleece production are competitive with those for other functions.

Wool production seems to have a role in the forseeable future, because suitable nonwool breeds are not available and their development through selection would require many years or generations. Thus, producers should seek to maximize the income obtained from the fleece. This can be done by producing finewool (64^s or finer) of adequate staple (2 3/4 in. or more) by giving special attention to packaging and merchandising the clip. How much emphasis should be placed on selection for fleece as contrasted to traits contributing to lamb production is questionable. Many producers and show-ring judges who practice selections based on visual appraisal may emphasize fleece because this trait is highly visible; whereas the animal's potential for lamb production is not evident on visual inspection. Also, feeding and management decisions based specifically on fleece production are questionable. If the sheep producer's position is based, instead, on improved lamb production, how can he become more efficient and competitive? This discussion, therefore, stresses increasing or improving the number of lambs produced.

For many years the major reproduction research emphasized endocrine manipulation or hormone stimulation. After decades of research and hundreds of experiments, these practices are little used by the industry. Other deterrents are the limited markets of the small industry and the difficulty and expense in securing FDA-approved drugs. The emphasis has now shifted from hormone stimulation to reproductive efficiency by genetic means. In earlier periods, animal breeders believed that reproductive heritability was low (about 10%). But studies by Shelton and Menzies (1968) have shown that reproductive heritability over a period of years is higher (about 20%). More recent experiments and experiences confirm reproductive efficiency can be improved by: 1) breed, 2) crossbreeding, and 3) selection within a breed.

BREED SELECTION

The most dramatic or immediate response theoretically is through breed selection; however, in practice the situation is often quite different. Although on a world basis there are literally hundreds or thousands of breeds or genotypes of sheep, producers are often restricted in owning or purchasing at a reasonable price those that are adapted to his environmental and production conditions. The producer, in fact, has very little choice in respect to breed.

Among traditional breeds the differences in lambing rate appear to be small. Some breed estimates taken from literature cited are shown in table 1.

TABLE 1. SOME BREED ESTIMATES OF LITTER SIZE

Breed	Litter Size	
Rambouillet	1.22[1]	1.38[3]
Targhee	1.29[1]	
Columbia	1.27[1]	
Corriedale	1.18[1]	1.45[3]
Hampshire	1.54[2]	
Shropshire	1.23[2]	
Southdown	1.26[2]	1.42[4]
Merino	1.30[2]	
Romnelet	1.35[3]	
Lincoln	1.56[4]	

[1]From Terrell and Stoehr (1939).
[2]From Sidwell, Everson and Terrell (1962).
[3]From Veseley and Peters (1965).
[4]From Wiener (1967).

Differences between flocks and seasons within a breed may well be greater than consistent or predictable differences between breeds. If man does not interfere, it seems likely that natural selection will tend to maximize net reproductive efficiency. In the writer's opinion, man has often tended to select against reproductive efficiency by emphasizing conformation, growth rate, breed type, fleece traits, etc. without simultaneously favoring those born or raised as twins. Among the available breeds in the U.S., breed selection, in most cases, must be made based on availability, adaptability, or personal preference. When these factors are considered, the breed options often are very limited. In the Southwest, for example many breeds may have been introduced and tried, but essentially all flocks continue to be Rambouillet (or related finewool types) or derivatives of this type.

On a world basis, known prolific breeds include Finnish Landrace, Romanov, Chios, Barbados Blackbelly, East Friesian, D-Man, and Booroola Merino. It is interesting to speculate on the origin of the high-fertility trait of these breeds. Most of the high fertility traits trace back to an earlier period and cannot be explained by conscious or man-imposed selection in modern times. In one case (Booroola Merino), the high fertility trait is thought to be due to the segregation of a single gene of unknown origin. The same may be true of some other breeds.

Of the various breeds listed, only Finnish Landrace has tested extensively in the U.S. The "Barbado" is found in substantial numbers, but the type found in the U.S. differs somewhat from the true Blackbellied Barbados from the Island of Barbados. Booroola, or sources of the Booroola gene, have recently been introduced in the U.S., but it will be a number of years before this breed or trait can be evaluated in our production systems. A reasonable assumption is that none of the highly prolific breeds, or at least those currently available in the U.S., will find favor for use as commercial breeds in pure form, thus they are discussed in connection with crossbreeding.

CROSSBREEDING

Exploitation of the phenomena of heterosis has been one of the major contributions of plant and animal breeding in modern times. Having acknowledged this contribution, it can be said that researchers or research data often oversell crossbreeding, especially in terms of sheep production. Some of the reasons are:

- Crossbreeding gains are not net gains because the increased production requires increased nutrients.
- Most of the world's sheep population produces fiber with little contribution of hybrid vigor to fiber production. Only the Merino has a highly developed fleece; almost any breed crossed on the Merino lowers the fleece value.
- The availability of suitable, unrelated, and adapted breeds within a given geographic area, environment, or production system often presents problems to the exploitation of crossbreeding.
- The exploitation of heterosis requires a program of controlled and systematic use of breeds that requires a minimum of three or more genetic groups. The maintenance or availability of these groups presents problems. For example, when F_1 females are crossed to a third or terminal sire breed, 5 types of animals exist: initial breeds used in crossing (breeds A & B), F_1 lambs and ewes, terminal sire (breed C) and three-breed-cross-market lambs. A failure to use these in a controlled system is a form of mongrelization.

Some of the most successful sheep industries of the world are based on a single, suitable, and well-adapted breed. Hybrid vigor tends to make the greatest contribution in traits that contribute to 1) fitness, 2) reproduction, or 3) breeds that complement each other in contributing to the desirable traits of the crossbred dam or market animal. Thus, crossbreeding tends to be utilized to maximize lamb production by using breeds that complement one another.

Leymaster (1982) summarized the expected gain from crossbreeding based on data from a number of studies (table 2). In simple programs based on two-breed crosses with the F_1 lamb being marketed, he suggests a 17.8% increase in production due largely to improved lamb survival and growth rate. This seems to be an overestimate when compared to the use of Suffolk rams or range Rambouillet ewes in Texas. Specific breed crosses might make greater--or fewer--contributions. The crossbred ewes indicated a net increase of 18.0% associated with conception rate and lamb growth. These two contributions should be somewhat additive, depending on the sire used on the F_1 ewes.

TABLE 2. ESTIMATES OF HETEROSIS

Division of effects	Trait	Individual heterosis	Maternal heterosis
Components	Conception	2.6	8.7
	Number born	2.8	3.2
	Postnatal survival	9.8	2.7
	Weaning weight	5.0	6.3
Composite	Weight of lambs weaned/ewe exposed	17.8	18.0

Individual breed crosses could result in more or less gain than shown in this generalized example.

These data make a strong case of the use of crossbreeding, which is not used universally by the sheep industry. The major limitations are the availability of suitable breeds, the problems of providing replacements, or managing the more complex systems. Individual breed crosses may well provide greater or less response than that indicated, due to breed complementarity. For a crossbreeding program in the U.S., the producer uses the breed that is both best adapted and available--such as the Merino types or derivatives. In terms of terminal sire breeds the sires are selected for market lamb production. For instance, Australian or New Zealand Southdowns are very popular because they sire lambs that will readily fatten on grass pastures at an early age. In the U.S., the Suffolk and Hampshire are the predominant sire breeds because they sire larger, faster-growing lambs with (possibly) superior carcass qualities.

A number of breeds are different in producing superior F_1 ewes--in particular the Border Leicester. Worldwide, they contribute good size, moderately high lambing rate, and significant hybrid vigor when crossed on unrelated types such as the finewool ewe. Border Leicester is of British origin and is poorly adapted to range conditions or hot, arid environments. The limited testing on (in the U.S.) them suggests they be placed outside the South or arid Southwest; Texas trials have not shown favorable results. The Dorset has received more crossing emphasis because of an extended breeding season and average or better lambing rate. Crossbred ewes sired by Dorset rams have given favorable results under farm conditions, but any superior traits found in this cross have not overcome their lack of adaptation to range conditions. Modest improvement in production conditions might change this conclusion.

Another breed of more recent introduction is the Finnish Landrace. Its high lambing rate is sufficiently dramatic that it is making an important contribution to the U.S. sheep industry (Dickerson, 1977) and is an apparent cause for the revival of the industry in the farm states. They have been used both in research and in practice in the Southwest under range conditions. Although limited, the research data has shown a modest gain from the F_1 ewes (Finn x Rambouillet) that adapted best to farm conditions or to improved feeding and management conditions on ranches. The available research to date (in Texas) has been with the 1/2 Finn ewes, but the 1/4 Finn is being investigated.

SELECTION WITHIN A LINE OR BREED

Selection within breeds is used because crossbreeding often limits the opportunities to enhance lamb production. One of the greatest mistakes of the sheep industry in the U.S. and other countries has been the failure to increase reproduction through selection. Contrary to earlier viewpoints, it is now well established that the reproductive trait responds to selection (Bradford, 1972), although perhaps not with the same ease of selection or at the same rate as many other traits. Partially offsetting low heritability is the great selection differential that can often be obtained and the economic importance of the trait. In the absence of emphasis on reproduction, producers may select against fertility by visually selecting for size and "bloom" on weaning-age lambs, thus favoring single born lambs. Selection for either size at later ages for wool production can adversely affect lamb production (Shelton and Menzies, 1968). Perhaps the greatest gain in the U.S. from selecting for fertility would be realized in the Rambouillet, or related finewool types, because 1) they are perhaps the single most important breed to the U.S. sheep industry, 2) they are concentrated in the South or Southwest where opportunities for crossbreeding are somewhat limited, and 3) perhaps more than other breeds they may have suffered from negative selection for fertility. Some ranchers still believe that twinning is not desirable, but the number is decreasing. Under adverse range conditions, it is difficult for a ewe to raise twin lambs, and so the economic advantage from a high lambing rate is decreased under these conditions. However, it is the writer's belief that it is feasible to provide supplemental feed to a ewe with twins; where or when this is not true, the U.S. sheep industry will not survive.

SELECTING FOR FERTILITY

Selecting for fertility, especially under range conditions, is very difficult. The indicated approach is to select sires from flock ewes or stud ewes that have had a good lamb production record. Early progress is likely to result if ranchers will use almost single-factor selection, but this seems unlikely. However, if the fertility trait were given 50% emphasis in selection, meaningful results could probably be obtained over a period of time. The ideal ewe produces a single lamb her first year, then twin lambs each year thereafter for five or more producing years. Less than 1% of range ewes actually do this, but a significant number of ewes should approach this level. (In the writer's opinion, triplet births should be tolerated but not favored. Larger litters should be selected against, unless the goal is to produce a super-fertile line to be exploited in a crossing program.)

There are some very real problems in selecting for the ideal ewe. A record system is necessary for keeping pedigree information, especially dam-offspring pairings. For range-lambing flocks of the Southwest, record keeping for dam-offspring pairings is nearly impossible. The suggested approach is to select a small stud flock on which the necessary records would be kept. This should automatically include all registered or purebred producers who are offering rams for sale. In the case of shed-lambing operations in the Northwest, dam-offspring records on larger numbers are possible. Basic to the record system is a cumulative record on ewe productivity; however, if this is to be maintained, a more extensive record system would be indicated.

For producers who are interested in selecting for lamb production but who wish to avoid paperwork, ewes may be ear tagged by year of birth so that the age of the ewe is instantly available. Then if each ewe receives an ear notch for each set of twins and a hole in the ear for each dry season, she carries her own production record. The problems with this approach are 1) the long period required when selection is based on the lifetime or a period of years for each ewe, and 2) the large number of ewes required to realize progress.

To overcome the problem of large numbers of ewes, several breeders could be coordinated. The <u>Group Breeding</u>

Schemes used in Australia and New Zealand are an example. Selection for improvement in fertility could make use of such a scheme, but it would be difficult to sell to a group of independent U.S. producers. Using a similar breeding scheme, sire selection, as well, could be improved. In experimental flocks selection for improved lamb production has often increased the lamb crop from 0-4% per year. There is good reason to believe this percentage could be increased if a number of flocks involving much larger numbers could be integrated into a common program.

An alternative to the long generation interval associated with ewe-productivity selection would be the identification of superior animals based on measurements made early in life. This would be useful if traits relating to ewe productivity could be identified in the young male. Early identification of superior ewes also would be useful. Two basic approaches have been tried on the male: 1) measuring the circulating levels of certain blood hormone levels such as the leutenizing hormone (LH) that is associated with ovulation in the ewe and reproduction in the male and is associated with reproduction. (Unfortunately, this hormone is released from the pituitary in an erratic or pulsatile manner, and a single measurement has little value. Numerous or repeat measures made over a period of time gives a better measure. Unfortunately, this would be expensive or difficult to accomplish, and has not been shown conclusively to be related to fertility of their female offspring.) Another possibility is the use of a single value obtained as a result of administration of a gonadotrophic-releasing hormone still in the experimental stage; 2) of making testicular measurements on the males. Ruling out orchidis due to injury or infection, it can be shown that rams with larger testes have greater sperm production capacity and, thus, greater total breeding potential. Earlier testicular development is associated with earlier sexual maturity and rams with greater testicular development might sire ewes with improved lambing rate since the hormone systems involved are the same. Comparisons between breeds tend to support this conclusion, but no research has been reported that shows a response in ewe fertility in the following generation as a result of selecting on testicular size of the male. Among rams within a breed of comparable age and treated alike, observed variation in testicular size is small. However, this may well parallel the situation in females in which true genetic differences in females are small. Still there is a likelihood of direct gain and potential for indirect gain from making use of testicular size in selection, and for these reasons producers would be encouraged to consider this as a selection tool.

REFERENCES

Bradford, G. E. 1972. Genetic control of litter size in sheep. J. Reprod. Fertil. Suppl. 15, pp 23-41.

Dickerson, G. E. 1977. Crossbreeding evaluation of Finn sheep and some U.S. breeds for market lamb production. North Central Publication No. 246, USDA and Univ. of Nebraska.

Leymaster, Kreg. 1982. Crossbreeding: Key to productive commercial ewes. Sheep Breeder and Sheepman Magazine 102:56.

Shelton, M. and J. W. Menzies. 1968. Genetic parameters of some performance characteristics of range finewool ewes. J. Anim. Sci. 27:1219.

Shelton, M. and J. W. Menzies. 1970. Repeatability and heritability of components of reproductive efficiency in finewool sheep. J. Anim. Sci. 30:1.

Shelton, M. and R. Kensing. 1980. The relative importance of meat and wool as a source of income to Texas sheep producers. Texas Agri. Expt. Sta. PR 3715.

Sidwell, G. M., D. O. Everson and C. E. Terrill. 1962. Fertility, prolificacy and lamb livability of some pure breeds and their crosses. J. Anim. Sci. 21:875.

Terrell, C. E. and J. A. Stoehr. 1939. Reproduction in range sheep. Proc. Am. Soc. Anim. Prod. p 369.

Veseley, J. A. and H. F. Peters. 1965. Fertility, prolificacy, weaned-lamb production and lamb survival in four range breeds of sheep. Can. J. Anim. Sci. 45:75.

Wiener, G. 1967. A comparison of body size, fleece weight, and maternal performance of five breeds of sheep kept in one environment. Anim. Prod. 9:177.

15

THE NEW ZEALAND NATIONAL FLOCK RECORDING SERVICE—"SHEEPLAN": PROVIDING FLEXIBILITY IN BREEDING OBJECTIVES AND RECORDING OPTIONS

R. L. Baker

INTRODUCTION

Performance recording has been defined by Owen (1971) "as the measurement of what are considered important traits, using simple unequivocal techniques that do not depend on the art of the skilled breeder in appreciating and integrating a number of rather ill-defined standards." Wide scale performance recording has been the basis of dramatic improvements in the efficiency of production in many livestock species in recent times and especially in poultry, dairy cattle, and pigs. Performance recording in sheep is very much in its infancy in most countries around the world, as has been comprehensively reviewed by Owen (1971). One exception is the Finnish Landrace breed (Finnsheep), for which a breed association was formed in 1918 that incorporated a selective registration scheme based on prolificacy records. The high level of prolificacy attained in this breed through performance recording and selection is now well-known all around the world, because germplasm was sought that would improve this important economic character in sheep.

The first national performance recording scheme for sheep in New Zealand was not initiated until 1968 and was aimed at the genetic improvement of the national flock. This original recording scheme produced a within-flock genetic ranking of animals for an index combining lamb production (weight of lamb weaned) and wool production (yearling fleece weight) in dual-purpose breeds, and for lamb growth (weaning weight) in meat breeds producing sires for crossbred lamb production (Clarke, 1967). This National Flock Recording Scheme (NFRS) operated with few major changes for 8 years (1968-1975) under the control of the Ministry of Agriculture and Fisheries.

During this time, comments and criticisms from breeders, the experience of field servicing provided by officers of the Advisory Services Division and of data processing by the Biometrics Division, along with further information from research, gave rise to a number of suggestions for its improvement. Broadly, these indicated that the NFRS was not sufficiently flexible to cope with the variety of needs of different breeds and breeders. Some regarded the scheme as too simple, others as too complex. In addition, the demands of many breeders for a wider range of traits to be recorded were to some extent supported by research findings, especially for live weights at various ages. There was also a strong demand from breeders for sire summaries of progeny performance and for a number of features to suit the special circumstances of integrated, large-scale breeding enterprises (open-nucleus breeding schemes) that have become an important and integral part of the New Zealand Sheep breeding scheme.

During 1974 and 1975, technical recommendations were made to establish and promote a modified and expanded recording service that was soundly based on past experience with the NFRS and current research knowledge. A revised national recording scheme called Sheeplan was then initiated in 1976 and continued to be run by the Ministry of Agriculture and Fisheries (Clarke and Rae, 1977). Like its predecessor, Sheeplan's prime objective was the genetic improvement of commercial sheep flocks.

Plans for genetic improvement on a national scale have several important ingredients:

- The measurement of traits of economic importance of individual animals.
- The processing and presentation of the records in a way that will assist in identifying genetically superior animals.
- The effective dissemination of genetic merit throughout the commercial population.

The technical background relating to the second of the objectives has been comprehensively presented by Clarke and Rae (1976, 1977) and Clarke (1979) and will be summarized here. In addition, some New Zealand examples of the sort of progress that can be made through selection for important economic traits in sheep will be presented along with current recommendations for effective dissemination of genetic improvement.

CHARACTERS RECORDED IN SHEEPLAN

An important feature of Sheeplan is its flexibility in terms of choice of recording options. The breeder is able to choose the characters he wishes to record and the output lists he requires (figure 1). This permits breeders to specify the objective that best applies to his breed and situation and to progress from a simple to a more complex breeding program.

Sheeplan enables breeders to identify, record objectively, rank, and select individual sheep within a flock for five basic characters:

- Number of lambs born (NLB) or reared (NLR).
- Lamb weaning weight (WWT).

- Live weights taken in the autumn (ALW) at 4 to 8 months of age, winter (WLW) at 8 to 12 months, or spring (SLW) at 12 to 15 months.
- Hogget (yearling sheep) fleece weight (HFW) at 10 to 15 months of age.
- Hogget fleece yield and fibre diameter (from laboratory testing).

Number of lambs born is the only character that is mandatory for all users (figure 1). It establishes pedigree information in the computer files and thereby allows pedigree relationships to be utilized to predict breeding values. Complete records of parentage permits NLB information on the dam to be used for genetic assessment of the progeny, especially ram progeny. Accurate identification of the sire of each lamb allows the processing and analysis of half-sister and sire progeny information.

INPUT LISTS		OUTPUT LISTS		IN ORDER OF
Lambing list	☑	Lambing summary	☐	
Weighing list (weaning wts)	☐	Ewe summary	☑	Ewe identif.
Weighing list (autumn wts.)	Rams ☐☐ Ewes	Ewe summary cross reference	☐	Lamb prod.
		Closed ewe file	☐	
Weighing list (winter wts)	Rams ☐☐ Ewes	Two-tooth selection list	Rams ☑☑ Ewes	Identification
Weighing list (spring wts.)	Rams ☐☐ Ewes	Two-tooth selection list	Rams ☐☐ Ewes	Index
Hogget shearing list	Rams ☐☐ Ewes	No. of lambs reared option	☐	
Fate lists	☐	Sire summaries	☐	

Figure 1. Input and output options in Sheeplan

From the point of view of genetic improvement from selection, the important common feature of most output listings is the calculation of breeding values for the characters recorded and for various combinations of them. Prime emphasis is given to the calculation of breeding values for the young ewes and rams that are candidates for entry into the flock. These are presented in the two-tooth (animals 15 to 20 months of age) selection lists prepared separately for ewes and rams. Breeding values for NLB are calculated for all sheep in, or available for inclusion in, the breeding flock. Both selection lists and ewe summary lists are distributed to all Sheeplan users (figure 1). On the ewe summary list, estimated breeding values for NLB are revised annually with current information from the lambing list to give an up-to-date assessment of ewe performance for culling purposes.

If requested, breeding values also are processed to summarize the performance of the progeny of the individual rams (sire summaries) used in the flock each year. Breeding values for the traits recorded are processed and presented sequentially as the information becomes available.

Quickly returning processed data in the form of breeding values for all animals in a breeder's flock encourages the use of this information for selection and culling decisions.

THE BREEDING VALUE

Breeding values constitute the technical basis underlying the processing and presentation of records to assist in making effective selection decisions. Selection for genetic improvement is concerned with how well the offspring of an animal will (on average) perform in the flock. The breeding value allows a prediction of the likely performance of an individual's progeny for the character concerned.

It is possible, although sometimes mathematically complex, to separate the observed (phenotypic) performance into a component due to average additive genetic merit (breeding value) and a residual component due to the environmental effect on performance. The heritability of a particular character is then defined as the fraction of total variation attributable to variation in breeding values.

The environmental contribution to total phenotypic variation is usually extremely complex, but in sheep it includes identifying effects such as birth-rearing rank, sex, age of dam, and age of lamb. Adjusting the characters for these recognizable environmental influences increases the effective heritability of the trait and the accuracy of the breeding value estimation. (For procedures used in Sheeplan see Clarke and Rae, 1976).

An important aspect in the prediction of breeding values in Sheeplan is that all available relevant information is used. This involves both direct and indirect pathways of genetic determination and again adds to the accuracy of breeding value estimation for each character assessed. Full details on this aspect of Sheeplan are given by Clarke and Rae (1976, 1977).

THE SELECTION INDEX

In sheep, productivity is usually made up of more than one trait. Thus, for dual-purpose breeds, the improvement objective will generally comprise not only the number of lambs produced and their growth rate, but also the weight and merit of the fleece. It is convenient to define the objective as an aggregate economic measure of overall merit that depends upon the relative economic values established for each of the component traits.

Once an aggregate breeding objective has been defined, it is possible to use the selection index method (Hazel, 1943) to balance the good features of an animal's performance against the bad.

The method used by Sheeplan for the prediction of breeding values for traits of the rising two-tooth ewes and rams of the dual-purpose breeds follows that given by Henderson (1963). This method is identical with, but more convenient than, the method of Hazel (1943) since it first uses methods of direct and indirect prediction to give breeding value estimates for each of the component traits and then combines these according to the economic values specified to estimate the breeding value for overall economic merit. This facilitates the processing of records using different sets of relative economic values for each breed or breed category and in the long term could permit the individual breeders to choose their own set of relative economic values for the traits they are recording.

The selection index predictions of breeding value are based on genetic and phenotypic parameters derived from several analyses of large amounts of New Zealand data (Clarke and Rae, 1977) for the Romney breed, which is the predominant sheep breed in New Zealand. Provision has been made to handle separate sets of parameters for different breed groups, should subsequent investigations indicate that this is necessary. At present, separate sets are used for only the broad categories of dual-purpose and meat breeds. The parameters currently used in Sheeplan for dual-purpose breeds are shown in table 1. The relative economic values (REV) in table 1 are based on gross rather than net returns, having been estimated from data from a variety of sources giving prices for wool, store and fat lambs over the five-year period from 1970 to 1975. At the present, live weight is regarded as having a zero REV on the grounds that the greater return from the larger animal when slaughtered is likely to be compensated by a higher maintenance cost

TABLE 1. PARAMETERS USED TO ESTIMATE BREEDING VALUES IN DUAL-PURPOSE BREEDS

Trait	REV (cents)	Standard Deviation	Herit-ability	Correlations[a]			
				NLB	WWT	SLW	HFW
NLB	554	0.57	0.10	--	0.12	0.20	0.05
WWT (kg)	24	3.0	0.20	0.12	--	0.70	0.20
SLW (kg)	0	4.5	0.35	0.15	0.15	--	0.30
HFW (kg)	92	0.45	0.30	0.00	0.30	0.40	--

[a]Genetic correlations above the diagonal and phenotypic correlations below. Repeatability of NLB taken as 0.15.

throughout its life. Some preliminary unpublished New Zealand analyses indicate that, when net relative economic values are considered, fleece weight should receive somewhat higher values and NLB lower values than those presently used. It is intended to continue work in this important, and somewhat neglected, area of research and make appropriate modifications when required. The big problem area here is to accurately predict feed intakes and maintenance costs in grazing animals.

The features of the selection index for aggregate merit derived from the dual-purpose parameters presented in table 1 are summarized in table 2.

The "value of each trait" shown in the first column of table 2 indicates the percentage reduction in the rate of improvement in aggregate breeding value that would result if that trait were excluded from the index but not from the aggregate breeding objective, i.e., selection is still aimed at improving the same overall objective. Thus the two traits that bear the major role are NLB and SLW, despite the fact that SLW does not contribute to the aggregate breeding value (because it has been given a zero economic value). This results from the contribution of indirect prediction because of the high heritability of SLW and the positive genetic correlations between SLW and NLB and SLW and WWT. For the same reasons both WWT and HFW tend to play a minor role. Nevertheless, these characters still show a positive response to selection using the index. This is shown by the column headed "rate of response" in table 2. These figures indicate the rate of genetic improvement in each trait per generation that can be expected from a selection differential of one standard deviation on the index (in this case, 43 cents). Thus, if the animals chosen for breeding average 43 cents above the average of the flock before selection, the expected rate of genetic progress would be 0.05 lambs born, plus 0.5 kg in WWT, 1.2 kg in SLW, and 0.03 kg in HFW for each generation of selection. For a typical breeding flock these figures should be divided by a factor of approximately 3 to 4 to convert them to expected annual rates of genetic gain. The last columns of table 2 indicate the contribution that the responses in the individual traits make to the overall response in aggregate breeding value. Because of its high economic value, improved NLB makes the greatest contribution to overall progress in economic terms.

TABLE 2. FEATURES OF THE SELECTION INDEX FOR DUAL-PURPOSE BREEDS

Trait	Value of each trait (%)	Rate of response [b]	REV (cents)	Contribution to economic response (cents)	
NLB[a]	24	0.05 lambs	554	28	(65%)
WWT	2	0.5 kg	24	12	(28%)
SLW	13	1.2 kg	0	0	
HFW	0	0.03 kg	92	3	(7%)
				43	

[a]Based on three lambings.
[b]Response per generation from one standard deviation of selection.

The index value for aggregate merit is presented on the two-tooth selection lists. But in addition, breeding values are presented for NLB (or NLR) and also for WWT, latest live weight (ALW, WLW or SLW) and HFW if these options have been selected by the breeder (figure 2).

The selection list shown in figure 2 allows breeders or buyers of their stock to:

- Choose rams that excel for any one particular character by referring to the individual breeding value for that character.
- Choose a ram that has high overall merit for a number of characters by referring to the selection index.

In figure 2 a breeding value of .064 lambs means that if this ram was mated to ewes of average breeding value, daughters would be expected to average .032 lambs more than daughters from an average ram (half the sire's genes are passed to his progeny). The breeding values for the other characters are interpreted in a similar manner so that the higher the breeding value the better the animal for that character. The breeding value for the index is expressed in cents.

SELECTION IN MEAT BREEDS

For the meat breeds, the selection lists supply breeding values for NLB, WWT and ALW (figure 3). The breeding values for WWT and ALW are the most important criteria and are relevant for early and heavy-weight lamb production, respectively. Breeding value for NLB is presented to cater to those breeders of sires for crossbred lamb production who wish to give emphasis to this component of the costs of their ram breeding enterprise. But because of the variation in emphasis that individual breeders wish to place on BV for NLB, no attempt is made to include NLB into an index along with breeding values for WWT or ALW. Breeding values for WWT and ALW are each predicted from any one, or from any combination of up to four live weights for which the breeder decides to record. The parameters on which these predictions are based are given in table 3.

Although using all four live weights gives the most accurate prediction of both weaning and autumn breeding values, combinations of any two weights achieve between 95 and 99% of this maximum accuracy. If the objective is improved weaning weight, then any one of the later weights adds substantially to the accuracy of predicting the BV, owing in part to the relatively low heritability of weaning weight and in part to the relatively high genetic correlation of weaning weight with later weights (ranging from 0.47 to 0.80).

EWE SUMMARIES

As in the selection lists, the ewe summary presents three types of information: pedigree information, individual performance data and progeny data relating to the ewe, and breeding values for the more important traits. Because of its importance in culling decisions, breeding values based on the number of lambs produced by a ewe are given major emphasis for dual-purpose breeds. By contrast, in the meat breeds where major attention attaches to the genes for meat production passed on through the sire to his slaughter lambs, breeding values for WWT are given most emphasis.

TABLE 3. PARAMETERS USED TO ESTIMATE BREEDING VALUES FOR MEAT BREEDS

	Standard	Herit-	Correlations[a]			
Trait	deviation	ability	WWT	ALW	WLW	SLW
WWT	3.1	0.10	--	0.70	0.60	0.50
ALW	3.4	0.22	0.62	--	0.67	0.47
WLW	3.8	0.36	0.53	0.78	--	0.85
SLW	4.6	0.45	0.50	0.63	0.81	--

[a]Genetic correlations above diagonal and phenotypic correlations below.

While all Sheeplan breeders receive breeding values for NLB in their ewe summary, when the weaning weight option is taken, breeders of dual-purpose sheep are also provided with a breeding value that reflects the mothering ability of

Fig. 2 SHEEPLAN TWO-TOOTH SELECTION LIST RAMS

FLOCK CODE 9109150   J T Smith   BREED OF SIRE -- DUAL-PURPOSE

TAG YEAR 1978   BREED OF EWE -- DUAL-PURPOSE

	Dam							Two-tooth											Remarks		
Sire identifi-cation No Yr	identifi-cation No Yr	Sire identifi-cation No Yr	Lambing Record Age at lambing (yrs) 1	2	3	4	No of lambs dev	Birth /rear rank	Indent No	Breeding value Index	No of lambs	Wean Wt	Latest Wt	Flce Wt	Adj Weanwt Dev	Adj Latest Wt Dev	Flce Wt Dev	Fibre Diam	Wool Type	Remarks Carried Forward
192/6	42/75	--		2	2		47	2/2	105	67	.064	1.16	2.02	0.4	5.0	5.4	0.26	52	6	1.7 8.0
144/76	238/74	82/73		2	2	3	68	3/2	106	60	.102	.45	1.36	-0.9	.5	3.0	-0.25	54/56	6	1.7 6.5

BV for total productivity in cents

Other Bvs in units measured--lambs and kgs.

The breeding value predicts how well a two-tooth, or its offspring, will perform relative to the average two-tooth.

Fig. 3 SHEEPLAN TWO-TOOTH SELECTION LIST RAMS

FLOCK CODE 9109152   A M MOORE   BREED OF SIRE -- MEAT BREED

TAG YEAR 1978   BREED OF EWE -- MEAT BREED

Dam							Two-tooth									Remarks
Sire identifi-cation No Yr	identifi-cation No Yr	Sirc identifi-cation No Yr	Lambing record Age at lambing (yrs) 1	2	3	4	Birth /rear rank	Breeding Value Ident No	No of lambs	Wean Wt	Autumn Wt	Adj Wean Wt Dev	Adj Autumn Wt Dev	Adj Winter Wt Dev	Adj Spring Wt Dev	Remarks Carried Forward
1191/76	169/74	--		2	2	2	2/2	9	12	68	80	6.75	4.72			11
4040/76	123/76	27/73		1			1/1	18	01	24	32	2.20	1.75			11

BV for fertility in lambs

BVs for weaning weight and autumn liveweight in kgs.

The breeding value predicts how well a two-tooth or its offspring, will perform relative to the average two-tooth.

their ewes. It is based upon an economic weighting of lamb survival to weaning and the average weaning weights of the lambs produced by the ewe each year and is predicted from these same traits. In this case, an overall measure of lamb production is calculated, combining both mothering ability (lamb survival and growth) and NLB (fertility and prolificacy). Fostering and lamb fate codes, indicating the breeder's assessment of whether lamb fosterings or deaths should be used to incriminate the ewe, are taken into account in calculating breeding values for mothering ability and lamb production.

All dual-purpose breeding values are based on each year's lambing information for both the ewe and her dam. They are updated annually with new information from the lambing and weaning lists. To encourage their use in culling and provide easier identification of high-ranking ewes, a ewe summary cross-reference list may be requested. It provides a list of tag numbers in order of breeding value for lamb production, or breeding value for NLB if the weaning weight option has not been taken.

Figure 4 outlines diagrammatically the procedures taken to derive the three breeding values summarizing ewe performance. Further details are given by Clarke and Rae (1976). On the ewe summary for dual-purpose breeds (table 4) the breeding values are expressed in cents, having been scaled according to their economic values during the process of calculating the different ewe indexes. This permits the different breeding values to be compared in a meaningful way.

SIRE SUMMARIES

The current approach to preparing sire summaries emphasizes breeding values based on adjusted data. The mean of the progeny of each sire is adjusted to take account of variation in numbers of progeny per sire. Retention ratios are calculated for each sire to highlight both the absolute and relative levels of culling taking place at various stages among the progeny. Studies are being made of how to interpret sire summaries as progeny test estimates of breeding value for the rams and of methods whereby these data may be included as half-sib information for the ranking of young flock replacements.

DOE SELECTION BASED ON PERFORMANCE RECORDS WORK

Consistent selection can bring about worthwhile improvement in important productive characters. Some of the most convincing evidence comes from a New Zealand study of the effectiveness of selecting on the basis of twinning in Romney ewes.

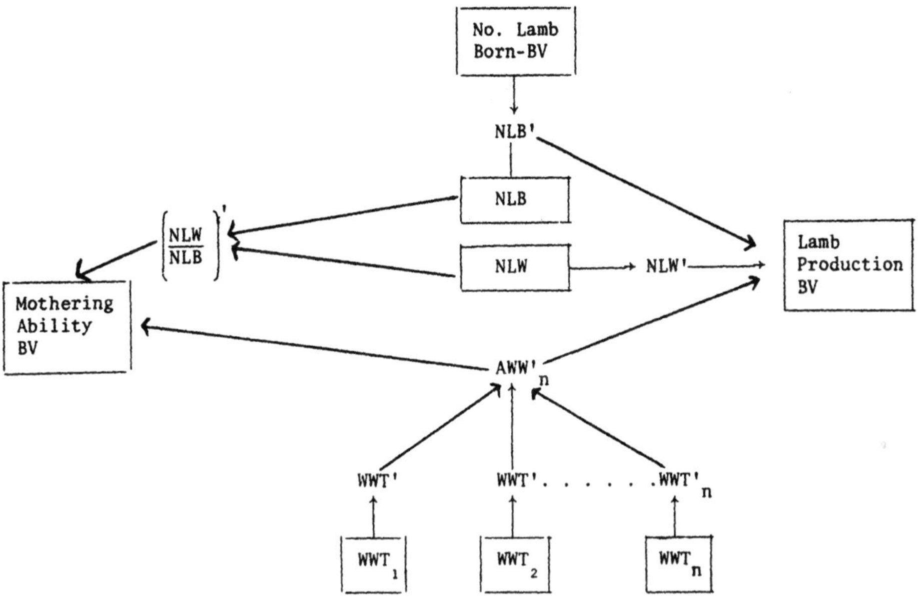

Figure 4. **Derivation of the different measures of ewe performance in Sheeplan. NLB = No. of lambs born; NLW = No. of lambs weaned; NLW/NLB = lamb survival; WWT = weaning weight; AWW = average weaning weight. The primes indicate the data is adjusted for environmental effects.**

TABLE 4. EWE SUMMARY LIST

EWE RECORDED					PROGENY RECORD					
	-Breeding value-									
Ident. No. Yr.	Lamb prod	No. lambs born	Moth abil	Sires No. Yr.	Ident. No. Yr.	SEX	ADJ WWT DEV	Dis-pos-al	Re-marks	Ewe fate
360/70				34/70	884/71	R	3.45			
					896/71	E	2.81			
				49/68	72					8
				26/69	17/73	R	9.25			
					18/73	R	3.25			
					115/73	E	3.03		F	
	31	-4	35	342/69	500/74	R		6		
					501/74	E	2.08			

In 1948, at Ruakura Dr. L. R. Wallace formed three flocks each of 100 mixed-aged ewes by selecting sheep from a recorded flock of 1,000 Romney ewes (Wallace, 1964). These consisted of a high-fertility (High) flock chosen on the ewes' or their dams' twinning records (Lambs born); a control flock (Control) selected without any attention to lambing records; and a low-fertility (Low) flock selected against twinning. The flocks were built up to 130 ewes each and have remained closed.

In the High flock, lambing percentage at birth showed a gradual but erratic increase over the Control or Low flocks over the period from 1948 to 1972 (Clarke, 1972). Over the five year period (1968-72), ewes in the High flock have produced an average of 52% more lambs at birth than those in the Low flock (table 5). The average performance of the Control flock indicates that more genetic response occurred in the High than the Low flock. This was expected from the selection pressures that could be applied in each.

TABLE 5. PERFORMANCE OF THE SELECTION FLOCKS (1968-72)*

	High	Control	Low	High-Low
Barrenness (%)	7	11	21	-14
Litter size (%)	163	117	111	52
Lamb deaths (%)	23	18	13	10
Docking (%)	116	85	77	39

*2 to 4 year old ewes.

The difference in docking percentage between the High and Control flocks (39%) indicates an average annual improvement rate of about 1 1/4 lambs docked/100 ewes mated for each of the 25 years of selection. On average, the High flock had less barrenness, considerably higher litter size (lambs born/ewe lambing), but disturbingly high lamb mortality to weaning. The Low flock demonstrated high barrenness as well as depressed litter size.

It appears that litter size, and probably ovulation rate, contributed most to the selection responses observed, although changes in ram fertility also may have been involved.

To investigate the response to selection more closely, free from inbreeding or culling biases among young breeding ewe replacements and in relation to a representative sample of present-day Romney sheep, an outcrossing experiment began at Ruakura in 1968.

In each of 3 consecutive years, 400 commercial Romney ewes were paddock mated in 3 balanced groups (4 rams/group) to either High or Low rams from the Ruakura flock, or to rams selected without attention to performance records from a wide sample of Romney ram-breeding flocks registered with the New Zealand Romney Sheep Breeders' Association (table 6).

All the ewe lambs were retained and after the collection of hogget performance (growth and fleece weights), they entered another breeding flock at Ruakura. The hogget performance data is not presented here, but the only major difference observed was 8% lower fleece weights in the progeny sired by High rams compared to those sired by industry rams. The outcross female progeny were run together (except during mating) and were retained without any culling on performance for 3 consecutive matings (2 to 4 year olds). They were mated to rams of various meat breeds in pedigreed matings, which were being tested for export lamb production.

TABLE 6. DESIGN OF THE RUAKURA OUTCROSSING EXPERIMENT

	Industry Romneys		
	Before 1948		1962-69
	Fertility flock rams		Industry rams
	H	L	I
	Commercial Romney ewe flock		
Outcross progeny	H(150)	L(150)	I(150)
1968-70 :	Ewe lambs retained		
1969-71 :	Hogget records		
1970-74 :	Performance as ewes (mated to export lamb sires)		

Progeny of High rams produced 21% more lambs docked than did those of Low rams (table 7). Since outcross progeny received only half their genes from the High and Low flocks, this indicates a genetic advantage of +42% for the High compared with the Low ram-source flocks at Ruakura. This estimate of the genetic superiority is close to the +39% average (table 5) in the selection flocks.

For litter size, the major component of improved ewe performance (table 7), the estimate of genetic superiority is the same as that given (52%) by the results in table 5 for the selection flocks.

The outcross results show a similar level of deaths among lambs from ewes derived from High and Low rams, despite the higher proportion of twin lambs in the former group. This suggests that inbreeding depression and/or ram effects may be responsible for the high lamb mortality in the High selection flock.

Inbreeding depression seems most likely because of the long period of intense selection practiced in the closed flocks. Both the High and Low flocks were small (100 to 130 ewes mated each year) and from 1959 onwards used only two new two-tooth rams each year. In the High flock, however, selection on dam's performance over several annual lambings has meant flock replacements are from common ancestors more often than is the case in the Low flock. In the Low flock, a greater proportion of dams exhibited comparable low levels of twinning. Accordingly, greater inbreeding would be expected in the High flock.

TABLE 7. REPRODUCTIVE PERFORMANCE OF OUTBRED PROGENY (1970-74)*

	High	Low	Industry	High-Low
Barrennes (%)	11	12	9	-1
Litter size (%)	152	126	136	26
Lamb deaths (%)	11	10	13	1
Docking (%)	121	100	107	21

*2 to 4 year old ewes

The similar incidence of barrenness among High and Low outcross progeny mated to the same rams (table 7) suggests that there has been little associated response in the ewes' contribution to barrenness from genetic improvement in twinning.

Components of overall ewe productivity for the three outcross groups are presented in table 8.

TABLE 8. EWE PRODUCTIVITY - OUTBRED PROGENY (1970-74)*

	High	Low	Industry
Premating wt (kg)	52.1	51.7	51.5
Lambs weaned (%)	119	97	104
Lamb weaning wt (kg at 16 week)	26.0	27.5	26.5
Wt of lamb weaned/ ewe (kg)	30.8 (112)	26.7 (97)	27.5 (100)
Ewe fleece wt (kg)	3.62 (94)	3.65 (95)	3.84 (100)
Productivity/ewe*	45.3 (106)	41.3 (96)	42.8 (100)

*Productivity/ewe = (lambs weaned x weaning wt) + (fleece wt x 4).

Lamb weaning weights refer to the crossbred lambs by a range of sire breeds. These have not been adjusted for variation in birth/rearing rank. Accordingly, weaning weights are lowest for lambs derived from outcross ewes sired by High rams, because of the greater number of twins. The wool production of these ewes also is depressed for the same reason. Weight of lamb weaned has been calculated as the product of average lamb weaning weight and lambs weaned/ewe.

The productivity index (last line, table 8) combines weight of lamb weaned with ewe fleece weight; it is assumed that 1 kg of wool is worth four times as much as 1 kg of weaning lamb live weight. The variation among the groups for this index is less than that for reproductive merit. This results from the opposing influences of reproduction and fleece production on overall productivity.

The similar premating live weights of each group suggest that the productivity ranking/ewe is likely to correspond to a ranking on productivity/ha. For the comparison between the High- and Low-sired ewes, this is supported by the ranking/ha observed for the High- and Low-fertility flocks in a stocking-rate trial (Joyce et al., 1976; Rattray et al., 1978).

DISSEMINATION OF GENETIC IMPROVEMENT--LESSONS FOR RAM BREEDERS AND RAM BUYERS

The long term selection experiment just described and other similar studies in New Zealand are starting to convince an increasing proportion of the ram breeding industry that performance recording and selection for economically important traits is worthwhile. The Breed Society umbrella covers 80% (3,500 flocks) of all ram breeders in New Zealand. The Federation of Livestock Breeding Groups represents the interests of a further 10%; most of the remainder breed rams just for their own use. For breed society registration of three breeds (Coopworth, Booroola, Bordersdale) it is mandatory to have performance recording (more specifically, membership of Sheeplan) and minimum performance levels.

Historically, breeders' selection decisions have been largely governed by breed society attitudes and showring standards, with the commercial ram buyers basically taking what was offered. Emphasis was on pedigree and sheep meeting breed-type standards. While the traditional breed societies (i.e., those not requiring performance recording) continue to have a major influence on breeders' and buyers' attitudes, they are becoming more aware of the mutual benefits available from a recording scheme like Sheeplan. Sheeplan is currently developing specific recording services for breed societies (e.g., pedigrees and flock books) that should further enhance this cooperation. The steadily increasing membership of Sheeplan confirms that it continues to meet the sheep industry's demands for a dynamic, effective national flock recording scheme. In March 1982, some 1300 flocks were recorded with Sheeplan; this represented about 30% of the ram breeding flocks in the country but contained 70% of the sire-producing ewes.

An important factor that has helped the success of sheep recording in New Zealand has been an active team of advisory specialists. When the first recording service was introduced in 1967, the practice of performance recording as an aid to ram breeding was very new. Not only was the industry inexperienced in this field but so were the advisers. In the initial years, therefore, there was emphasis on data collection and processing, somewhat to the detriment of emphasis on interpretation of processed records and putting those records to use in decision making.

With the introduction of Sheeplan some 8 years later, advisers and many breeders were by now familiar with performance recording. Furthermore, since Sheeplan now offered a variety of recording options, this placed pressures on breeders and their advisers to clearly define their breeding objectives and then choose the most suitable breeding plan and associated combinations of input and output options. Through this there is now a clearer realization that the recording scheme is simply a tool to assist the decision-making process; nothing will be achieved, unless the records are actually used rather than just collected.

There are currently 30 specialist advisers of the Ministry of Agriculture and Fisheries with animal breeding and extension training and skills. They are the principal people responsible for advising on Sheeplan in the field and its counterpart for the breed industry - Beefplan. They are also responsible for breeding advice to commercial farmers.

Another powerful force driving the breeder towards performance recording has been the purchasing power of the commercial producers. Once some of the early-performance-recording participants began seeing progress and having success selling performance recorded rams, the commercial ram buyer began to accept the philosophy of performance recording. This development received considerable encouragement from the Sheeplan advisory team. This has developed to the point where a high proportion of ram buyers now demand that the rams they buy be from performance-recording flocks and that the records be made available to them to assist in their choice of rams. Sheeplan assists advisers in this area by providing an annual membership list of Sheeplan participants, with details of breed, location, flock size, years on Sheeplan, and characters recorded. The design of the Sheeplan computer output recognizes that ram buyers and their agents, as well as breeders, will want to read them. For example, the important breeding values are grouped for easy reference, and cross-reference listings in breeding-value order are available.

ACKNOWLEDGEMENTS

Professor A. L. Rae and his associates over the years at Massey University were primarily responsible for the inauguration of performance recording with sheep in New Zealand. Professor Rae and Dr. J. N. Clarke together were mainly responsible for the technical aspects of developing

Sheeplan. This paper draws very heavily on their published work. Clare Callow kindly provided me with up-to-date details of recording options and input and output forms in her role as coordinator of Sheeplan activities.

REFERENCES

Clarke, E. A. 1967. Performance recording of sheep. Proc. of N.Z. Society of Animal Production 27:29.

Clarke, J. N. 1972. Current levels of performance in the Ruakura fertility flock of Romney sheep. Proc. of N.Z. Society of Animal Production 32:99.

Clarke, J. N. 1979. Providing flexibility in the choice of breeding objectives and recording options in centralized recording programs for breeders of dual-purpose sheep. Proc. of the Inauguaral Conference of the Australian Association of Animal Breeding and Genetics. p 397.

Clarke, J. N. and A. L. Rae. 1976. Sheeplan Advisers Manual. 78 pp.

Clarke, J. N. and A. L. Rae. 1977. Technical aspects of the present national sheep recording scheme (Sheeplan). Proc. of N.Z. Society of Animal Production 37:183.

Joyce, J. P., J. N. Clarke, K. S. Maclean and E. H. Cox. 1976. Proc. Ruakura Farmers Conference. p 34-38.

Owen, J. B. 1971. Performance Recording in Sheep. Commonwealth Agricultural Bureaux, Slough, England.

Rattray, P. V., K. T. Jagush, J. N. Clarke and K. S. Maclean. 1978. Proc. Ruakura Farmers Conference.

Wallace, L. R. 1964. Proc. Ruakura Farmers Conference p 1-12.

16

ECONOMICAL PERFORMANCES OF SUFFOLK SHEEP FOR MUTTON, WOOL, AND BREEDING IN MEXICO

Miguel A. Galina, M. Guerrero,
J. Gutierrez, J. J. Salas

Lamb production traditionally has been an important part of agriculture in Mexico (Perez, 1976). However, since the early 1900s sheep production has decreased from 16 million head to about 5 million in 1979, with a leveling off in the last ten years because of low profits and agricultural reforms (Galina et al., 1981; 1982) (figure 1).

FIGURE 1. OVINE FLOCK IN MEXICO FROM 1970 TO 1979

1970	1971	1972	1973	1974	1975	1976	1977	1978	1979
7.87*	7.95	8.09	8.18	7.83	8.29	7.86	7.86	4.79	5.84

*Numbers are in millions.
Source: Galina et al. (1981).

In 1980 74,662 sheep, valued at $2 million (U.S.), were imported for slaughter (mainly from North America) (Galina et al., 1981). The biggest Mexican expenditure for small ruminants was in the purchase of wool (mainly from Australia and Argentina who supply 88% of the wool imported by the Mexican textile industry) with a value of $25 million (U.S.) (Galina et al., 1981).

It has been well documented that sheep and sheep products are well accepted in Mexico and that some of the Mexican dietary customs rely heavily on mutton and that the artesans rely on the wool (Galina, 1980).

Mexicans have imported sheep from the U.S. for some time and some of the most popular breeds in Mexico come from U.S. livestock. However, U.S. sheep often have not fulfilled breeders' expectations due to miscalculations of sheep production capabilities under Mexican conditions. Thus, Mexico recently has developed a large program of sheep repopulation with Corriedale sheep imported from Australia and from New Zealand. This program is valued at more than $10 million (U.S.). These sheep will endure severe grazing conditions (Perez, 1982) and can be a viable alternative. However, the results of this program have not yet been evaluated and economic returns must be carefully analyzed before any conclusions are drawn from such a large program.

Because U.S. sheep often have not performed well in Mexico, several studies have been conducted by Mexican researchers, particularly with regard to the U.S. Suffolk. This breed has developed well in one of the largest sheep areas in Mexico. It is used for mutton, which is a very popular meat among Mexicans, and for wool that (in spite of poor quality) is used by Mexican artesans whose handicrafts are a source of income for the "campesinos" (Guerrero, 1982; Gutierrez, 1982).

MATERIAL AND METHODS

A series of experiments was developed to observe production capabilities of Suffolk sheep imported from the U.S. and Canada, as well as of sheep bred on Mexican farms. The flocks were purchased and located in a government breeding center and/or in a private enterprise. The objective was to produce ewes and sires of high quality for breeding purposes. The return of both enterprises was 70% to 80% of the ewes and sires sold as breeding stock and 20% to 30% sold for mutton. The objective was to determine production performance on these two farms as compared to that expected in North America in terms of costs and profits from such agricultural enterprises. The farms were located in the State of Mexico. The government farm is in Chapa de Mota, at 19°48' N and 99°37' W, 2800 m above sea level. The climate is temperate, with a rainy season during the summer months. The average rainfall is 800 mm, and the average temperature is 13.3°C (±10°C). The total surface area of the farm is 270 ha. The private farm is located in Huehuetoca at 19°33' N and 99°11' W. The farm's 292 ha are 2350 m above sea level with 600 mm of rainfall. Average temperatures are 14.4°C (± 11°C). Both farms were managed in a semitabulated system--with grazing during the summer and the fall and with zero grazing in the winter and spring. Both enterprises were studied from 1979 to 1981. In Chapa de Mota, 338 ewes and 22 Suffolk sires were imported from the U.S. or Canada, and 100 ewes were purchased from Mexican farmers. In Huehuetoca, 250 ewes were Mexican Suffolk and 250 were imported from the U.S. All sires were imported from the U.S.

Breeding season in Chapa de Mota was from November 15 to January 15. In Huehuetoca, breeding was from August 18 to September 18th. All of the experiment was recorded by the method described by Sidwell and Miller (1971 a; 1971b; 1971c; 1971d) and corrected by Notter and Copenhaver (1980a; 1980b) and Magid et al. (1981). The analyses and results were compared with the nutritional requirements recommended by Pope et al. (1974) in the National Research council (NRC) publication covering dry matter, protein, and energy. All traits were analyzed by least-square procedures for data with unequal class numbers, as described by Magid et al. (1981) after Harvey (1975). The statistical model of fixed effects used for the analysis of birth and weaning effects included discrete classes for age and birth year of ewe; sex; type of birth or rearing of lambs; sire breed of ewes; and dam breed. Weaning age of lambs was included as a constant variable.

RESULTS

Reproduction

The results obtained on both farms are summarized in figure 2.

FIGURE 2. REPRODUCTION PERFORMANCE OF SUFFOLK SHEEP

	Chapa de Mota (%)	Huehuetoca (%)
Fertility	67.6	80.0
Prolificacy	108.0	128.7
Lamb survival at 15 days	87.4	93.4
Lamb survival at weaning	98.0	92.6
Age at first breeding	18 months	9 months

Source: Galina et al. (1982).

As figure 2 shows, fertility was 67.6% for Chapa de Mota and 80.9% for Huehuetoca. Births occurred from April to June in Chapa de Mota and from January to March in Hueuetoca, with 80% of those births recorded in January or early February in Huehuetoca.

Average age per ewe breeding was 41 months (± 7) in Chapa de Mota and 48 months (± 10) in Huehuetoca. Age of the ewe at first breeding was 18 months for Chapa de Mota and 9 months for Huehuetoca. The average weight per ewe at breeding was 58.3 kg (± 5) in Chapa de Mota and 60 kg (± 4) in Huehuetoca. The average weight at first breeding was 40 kg (± 5) in Chapa de Mota and 42 kg (± 4) in Huehuetoca. The percentages of single births was 91.6% in Chapa and 62.4% in Huehuetoca; twin births were 8.4% and 37%, respectively. The average weight of single births was 4.8 kg in Chapa and 4.5 kg in Huehuetoca; average weights for multiple births were 3.5 kg and 3.4 kg, respectively. Average daily gain for each lamb was 174 g in Chapa de Mota and 221 for Huehuetoca . Figure 3 shows the nutrition management data from the farms compared with the NRC data.

The average cost per Mcal in Mexico in 1982 was about $.06 (U.S.).

In the least-square analyses, comparisons of the U.S. import Suffolk vs the Mexican Suffolk showed the U.S. Suffolks to be significantly ($P<.01$) better. However, in comparing performance on low-energy diets or with extensive grazing, Mexican sheep consistantly achieved better results ($P<.01$).

FIGURE 3. NUTRITION

Maintenance	Dry matter kg	Protein g	Energy Mcal
NRC	1.0	89	2.8
Huehuetoca	.86	108	3.61
Chapa de Mota	.80	75	2.4
Gestation			
First 15 weeks			
NRC	1.1	99	2.64
Huehuetoca	1.4	108	3.26
Chapa de Mota	1.2	90	2.04
Last 8 weeks			
NRC	1.7	158	4.23
Huehuetoca	1.0	127	4.23
Chapa de Mota	1.0	110	3.65
Lactation			
First 42 days			
NRC	1.7	158	4.36
Huehuetoca	1.6	180	4.20
Chapa de Mota	1.4	130	3.70
Last 46 days			
NRC	2.1	218	5.98
Huehuetoca	1.8	152	5.12
Chapa de Mota	1.5	130	3.90

Source: Galina et al. (1982).

DISCUSSION

The study data indicate a fertility rate of 68% in Chapa de Mota and 80% in Huehuetoca. These last results were in accordance with the fertility levels reported by Sidwell and Miller for Suffolk in the U.S. (1971 a; 1971b). Age, management, and nutrition were among the factors that negatively affected the fertility in Chapa de Mota. Similar conclusions were demonstrated by Magid et al. (1981). It has been reported that the low supply of nutritional requirements recommended by the NRC affects fertility due to the direct effect of nutrition on ovulation (Hulet et al., 1979; Scoot, 1977; Guerrero, 1982). Age apparently does not have a direct effect on the reproductive performances of

Suffolk when their average age at lambing is from 48 to 60 months. A breeding season delayed to December (instead of August) had a direct effect ($P<.05$) on fertility, which was perhaps related to previous management of the Suffolk in the U.S. where breeding management is similar to that at Huehuetoca.

Galina et al. (1982) reported on the direct effect of nutritional management on sheep production. In the present study, it was shown that the cost per ewe in Chapa de Mota was $100 (U.S.) per year and $132 (U.S) in Huehuetoca, due largely to the high cost of the Mcal in Mexico. However, because of the average cost ($200 [U.S.]) of the ewe at 6 months of age, the greater cost in Huehuetoca is economically justifiable. Prolificacy of the Suffolk in Chapa de Mota was extremely low and, in Huehuetoca, prolificacy was less than that reported in the U.S. for Suffolk (Sidwell and Miller, 1971c). It has been previously demonstrated that heritability of this factor must be improved by selection and no such effort was made in Chapa de Mota (Scott, 1977). Lamb survival at 15 days and at weaning has justified economically productive management measurements in both farms, as was demonstrated by Guerrero (1982); Gutierrez (1982); and Galina et al. (1982).

Birth weight had a direct influence on the age at first breeding, a finding that is in accord with previous reports (Notter and Copenhaver, 1980a; 1980b). Cost at first breeding was $150 (U.S.) for Chapa de Mota and $100 (U.S.) in Huehuetoca. These results justified the large initial investment due to faster return in profits. Average daily gain in lactating lambs of 171 g per day in Chapa and 221 g in Huehuetoca indicated that management was better in Huehuetoca. One of the more debatable issues regarding the use of U.S. sheep on Mexican farms is that U.S. Suffolk were developed to gain rapidly on high energy diets available because of the low cost of grain in the U.S. and Canada.

In this study, the data suggest that when a farm does not have the resources that will guarantee conditions similar to those in the U.S. regarding feed and sanitation, Mexican Suffolk will perform better than those imported from the U.S. and Canada (Galina et al., 1982).

Similar performances are generally difficult to obtain in Mexico because of high cost of grains and concentrates and because of exposure to environments commonly infested with parasites to which the sheep are not resistant (Galina, 1981). When a sheep producer invests $300 to $400 (U.S.) in U.S. sheep imported from Texas, which have shown high performances and growth, and then introduces these sheep to extensive grazing in a foreign environment with very low energy supplement and poor sanitation, he will probably be discouraged by low performances. However, a large proportion of sheep farmers seem to have developed a pride by owning U.S. livestock without evaluating their production performance under local conditions.

REFERENCES

Galina, M. 1981. Enfermedades mas frecuentes en ovinos en el valle de Mexico. Memorias ovinas. I Encuentro nacional sobre la produccion de ovinos y caprinos, Facultad de Estudios Superiores, Cuautitlan, National Autonomous University of Mexico (UNAM), Mexico.

Galina, M. 1980. Proyecto para la creacion de la maestria en produccion animal (ovinos y caprinos) Facultad de Estudios Superiores, Cuautitlan, National Autonomous University of Mexico (UNAM), Mexico.

Galina, M. J. Elizalde, M. Guerrero, J. Gutierrez and J. J. Salas. 1981. Comportamiento productivo del ovino Suffolk en el altiplano del valle de Mexico. Memorias Buriatria, Mexico. (In press.)

Galina, M., O. Rojas and J. Hummel. 1981. Diagnostico y perspectivas de la produccion ovina en Mexico. Memorias Ovinas. I Encuentro Nacional sobre produccion de ovinos y caprinos. Facultad de Estudios Superiores, Cuautitlan, National University of Mexico (UNAM), Mexico.

Guerrero, M. 1982. Evaluacion de la eficiencia productiva del rebano Suffolk del Centro Nacional de Fomento ovino en Chapa de Mota estado de Mexico (SARH de 1979-1981). Thesis, Facultad de Estudios Superiores, Cuautitlan, National Autonomous University of Mexico (UNAM), Mexico.

Gutierrez, J. 1982. Evaluacion de la eficiencia productiva de un rebano Suffolk en Huehuetoca, Estado de Mexico, de 1980-1981. Thesis. Facultad de Estudios Superiores, Cuautitlan, National Autonomous University of Mexico (UNAM), Mexico.

Harvey, W. R. 1975. Least-squares analysis of data with unequal subclass numbers. USDA.

Hulet, C.V., T. Tueller and D. Knight. 1979. Semiconfinement sheep production. USDA Exp. Sta. Bull., Dubois, Idaho.

Magid, A. F., V. B. Swanson, J. S. Brinks, G. E. Dickerson and G. M. Smith. 1981. Border Leicester and Finnsheep crosses II. Productivity of Fl Ewes. J. Anim. Sci. 52:1262.

Notter, D. R. and J. Copenhaver. 1980 Performances of Finnish Landrace crossbred ewe under accelerated lambing. I. Fertility, prolificacy and ewe productivity. J. Anim. Sci. 51:1033.

Notter, D. R. and J. S. Copenhaver. 1980. Performances of Finnish Landrace crossbred ewe under accelerated lambing. II. Lamb growth and survival. J. Anim. Sci. 51:1043.

Perez, I.M.A. 1982. Personal communication. Direccion General de Ganaderia SARH. Hamburgo 45, Mexico.

Perez, I.M.A. 1976. Analisis evolutivo de la ganaderia ovina nacional de 1940-1976. Thesis. Facultad de Medicina Veterinaria y Zootecnia National Autonomous University of Mexico (UNAM), Mexico.

Pope, L. A., E. J. Butcher, E. Dinusson, S. Garringus, E. D. Hogue, C. W. Weir. 1975. Requirements of sheep. National Academy of Sciences, National Research Council (NRC).

Scot, G. E. 1977. The sheepman's production handbook. Sheep Industry Development Program, Denver, Colorado.

Sidwell, G. M. and L. Miller. 1971a. Production in some pure breeds of sheep and their crosses. I. Reproduction efficiency in ewes. J. Anim. Sci. 32:1084.

Sidwell, G. M. and L. Miller. 1971b. Fertility, prolificacy and lamb livability of some pure breeds and their crosses. II. Birth weight and weaning weight of lambs. J. Anim. Sci. 32:1090.

Sidwell, G. M. and L. Miller. 1971c. Fertility, prolificacy and lamb livability of some pure breeds and their crosses. J. Anim. Sci. 32:1079.

Sidwell, G. M. and L. Miller. 1971d. Production in some pure breeds of sheep and their crosses. III. Production of ewes. J. Anim. Sci. 32:1095.

17

DEVELOPMENT OF THE POLYPAY BREED OF SHEEP

Clarence V. Hulet, S. K. Ercanbrack,
A. D. Knight

INTRODUCTION

The U.S. Department of Agriculture reported an estimated 1981 lamb crop of 101 lambs saved for every 100 ewes on hand one year of age or older (lamb crop defined as those born in the "native" states and lambs docked or branded in the western states). This is the highest U.S. lamb crop ever reported and is the first time the national average lamb crop has exceeded 100%. Although the improvement is encouraging, sheep are not efficient converters of feed to meat and fiber at such levels of reproductive performance (Byerly, 1967). However, there is a great opportunity for major improvement in production efficiency if reproductive rate can be markedly increased (Dickerson, 1970).

Achievement of improved reproductive performance in most domestic sheep breeds of the U.S. has been regarded with considerable caution because of low heritabilities for reproduction traits. However, this view is changing with the increasing availability of exotic breeds that have high reproductive capacity. The characterization of these exotic breeds in crosses with domestic breeds has shown that major improvement in reproductive rate is attainable. Crossbreeding has been demonstrated to be an effective method for exploiting the commercial value of highly prolific breeds (Carter, 1976; Dickerson, 1977; Jakubec, 1977; Meyer et al., 1977; Turner, 1969). An important extension from crossbreeding for important and permanent increases in production efficiency is the development of synthetic breeds that combine into a purebred type the potential for greatly increased reproductive rate along with the desired growth, carcass, and fleece attributes. (Carter, 1976; Hulet et al., 1981; Smith et al., 1979).

Development of the "Polypay" breed (the name was coined to suggest more than two paying crops per year, i.e., one wool and two lamb crops) was begun with matings made at the U.S. Sheep Experiment Station in 1968. The objective was to develop a breed with a reproductive capacity markedly superior to that of domestic western-range breeds.

The purpose of this paper is to compare some of the reproductive and body trait records of the Polypay during the first few years of its development, with those of the contemporary breeds and crosses from which the Polypay was derived.

MATERIAL AND METHODS

Five primary goals were adopted for the proposed new breed:

- High lifetime prolificacy.
- A good lamb crop at 1 year of age.
- Ability to lamb more frequently than once per year.
- Good growth rate of lambs.
- Good carcass quality.

A twice-a-year lambing schedule (conventional spring lambing plus a fall lambing) was adopted because it is more adaptable to range operation management than was a schedule of three lambings at approximately equal intervals during a two-year period.

Four breeds were selected for the foundation of the new breed -- the Rambouillet and the Targhee because of their hardiness, size, long breeding season, herding instinct, and fleece characteristics; the Dorset because of its carcass quality, milking ability, long breeding season, and white fleece; and the Finnsheep because of its early puberty, early postpartum fertility, and high lambing rate.

Four unrelated Dorset rams from widely separated areas of the U.S. (Oregon, California, Montana, and North Carolina) and a Finnsheep ram (one from among the initial group imported into the United States in 1968 [Oltenacu and Boylan, 1981]) were purchased to cross on Rambouillet and Targhee ewes from the U.S. Sheep Experiment Station. Initial Dorset x Targhee and Finn x Rambouillet crosses were made in 1968. These were followed in 1969 by Dorset-Targhee x Finn-Rambouillet crosses. Crossbreds resulting from the two-and four-breed foundation crosses were then inter se mated and selected according to established criteria. Between 1969 and 1972, four additional Finnsheep rams were purchased. During years subsequent to the initial crossings, additional two- and four-breed crosses were produced, top-crossed with inter se mated sires of the same respective foundation cross, and their selected offspring included in the inter se mated groups.

Ewes were managed under fenced range conditions during most of the grazing season but were herded on open sagebrush-wheatgrass range during the late fall. When grazing was not available, ewes were kept in dry lot and fed pelleted alfalfa hay. Hormone therapy was used for the spring breeding (Hulet and Stromshak, 1972) through 1978, and lambs were born in April and October of each year. Lambs were weaned at about 80 days of age and put on pasture with creep or in dry lot (depending upon the availability of grass) until weanlings averaged about 120 days of age. Weanling average daily gain was determined for the interval between birth and 120 days of age.

The use of hormones as an aid to achieving twice-a-year lambing was discontinued in 1978 because of uncertain availability of exogenous hormones and possible future problems with regulatory agency approval. After 1978, selection was combined with management procedures to lengthen the natural breeding season to enhance the achievement of twice-a-year lambing. Adjustments in lambing dates and weaning ages were a part of the change in management procedures.

Ewes started lambing on January 16, 1979. During each subsequent year, winter lambing was scheduled to start about one week earlier than during the previous year. Winter-born lambs generally were early-weaned at about 31 days of age to optimize early postpartum rebreeding. Early weaned lambs were kept in dry lot and fed to achieve a rapid growth rate. Summer-born lambs were generally not early weaned. Collection of weanling data since 1979 continued to be at about 120 days of age for both winter-born and summer-born lambs.

Weanling and yearling scored traits were estimated by three independent scorers using five-point system with plus and minus scores for each point, thus providing 15 scoring units. Lower scores indicate superior merit. To facilitate analysis, scores were coded to a scale ranging from 2 through 16 with codes 3, 6, 9, 12, and 15 corresponding to score values of 1, 2, 3, 4, and 5, respectively. Wool grade is a code of spinning count and ranges from 1 (70s) through 9 (48s). Thigh score is a measure of uniformity between side-and thigh-wool grades. A thigh score of 1 indicates that side grade and thigh grade are the same. The score increases by 1 for each grade that thigh grade is coarser than side grade. Yearling data included in the study were obtained only from spring-born ewes and collected at shearing time (late May) when the average age was near 400 days. Yearling staple length, wool grade, and face score were determined just prior to shearing. Fleeces were weighed on the shearing floor and body weights were determined immediately after shearing. Yearling average daily gain was determined for the period between weaning (120 days) and shearing (400 days).

Selection of replacement-ewe lambs was made after weanling data were obtained. Culling occurred in late summer of each year before ewes were put into breeding. Ewes put into breeding generally completed the annual production cycle, and no additional culling took place until the next summer. Primary emphasis in selection was on lambing performance of dams when given two opportunities to lamb per year and also on each ewe's own lambing performance at about 12 to 14 months of age. Sires were selected on the basis of dam lifetime-lamb-production record plus his own growth rate from birth to weaning.

Because of the nature of the breeding program, the age profiles of the various breeding groups were similar but not characteristic of a typical flock. During the first five years of the breeding program (1969 through 1973) a chi-square analysis was made of the respective age group data on a within-year basis. Data for 1974 and 1975 were obtained from a reasonable consistent population of different-aged ewes (2 through 5 years old) within each breed group and were analyzed by least-squares procedures. The 1979-1981 reproductive performance data have been included for the Polypay and some contemporary breeds and crosses; however, statistical comparisons are not appropriate because of differences in management practices among the breeds and crosses.

RESULTS AND DISCUSSION

Initial Comparisons Among Breeds and Crosses

Superior lambing performance at yearling age was, from the beginning, regarded as one of the important characteristics to be possessed by the new breed. The first yearlings of the newly established Dorset x Targhee and Finn x Rambouillet crosses lambed in April 1969. Comparisons made among Rambouillet, Targhee, and Columbia purebreds and the Dorset x Targhee and Finn x Rambouillet crosses showed that lambing performance of the Finn x Rambouillets was outstanding. performance of the Columbias was markedly inferior to the other breed groups, thus they were not included in future trials.

Reciprocal Dorset-Targhee x Finn-Rambouillet matings were first made in the fall of 1969. Beginning in 1970, the respective two-breed crosses and the Dorset-Targhee x Finn-Rambouillet cross were each mated inter se and selected, along with purebred Rambouillets and Targhees, for best lamb production when given two opportunities to lamb each year. Spring lambing performance of yearling ewes over the three-year period from 1971 through 1973 is shown in table 1. Performance of Finn-Rambouillet and Dorset-Targhee x Finn-Rambouillet (Polypay) was consistently at a high level. Targhee and Dorset-Targhee generally were poorer performers than were Finn-Rambouillet and Polypay but were superior to the slower maturing Rambouillet.

Early indications of response to twice-a-year lambing are shown in table 2. Of all yearlings exposed to breeding during the spring of 1972, 52% of the Polypay lambed that fall, a much higher percentage (P<.01) than of Rambouillet (27%), Targhee (29%), Dorset-Targhee (26%) or Finn-Rambouillet (22%). Ewes that lambed during the fall of 1972 were exposed (while lactating) to rams for about 6 weeks, beginning within 2 weeks of lambing. During the next spring lambing (as 2-year-olds) the lambing performance (fertility) of Polypay (100%) was superior (P<.01) to that of Rambouil-

TABLE 1. FERTILITY OF PUREBRED AND INTER SE MATED EWE LAMBS BRED IN NOVEMBER AND DECEMBER AT 7 TO 8 MONTHS OF AGE

	No. ewes exposed			Percent of ewes lambing		
Breed or cross	1971	1972	1973	1971	1972	1973
Rambouillet	27	38	30	4[a]	16[a]	37[a]
Targhee	31	35	37	10[ab]	57[b]	73[b]
Dorset-Targhee	20	39	42	35[b]	64[b]	86[bc]
Finn-Rambouillet	39	32	31	92[c]	94[c]	87[bc]
Polypay[2]	35	65	33	77[c]	94[c]	97[c]

[1] Feeding, management, and weather conditions varied in some respects among years, but all breeds and crosses were treated similarly within years.
[2] Polypay are (Dorset-Targhee) x (Finn-Rambouillet).
[a, b, c] Means within columns not followed by the same superscript differ significantly ($P<.05$).

TABLE 2. EARLY POSTPARTUM FERTILITY OF FALL-LAMBING PUREBRED AND INTER SE MATED EWES AS INDICATED BY NUMBER OF EWES LAMBING THE FOLLOWING SPRING

	Ewes lambing					
	Fall	1972	Spring	1973	Spring 1973 by April 24[1]	
Breed or cross	No.	%[2]	No.	%[3]	No.	%[4]
Rambouillet	33	27[a]	25	76[a]	8	32[a]
Targhee	21	29[a]	17	81[a]	11	65[b]
Dorset-Targhee	19	26[a]	18	95[ab]	14	78[b]
Finn-Rambouillet	17	22[a]	14	82[a]	11	79[b]
Polypay[5]	14	52[b]	14	100[b]	11	79[b]

[1] Cut-off date for hormone treatment of spring lambing ewes.
[2] Percentage of 122 Rambouillet, 72 Targhee, 73 Dorset-Targhee, 77 Finn-Rambouillet, and 27 Polypay ewes treated and exposed to breeding during the spring of 1972.
[3] Percentage of 1972 fall-lambing ewes that lambed during the spring of 1973.
[4] Percentage of 1972 fall-lambing ewes that lambed before April 24, 1973.
[5] Polypay are (Dorset-Targhee) x (Finn-Rambouillet).
[a,b] Means within columns not followed by the same superscript differ significantly ($P<.05$).

TABLE 3. LEAST-SQUARES ESTIMATED MEANS AND STANDARD ERRORS FOR 1974-1975 REPRODUCTIVE PERFORMANCE OF TARGHEE, DORSET-TARGHEE, FINN-RAMBOUILLET, AND POLYPAY

	Winter lambing					Summer lambing			
		Lambs born	Lambs weaned	Weight (kg) weaned		Lambs born	Lambs weaned	Weight (kg) weaned
Breed group	Ewes exposed	ewes exposed	ewes exposed	ewes exposed	Ewes exposed	ewes exposed	ewes exposed	ewes exposed
Targhee	155	1.37 ± .05[a]	1.07 ± .06	35.96 ± 2.02	116	.48 ± .07[a]	.32 ± .07[a]	9.97 ± 2.06
Dorset-Targhee	168	1.28 ± .05[a]	.98 ± .06	33.39 ± 1.99	131	.60 ± .07[a]	.41 ± .07[a]	12.90 ± 2.04
Finn-Rambouillet	202	1.72 ± .04[b]	1.11 ± .06	34.30 ± 1.87	146	.55 ± .06[a]	.42 ± .06[a]	12.08 ± 1.93
Polypay	175	1.43 ± .05[a]	1.10 ± .07	36.22 ± 2.27	141	.79 ± .08[b]	.62 ± .08[b]	18.52 ± 2.31
Age								
2	228	1.22 ± .04[a]	.82 ± .05[a]	26.46 ± 1.72[a]	180	.77 ± .06[a]	.52 ± .05	15.31 ± 1.70
3	205	1.55 ± .04[b]	1.16 ± .06[b]	38.09 ± 1.77[b]	152	.58 ± .05[b]	.43 ± .06	13.07 ± 1.85
4+[c]	267	1.58 ± .05[b]	1.22 ± .06[b]	40.35 ± 1.83[b]	202	.46 ± .05[b]	.37 ± .06	11.73 ± 1.87

[a,b] Means in the same column (within breed group or within age group) with different superscripts differ significantly (P<.05).

[c] Age group includes all ewes 4 years of age and older.

let (76%), Targhee (81%), and Finn-Rambouillet (82%), but not to Dorset-Targhee (94%). Although 76% of the Rambouillet lambed in the spring, only 32% of those that lambed did so sufficiently early to be hormone-treated and rebred for the next fall lambing. This percentage was somewhat lower ($P<.05$) than the other breed groups (65%, 78%, 79%, and 79% for Targhee, Dorset-Targhee, Finn-Rambouillet, and Polypay, respectively). Estrous observations indicated that the poor performance of the Rambouillet was due in part to delayed estrus after parturition. the longer gestation in the Rambouillet probably also contributed to poorer performance. Because of continued poor performance of Rambouillet as yearlings, and their poor response to a twice-a-year lambing schedule, they were dropped from the trials in 1975 to provide additional space and facilities for increasing numbers in the other groups.

Reproductive Performance of Young Mature Ewes During Spring and Fall Lambings

By 1974 some of the early two-breed and four-breed-cross ewes were 4- and 5-year-olds. Advanced generation inter se cross ewes, as well as some additional F_1 generation crosses that had been added to the breed groups each year, were 2- and 3-year-olds. Reproductive performance traits of these young mature ewes are shown in table 3.

Winter lambing prolificacy of Finn-Rambouillet was superior ($P<.05$) to other breed groups, but net reproductive rate and weight of lamb weaned was comparable among all breed groups. During the summer lambing season, performance of Polypays was superior ($P<.05$) to each of the other breed groups for fertility, net reproductive rate, and weight of lamb weaned. When total winter and summer production was expressed on the basis of number of ewes put into breeding for the annual winter lambing, Polypay ewes weaned 1.60 lambs and 51.2 kg of lamb per year (number of lambs alive at 120 days of age and weight of lamb adjusted to 120 days of age). The corresponding number of lambs and kilograms of lamb, respectively, at 120 days of age was 1.31 and 43.2 for Targhee, 1.30 and 43.5 for Dorset-Targhee, and 1.42 and 43.0 for Finn-Rambouillet.

Weanling and Yearling Growth, Body and Fleece Traits

Least-squares means for 1974-1975 weanling traits are presented in table 4. Breeding was a significant source of variation for all traits except staple length (table 4). Dorset-Targhee were superior ($P<.01$) to the other breed groups for type score and condition score. Targhee and Polypay were superior ($P<.01$) to Finn-Rambouillet for both type and condition scores, but Polypay also were superior ($P<.01$) to Targhee for type score. Targhee wool was finer than the wool of the other three breed groups. Wool of Finn-Rambouillet was slightly finer ($P<.01$) than that of

TABLE 4. LEAST-SQUARES ESTIMATED MEANS AND STANDARD ERRORS BY BREEDING GROUP FOR BIRTH WEIGHT AND WEANLING TRAITS

Breed or cross[1]	No.	Birth weight kg	Weaning weight, kg	Face score[2]	Type score[2]	Condi-tion score[2]	Staple length, cm	Wool side grade[3]	Wool thigh grade [3]	Wool thigh score[4]	Weaning ADG, kg
T	217	4.40 ± .05^{a}	34.29 ± .39^{a}	10.17 ± .16^{a}	8.63 ± .10^{a}	7.58 ± .10^{a}	4.55 ± .05^{a}	3.11 ± .07^{a}	4.12 ± .09^{a}	2.00 ± .05^{a}	.236 ± .003^{b}
DT	222	4.10 ± .06^{b}	34.37 ± .41^{a}	10.55 ± 17^{a}	7.45 ± .11^{b}	6.84 ± .10^{b}	4.64 ± .05^{a}	3.59 ± .08^{b}	4.79 ± .09^{b}	2.20 ± .05^{b}	.239 ± .003^{a}
FR	290	3.51 ± .05^{c}	32.21 ± .38^{b}	6.19 ± .15^{b}	9.24 ± .10^{c}	8.27 ± .10^{c}	4.57 ± .04^{a}	3.36 ± .07^{c}	4.49 ± .08^{c}	2.14 ± .05^{b}	.226 ± .003^{b}
P	277	3.78 ± .05^{d}	33.70 ± .36^{a}	8.91 ± .15^{c}	8.22 ± .10^{d}	7.44 ± .09^{a}	4.63 ± .04^{a}	3.64 ± .07^{b}	4.84 ± .08^{b}	2.19 ± .05^{b}	.236 ± .003^{a}

[1] T = Targhee, DT = inter se mated Dorset-Targhee, FR = inter se mated Finn-Rambouillet, P = Polypay.
[2] Scored traits are ranked according to merit on a 15-point scale ranging from 2 through 16; lower score values indicate superior merit.
[3] Wool grade (spinning count) coded on a scale from 1 (=70s) through 9 (=48s).
[4] A thigh score of 1 indicates side and thigh grades are equal. The score increases by 1 for each grade thigh is coarser than side.
[a,b,c,d] Means within columns not followed by the same superscript differ significantly ($P<.05$).

Dorset-Targhee and Polypay, but none of the four breed groups differed in thigh score.

Least-squares means for yearling traits are presented in table 5. Although results reflect the selection that occurred between weaning and yearling ages, it is evident that some loss in fleece weight was associated with the influence of Finnsheep and Dorset breeding. Targhee grease-fleece weight was heavier (P<.01) that that of the Finn-Rambouillet (by 38%), that of .Polypay (by 24%), and that of Dorset-Targhee (by 11%). Targhee also had finer wool (P<01) than the other three breed groups. Targhee wool grade averaged from 60s to 62s, Polypay and Finn-Rambouillet--60s, and Dorset-Targhee 60s to 58s.

Differences in face score (Finn-Rambouillet most open, followed by Polypay, Targhee, and Dorset-Targhee) reflected a breed-type characteristic, which was probably only minimally affected by selection. But scores indicated that average face covering of yearling ewes was less than that of weanling-age lambs. Dorset-Targhee at yearling age continued to have a slight advantage over the other breed groups for type score, but condition scores of Targhee, Dorset-Targhee, and Polypay were very similar and all were superior (P<.01) to Finn-Rambouillet.

By 1976 it was evident that the inter se mated Polypay offered good balance of body growth and type traits, acceptable fleece characteristics, and an excellent opportunity for achieving the reproductive performance goals that had been established for the new synthetic breed. Dorset-Targhee and Finn-Rambouillet inter se matings were discontinued as a part of the accelerated lambing program. Polypay numbers were increased, but intensive selection was continued for twice-a-year lambing ability.

The 1978 decision to discontinue hormone therapy during the winter breeding season was accompanied by major management adjustments. A series of experiments was initiated to study such factors as lactation, nutritional status, breeding season of year, the sudden introduction of the ram (ram effect), and other management considerations for achieving early postpartum rebreeding (Hulet et al., 1982) Stellflug and Hulet, 1982). Under these new management programs, Polypay continued to perform very satisfactorily and exhibited a clear potential for superior reproductive performance on an accelerated lambing schedule. Additionally, two Polypay lines were initiated in 1978 to be selected for once-a-year lamb production when managed under typical range conditions. Current performance of the once-a-year lambing Polypay is very competitive with the of one-half Finn crosses and superior to that of Rambouillet. Targhee, and one-fourth Finn crosses (table 6). Contemporary reproductive performance means of Polypays on the twice-a-year lambing schedule are also included in table 6. The current annual production of Polypay on the twice-a-year lambing schedule was about 14% more lambs weaned and 19% more weight of lamb weaned than Polypay on the once-a-year lambing

TABLE 5. LEAST-SQUARES ESTIMATED MEANS AND STANDARD ERRORS BY BREEDING GROUPS FOR YEARLING EWE TRAITS

Breed or cross[1]	No.	Yearling weight, kg	Yearling ADG, kg	Face Score[2]	Type score[2]	Condition score[2]	Staple length cm	Grease fleece weight, kg	Fleece grade[3]
T	62	50.13 ± .88^{a}	.048 ± .003^{a}	8.56 ± .24^{a}	6.90 ± 16^{ab}	7.37 ± 11^{a}	9.12 ± .15^{a}	4.18 ± .08^{a}	3.76 ± .19^{a}
DT	47	48.84 ± 1.02^{a}	.042 ± .003^{a}	8.78 ± .28^{a}	6.59 ± .19^{b}	7.34 ± .13^{a}	8.73 ± .17^{a}	3.74 ± .10^{b}	4.28 ± .23^{a}
FR	55	48.41 ± 1.12^{a}	.045 ± .003^{a}	4.28 ± .31^{b}	7.65 ± .20^{c}	7.98 ± .15^{b}	8.74 ± .19^{a}	3.03 ± .11^{c}	4.04 ± .25^{b}
P	53	47.78 ± 1.31^{a}	.042 ± .004^{a}	7.16 ± .36^{c}	7.26 ± .24^{ac}	7.45 ± .17^{a}	8.68 ± .22^{a}	3.37 ± .12^{d}	4.02 ± .29^{b}

1 T = Targhe, DT = inter se mated Dorset-Targhee, FR = inter se mated Finn-Rambouillet, P = Polypay.
2 Scored traits are ranked according to merit on a 15-point scale ranging from 2 thourgh 16; lower score values indicate superior merit.
3 Wool grade (spinning count) coded on a scale from 1(=70s) through 9(=48s).
a,b,c,d Means in the same column (within breed group or within age group) with different superscripts differ significantly ($P<.05$).

TABLE 6. AVERAGE REPRODUCTIVE PERFORMANCE OF ONCE- AND TWICE-A-YEAR LAMBING GROUPS, 1979-1980

			Lambs born	Lambs weaned	Weight (kg) weaned
Breed	Age	Ewes exposed	Ewes exposed	Ewes exposed	Ewes exposed
		Once-a-year lambing groups			
Rambouillet	1	609	.69	.47	14.2
	2	400	1.27	.93	30.4
	3	390	1.47	1.18	36.7
	mature	1007	1.59	1.32	40.6
Targhee	1	623	.54	.36	10.5
	2	407	1.18	.86	29.0
	3	435	1.29	1.00	32.8
	mature	1069	1.48	1.17	37.8
1/4 Finn	1	458	.91	.53	15.8
	2	278	1.43	1.01	32.5
	3	163	1.68	1.30	42.0
	mature	480	1.72	1.43	46.3
1/2 Finn	1	532	1.23	.76	21.1
	2	331	1.81	1.27	38.6
	3	212	2.01	1.54	49.2
	mature	485	2.09	1.59	50.7
Polypay	1	312	1.11	.73	21.5
	2	196	1.66	1.34	42.2
	3	171	1.89	1.61	49.3
	mature	165	1.83	1.49	46.5
		Twice-a-year lambing Polypay			
Polypay (winter lambing)	1	362	1.12	.85	25.4
	2	268	1.45	1.24	40.6
	3	204	1.76	1.50	49.2
	mature	242	1.74	1.41	47.9
Polypay (summer lambing)	1	-	-	-	-
	2	233	.36	.33	10.7
	3	164	.56	.43	13.4
	mature	190	.47	.37	11.7
Polypay (annual[1])	1	362	1.12	.85	25.4
	2	268	1.76	1.52	49.9
	3	204	2.21	1.85	60.0
	mature	242	2.11	1.70	57.1

[1] Annual reproductive performance of Polypay ewes is based on the number of ewes exposed for the winter lambing.

schedule. Note, however, that caution must be exercised in comparing once-a-year and twice-a-year lambing performance because of management differences between these groups.

General Conclusions

The challenge in development of the Polypay has been to achieve markedly improved reproductive performance in such a way that adaptability, growth rate, carcass quality, and wool characteristics are not unacceptably compromised. Polypay lamb growth rate, body type, and body-condition scores are comparable to or superior to those of other breeds and crosses at the U.S. Sheep Experiment Station. The typical mature Polypay ewe (figure 1) weighs about 65 kg and produces 4.2 kg of 58s spinning count wool (table 7). Although fleece weight and wool grade characteristics show some influence from the Finnsheep in the foundation cross, the compromise is relatively economically unimportant when considered in conjunction with the Polypay's superior reproductive performance.

ACKNOWLEDGMENT

The authors acknowledge the cooperation of the University of Idaho, Moscow, Idaho.

TABLE 7. LEAST-SQUARES MEANS AND STANDARD ERRORS FOR BODY WEIGHT, FLEECE WEIGHT, AND FLEECE GRADE OF POLYPAY EWES AND RAMS

Sex	Age group	No.[a]	Body weight, kg	Fleece weight, kg	Fleece grade score[b]
Ewes	1	338	45.8 ± .42	3.8 ± .05	4.7 ± .03
	2	269	57.7 ± .46	4.1 ± .06	5.1 ± .04
	3	188	64.0 ± .56	4.3 ± .07	5.2 ± .04
	4[c]	215	65.3 ± .52	4.2 ± .06	5.3 ± .04
Rams	1	159	65.0 ± .61	5.2 ± .10	5.0 ± .09

[a] Includes weights and fleece grade scores obtained during the two-year period, 1980-1981. Animals and fleeces were weighed immediately after shearing.
[b] The numerical score for fleece grade (spinning count) ranges from 1 (=70s) through 9 (=48s).
[c] Age group 4 includes 4-, 5-, and 6-year olds.

REFERENCES

Byerly, T. C. 1967. Efficiency of feed conversion. Science 157: 890-895.

Carter, A. H. 1976. Exploitation of exotic genotypes. In: Sheep Breeding. Proceedings of the 1976 International Congress, Muresk and Perth, Western Australia. pp 159-170.

Dickerson, Gordon. 1970. Efficiency of animal production-molding the biological components. J. Anim. Sci. 30-849.

Dickerson, Gordon E. 1977. Crossbreeding evaluation of Finnsheep and some U.S. breeds for market lamb production. North Central Regional Publication No. 246, pp 1-30.

Dufour, J. J. 1975. Effects of seasons on postpartum characteristics of sheep being selected for year-round breeding and on puberty of their female progeny. Can. J. Anim. Sci. 55:487.

Gabriel, K. R. 1966. Simultaneous test procedures for multiple comparisons on categorical data. American Statistical Association Journal 61:1081.

Harvey, W. R. 1975. Least-squares analyses of data with unequal subclass numbers. U.S. Department of Agriculture. ARS H-4.

Hulet, C. V. and F. Stormshak. 1972. Some factors affecting response of anestrous ewes to hormone treatment. J. Anim. Sci. 34:1011.

Hulet, C. V., A. D. Knight and S.K. Ercanbrack. 1981. The future for new or synthetic breeds or types of sheep. International Goat and Sheep Research. In press.

Hulet, C. V., J. N. Stellflug and A. D. Knight. 1982. Effect of time of early weaning and time of lambing on accelerated lambing in Polypay sheep. J. Anim. Sci. (In press).

Jacubec, V. 1977. Productivity of crosses based on prolific breeds of sheep. Livestock Prod. Sci. 4:379.

Kramer, C. Y. 1957. Extension of multiple range tests to group correlated adjusted means. Biometrics 13:13.

Meyer, H. H., J. N. Clarke, M. L. Bigham and H. H. Carter. 1977. Reproductive performance, growth, and wool production of exotic sheep and their crosses with the Romney. Proceedings of the New Zealand Society of Animal Production 37:220.

Oltenacu, E. A. Branford and W. J. Boylan. 1981. Productivity of purebred and crossbred Finnsheep. I. Reproductive traits of ewes and lamb survival. J. Anim. Sci. 52:989.

Smith, C., J. W. B. King, D. Nicholson, B. T. Wolf and D. R. Brampton. 1979. Performance of crossbred sheep from a synthetic dam line. Anim. Prod. 29:1.

Stellflug, J. N. and C. V. Hulet. 1982. Photoperiod control and management for improving reproductive efficiency in sheep. J. Anim. Sci. (In press).

Turner, Helen Newton. 1969. Genetic improvement of reproduction rate in sheep. Animal Breeding Abstracts 37:545.

18

INDIVIDUAL RECORDS: THEIR EFFECT ON PRODUCTIVITY

Arthur Christensen

All sheep producers are looking for ways to make their operation more efficient and profitable. To make programs, ewes must be culled and replacements selected. Management practices and breeding stock must be evaluated. There are two ways to make these decisions: visual appraisal and production records. Each of these methods has advantages and disadvantages and each has its place. "Eyeballing" by a trained person is very accurate in grading wool or determining body condition and conformation. A good observer can spot health problems quickly and evaluate feed and management programs. However, when it comes to culling sound ewes and selecting replacements for lamb production, the visual system breaks down. We tend to sell the thin ewes that raised twins and to keep the beautiful ewe that didn't raise a lamb. Although research shows that twins have the potential for higher production, twin lambs never seem to have the "quality" of a nice big single. It's a rare breeder that isn't fooled by the saying, "Fat is a beautiful color." Production records are the only accurate means of measuring unseen reproductive traits. The dairy industry is a good example of the progress that can be made by production testing.

The big disadvantage of keeping records is the time and cost involved. In our operation, we estimate that the extra cost is about $5.00 per head. Although individual records are better than eyeballing, they are no more than 80% accurate in predicting the productivity of an animal.

LESSONS LEARNED FROM NEARLY 30 YEARS OF RECORD KEEPING

- Records are of little value unless all other management aspects are under control. A high-producing ewe is no better than a cull if she aborts her lambs, loses them to coyotes, or is improperly fed. Progress is not always a straight line upwand and onward. Sometimes there are dips. Our 1982 production is lower than it was 3 years ago. We know that we must go back to the "drawing board" to find a way to control enzootic abortion and mastitis. Two weeks of heavy snow in the middle of lambing this year did not help.

- Ewes that are nursing three or more lambs require much more feed than most authorities recommend.
- Progress comes quickly when production testing starts and the best producing are selected. Records on each ewe eliminates the "born losers" that can't be detected by eyeballing.
- Record keeping can be drudgery, or fun--even fascinating. We spend about 1 hour per ewe a year at this activity. On the other hand, feeding and caring for animals that are not producing is also expensive and time consuming. A good set of ear tags and number brands save time and labor not only for keeping a good set of records but also when sheep must be segregated or handled.
- Records should be simplified, standardized, and include only vital information. Abbreviations and color codes help. Selecting for too many traits is a mistake. Since we get paid only for the pounds of lamb and wool we sell in the fall, those are the traits we select for.

Father Time would overtake a horse breeder before the breeder could develop one animal that could serve as a kid's pony, pull a beer wagon, and win the Kentucky Derby.

Just as horses serve different purposes, sheep are not all alike. We have several ewes that consistently raise four lambs and about 100 ewes that raise three. One ewe has raised five, and a few haven't raised any. One ewe that gave birth to four lambs lost all of them, plus three grafts. She did a better job in the freezer.

The record system we use was developed over many years, mostly with the help of Agricultural Extension personnel.

<u>Identification: Identify animals with ear tags.</u> Shortly before breeding season, replacement ewe lambs are assigned a lifetime number. A large plastic tag with 1 1/2" numbers is placed in the right ear and a small metal tag with the same number is attached to the left ear. Each age group has a colored-coded plastic tag. Since some tags are always lost, no identification system will work unless there are at least two tags or marks.

Shortly after shearing, the ewe's lifetime number is painted high on her right side in large (4") black numerals. These numbers are readable for about three months and are invaluable from prelambing until the herd goes to the mountains in June. When ewes lamb the lambing date is painted beneath the lifetime no. in a color code to show the type of rearing. Newborn lambs are assigned their mother's number in an eartag and one side is painted with 2" numbers in a color that indicates rearing type: black for singles; green, twins; red, triplets; blue, quads. The eartag has their identification number plus a letter: A for 1st lamb; B for 2nd lamb; and C for 3rd lamb.

Record Cards. Each replacement ewe is assigned a 5 x 8 record card that shows her lifetime production. Figures 1 and 2 show two production records and the wide variation in performance. Few ewes have a bad record since none are kept if they haven't raised twins or better after two years.

In the following example (figure 2) 93X raised 26 lambs in 10 years while no. 211 weaned one lamb in three years.

Explanation of Record Cards

Section A of figure 1 illustrates the lifetime record for a ewe.

Column No.	Information Recorded
1	Permanent no.
2, 3, 4	Self-explanatory
5	No. of lambs born: S = single; tw = twin; tr = triplet; quad = quadruplet; quint = quintuplet
6	Weight at approximately 5 mo. of age
7	Grade of the wool: 1/4", 3/8", 1/2", or fine
8	Breed type

Section B of figure 1 is an example of records for:

Column No.	Information Recorded
1	Year involved
2	Sire of lamb
3	Date of birth
4	Alphabetical letter prefix to identify each lamb of multiple birth; example, no. 10s triplet lambs would be A10, B10, and C10
5	E = ewe, W = wether, and B = buck
6	Individual weight
7	Total weight of lambs
8	Weight wool
9	Comments on health, disposition, lamb mortality, etc.
10	Point system for total production: one lb lamb = 1 point and one lb wool = 2 points
11	Group I = highest producers. (Our replacements come from this group.) Group II = good, average ewes. Group III = unproven ewes or candidates for culling. (These ewes are bred to Suffolk rams and usually not sold for breeding.)

Section A

No.	Sire	Dam	Date Lambed	Birth Type	Wean Wt.	Wool Grd.	Breed Description	Remarks Ewe
1	2	3	4	5	6	7	8	

Section B

Year	Sire	LAMBING RECORD Date		Sex		Mkt. Wt.	Ttl. Lamb Wt.	Wool Wt.			Yearly Remarks	Points	Breed Group
1	2	3	4	5		6	7	8			9	10	11

No.	Sire	Dam	Date Lambed	Birth Type	Wean Wt.	Wool Grd.	Breed Description	Remarks Ewe
211	Targhee	392 Targ	3-27-72	Tw.	75	½	Targhee	

Year	Sire	LAMBING RECORD Date		Sex		Mkt. Wt.	Ttl. Lamb Wt.	Wool Wt.			Yearly Remarks	Points	Breed Group
73	Suffolk	Dry						11					III
74	Suffolk	3/19		w		70		16					III
75	Suffolk			Dry				15			Sell		III

Figure 1.

No.	Sire	Dam	Date Lambed	Birth Type	~~Wean~~ Wt.	Wool Grd.	Breed Description	Remarks Ewe
93X	169-170 Finn	4028-01 Ramb	4-16-70	Tw.	135	3/8	Finn-Ramb	Purchased from Dubois Sheep Experiment Station in fall 1972

Year	Sire	LAMBING RECORD Date		Sex		Mkt. Wt.	Tl. Lamb Wt.	Wool Wt.			Yearly Remarks	Points	Breed Group
73	Finn	3/6		B B E		102 107 83	292	6				304	I
74	Finn	2/26		B E E E		60 65 65 60	250	9				268	I
75	Finn	2/27		B E		75 75	150	8			Had 3 1 died 3/20	166	I
76	0718 Finn	3/17		W W E		65 60 60	185	10				205	I
77	½ Finn	3/20		B E E		100 65 80	245	8				261	I
78	Ramb	3/20		W E E		80 80 85	245	8			Ewe very sick Pneumonia 5/10	261	I
79	½ Finn	3/10		E E		100 95	195	7				209	I
80	½ Finn	3/1		W E		100 95	195	7			W-Single E-graft from 630-Quint	209	I
81	½ Finn	3/19	A B	E W		75 80	155	9			also had a monster by Caesarean	173	I
82	½ Finn	3/25	A B	E B				8			1 killed by Coyote		

Figure 2.

August evaluation. In mid-August, ewes are evaluated and culled. Lambs are individually weighted in a crate on a platform scale at the end of a chute. Five people can process 150 head per hour. Any gains made after this date are not credited to the ewe. Lambs are classified and paint branded: wether lambs and single ewe lambs over 95 lb are sold; light wethers and any ewe lambs with a visual weakness are marked for feeders; replacements are selected from 80 lb lambs of multiple birth whose mother's production was in the top 1/3 for at least 2 years.

Wool evaluation. All fleeces are individually weighted at shearing time. No replacements are kept from a ewe with an inferior fleece.

SUMMARY

Although record keeping is a lot of work, it is also very interesting. It's no more work to keep records and care for 100 ewes than it is to care for 200 without production information. The net return would be about the same.

Figure 3 shows our sheep production by breed and by age within each breed.

Although we don't have many Targhees left, this shows what can be done within the breed by keeping records and selecting replacements from top producers.

Adding the Finn breeding does increase the production but lamb mortality is higher and the lambs are lighter at weaning.

Records are cheaper than hay. There are hundreds of management techniques that must be used on a successful sheep operation. After working with sheep for a lifetime, if I were asked to select the two most valuable tools, I would choose a good set of production records and a lead pencil.

Targhee Ewes

Year born	Age	No. in group	No. lambs born	No. lambs raised	% lambs born	% lambs raised	Lb. lamb per ewe	Lb. per lamb
1972	8	8	14	12	175	150	150	100
1973	7	14	29	27	207	193	183	95
1978	2	14	26	23	185	164	163	99
Breed total		36	69	62	191	172	168	98

1/2 Finn x 1/2 Targhee

Year born	Age	No. in group	No. lambs born	No. lambs raised	% lambs born	% lambs raised	Lb. lamb per ewe	Lb. per lamb
1970	10	1	1	2	100	200	195	97
1973	7	7	26	16	371	229	207	90
1974	6	95	265	204	279	215	190	89
1975	5	19	60	50	316	263	244	93
1976	4	21	57	45	271	214	192	90
1977	3	79	216	188	273	238	213	90
1978	2	58	131	103	226	178	156	88
Total Mat. 1/2 Finn		280	756	608	270	217		
1979	1	161	286	217	178	135	119	88
Breed Total		441	1042	825	236	187	166	89
Grand Total Mat. Ewes		316	825	670	261	212		
Grand Total All Ewes		477	1111	887	233	198		

Figure 3. Christensen Ranch 1980 Ewe Production

Part 4

PHYSIOLOGY AND REPRODUCTION OF SHEEP

19

HORMONAL REGULATION OF THE ESTROUS CYCLE

Roy L. Ax

INTRODUCTION

All reproductive events are regulated by hormones. In simple terms, if the organs of reproduction correspond to the "plumbing," the endocrine system can be called the "wiring." A delicate balance exists between the nervous system and the endocrine system. We are entering an era in which artificial control of the estrous cycle with hormones promises to become more commonplace. Thus, hormones can be considered valuable management tools. If producers are to use the tools effectively, we must develop a better understanding of the complex hormonal interrelationships between the hypothalamus, pituitary, and ovary.

HORMONES DEFINED

A classical definition for a hormone is that it is a substance produced in one tissue that is transported to another tissue to exert a specific effect. (Some of the confusion about hormone actions should be clarified in the next section). Hormones have many chemical classifications; some of the most common reproductive hormones are briefly described here. Gonadotropin-releasing hormone (GnRH) is composed of amino acids and is thus a polypeptide in nature. The follicle-stimulating hormone (FSH) and luteinizing hormone (LH) are glycoproteins. This means they are composed mostly of protein, with some carbohydrate attached to the protein. Estrogen and progesterone are steroids that are synthesized from cholesterol. Prostaglandins are produced from a fatty acid--arachidonic acid. The diversity of the composition of hormones leads to the variation in their biological functions. Most hormone concentrations are in billionths or trillionths of a gram per milliliter of blood.

HOW DO HORMONES WORK?

The fact that a hormone is produced by a tissue does not necessarily imply that it will exert a physiological effect somewhere else. The ability of one tissue to respond to a particular hormone rests in whether that tissue possesses a receptor to the particular hormone. A receptor functions as the lock, and the hormone functions as the key that fits the lock. Therefore, as an example, if a tissue is going to respond to estrogen, its cells must possess estrogen receptors. After a hormone is bound to its receptor, a cellular response is initiated in the target tissue. A target tissue may possess receptors for several different hormones, and exposure to the various hormones can modulate the final response.

THE HYPOTHALAMUS

The hypothalamus is located at the base of the brain. It contains nerve endings to integrate sensory information and sorts out hormonal signals as well. The major reproductive hormone of the hypothalamus is gonadotropin-releasing hormone (GnRH) that is sometimes called luteinizing-hormone-releasing hormone (LHRH). For purposes of this discussion we will use GnRH nomenclature. GnRH is transported in blood vessels to the pituitary gland to regulate secretion of FSH and LH from the pituitary.

THE PITUITARY

The pituitary is positioned underneath the hypothalamus directly above the roof of the mouth. The major reproductive hormones produced in the anterior lobe of the pituitary are called gonadotropins, which means to stimulate the gonads. Follicle-stimulating hormone (FSH) and luteinizing hormone (LH) are the two gonadotropins that regulate the ovary. They are secreted by the pituitary and transported in the circulation to the ovary where they interact with their respective receptors to affect ovarian functions. The main action of FSH is to initiate growth of follicles on the ovary. Continued follicle growth depends on the presence of both FSH and LH. The major effect of LH is to promote ovulation, but there is increasing evidence that FSH can exert a major influence to facilitate ovulation.

THE OVARY

The ovary has two biological functions: (1) to provide the eggs (ova) for the female genetic contribution to the next generation, and (2) to produce hormones to coordinate behavioral changes with ovulation and prepare the reproductive tract for pregnancy.

Estrogen is the hormone produced by follicles as they develop on the ovary. As the predominant follicle or follicles approach ovulatory size, the increased amounts of estrogen are transported to the hypothalamus to cause behavioral heat. The pituitary also responds to the elevated estrogen by releasing a surge of LH which leads to ovulation. Thus, estrogen coordinates behavioral acceptance of a male when the egg will be released into the female tract. This is Mother Nature's attempt to ensure that the probability of fertilization occurring is maximized.

After ovulation has occurred, the tissue that a moment ago was a follicle starts a dramatic change into becoming a corpus luteum (yellow body). The corpus luteum produces progesterone to prepare the female tract for a possible ensuing pregnancy. The corpus luteum forms, regardless of whether or not mating occurs in farm animals. If the corpus luteum remains functional due to pregnancy occurring, the sustained production of progesterone prevents cyclicity. Until the corpus luteum regresses, the typical pattern of cyclic hormonal changes is absent.

THE FOLLICLE

In a simple sense, the follicle is the dwelling of the ovum. Ovulation of a follicle is the exception rather than the rule because over 99% of all potential oocytes are never shed from the ovary. This loss is called atresia. Atresia can occur at any time during follicle growth. When animals are injected with gonadotropins to induce superovulation, some follicles that would have undergone atresia are rescued. This supports the hypothesis that continued follicle growth is dependent upon continued exposure to gonadotropins and the presence of gonadotropin receptors in the follicle.

OVULATION

Once the LH surge has been elicited from the pituitary, the follicle starts to undergo a series of changes to prepare for impending ovulation. The cells lining the inside of the follicle begin to luteinize and secrete progesterone as the major steroid rather than estrogen. The oocyte commences its maturational steps to get it in the proper meiotic configuration for chromosome pairing with the meiotic contribution from a sperm.

Enzymes are activated to degrade the follicle wall and permit the egg to pass into the oviduct. Biochemical studies have pointed to FSH being responsible for stimulating production of those enzymes. FSH also promotes the spreading apart of cells that are tightly surrounding the egg, which then leads to some of the subsequent maturational changes in the egg. Prostaglandins are required for normal ovulation. Substances known to inhibit prostaglandin formation prevent ovulation. Due to the enzyme, steroid, and prostaglandin effects, a hypothesis was formulated comparing the ovulatory process to an inflammatory reaction.

THE CORPUS LUTEUM

The scar tissue remaining after ovulation becomes the corpus luteum; this endocrine tissue has been studied extensively. Low amounts of LH from the pituitary are essential for establishment and continued function of the corpus luteum. In all livestock species, the corpus luteum functions to maintain early pregnancy by secreting progesterone. Progesterone prevents cyclicity. The placenta of the developing fetus eventually sustains the pregnancy by producing progesterone in the bovine, equine, and ovine. In the porcine, the corpus luteum is required to support the entire gestational period.

If cyclicity is to resume, the corpus luteum must regress and cease progesterone production. In pregnant animals, an embryonic signal leads to maintenance of the corpus luteum, (discussed in the next section). In nonpregnant animals, the uterus recognizes the absence of an embryo and secretes prostaglandins. Those prostaglandins are transported to the corpus luteum and cause it to regress. Thus, the inhibitory effects of progesterone are removed, new follicles start to develop, and heat occurs in a few days.

PREGNANCY RECOGNITION

The corpus luteum continues to function to provide a signal from the embryo to the dam. These signals are hormones identified in the human as chorionic gonadotropin (hCG) and in the mare as pregnant mare serum gonadotropin (PMSG). These two gonadotropins can be used to regulate the estrous cycle of animals. PMSG is biologically similar to FSH, and hCG exerts an action similar to LH. Thus, PMSG can be used to induce follicle growth and hCG will promote ovulation of these follicles. However, both PMSG and hCG are recognized as foreign proteins by livestock, and the animals build up antibody resistance to them, if they are used too frequently.

An ideal pregnancy test would be the identification of the embryonic signal in the bovine, ovine, and porcine. Experiments have shown that embryonic extracts will maintain a corpus luteum in an animal that has not been bred. However, even with sophisticated biochemical tests, specific signals from embryos in the dam's circulation have not been detected. The livestock industry could benefit significantly from pregnancy tests of these types if they are ever developed successfully. Since an embryonic signal would have to be apparent to prevent a subsequent heat in the dam, a pregnancy test would also pinpoint which animals would be returning to heat within a few days.

HORMONAL REGULATION OF THE CYCLE

The preceding sections indicate that follicle growth, ovulation, and corpus luteum formation are a dynamic process of sequential steps in an intricate balance. Administration of a hormone to mimic the effect of that hormone in the animal can be used to regulate the cycle. If hormone administration is to produce the desired result, it must be given at a time that is physiologically compatible with the cycle. The common hormones that have been used experimentally or commercially are progesterone-like drugs (progestins), GnRH, and prostaglandins.

Progestins

These compounds were the first to be experimentally employed to regulate the cycle. Progestins have been injected, fed, implanted, or administered via vaginal sponges. Regardless of what stage of the cycle an animal is in when the progestin commences, cyclic fluctuations in other hormones are arrested. As long as progestin is administered, cyclicity ceases. Removal of the source of the progestin results in renewed follicle growth, and estrus, within a few days. Field trial data suggest that fertility at the first estrus after progestin withdrawal is lowered. Thus, it is usually recommended that breeding be done at the second estrus, since the cycles of the animals will still be in close synchrony.

Prostaglandins

These compounds have largely replaced the use of progestins because (1) only one or two injections are required, and (2) fertility is not affected by use of prostaglandins. Prostaglandins are only effective if the animal possesses a functional corpus luteum. Contrary to some opinions, prostaglandins are not a heat-inducing drug. Rather, they cause a corpus luteum to regress, and the animal secretes her own gonadotropins to regulate the ovary and cause a physiological heat. Success has occurred regularly by breeding animals at a predetermined time after prostaglandin injection. Greater success in conception rates can occur if animals are watched for estrual behavior after receiving prostaglandins and are bred in relation to standing heat. Care must be exercised with prostaglandins, because injections into an animal with a functional corpus luteum sustaining a pregnancy could induce an abortion.

GnRH

GnRH is composed of 10 amino acids. It can now be chemically synthesized in a laboratory, and this has permitted chemists to develop some powerful analogs. There are no noticeable ill effects from administering GnRH; its action is to promote a release of gonadotropins from the pituitary. Maximum gonadotropin output occurs approximately 2 to 4 hours after GnRH injection.

A common use for GnRH is to initiate cyclicity in animals with anestrous. GnRH is the most widely used therapy for treating cystic ovarian degeneration. Cystic ovaries usually result from inadequate gonadotropin production. Thus, GnRH triggers release of gonadotropins to restore ovarian function.

A new use for GnRH is for injection after prostaglandin administration; the interval to the gonadotropin surge, and hence, ovulation, can be coordinated more closely. This reduces the variation in time between prostaglandin injection and standing heat that is ordinarily seen among animals.

We have an ongoing study at the University of Wisconsin to evaluate the efficacy of GnRH injections at the time of insemination in dairy cattle. The heifers receiving GnRH have shown no advantages over heifers receiving the saline control. In lactating cows, administration of GnRH 14 days postpartum or at the first artificial insemination has improved first-service conception rates by 15% to 19%. In cows presented for third service (and thus classified as "repeat" breeders in commercial herds) conception rates were about 30% higher for cows that received the GnRH. The physiological effect elicited by GnRH has yet to be experimentally established. We have postulated that gonadotropins produced in response to GnRH cause a corpus luteum to form that may have otherwise been deficient and led to early embryonic death. GnRH could also promote what would have been a delayed ovulation to occur sooner or have a direct effect on the ovary. The lactational stress imposed on a dairy cow may make her unique to respond to GnRh in this manner. Experiments with other farm animals are needed to determine if similar effects result.

SUMMARY

The reproductive cycle is regulated by fluctuations in different hormones. The cycle can be regulated by administering hormones to mimic the effect that would occur in the animal. Therefore, producers have endocrine tools to assist them in managing their animals. For maximum success the producers must understand how the hormones work biologically and realize that they are powerful drugs. We will see an increasing frequency of producers regulating the reproductive cycle to maximize reproductive efficiency.

REFERENCES

Britt, J. H. 1979. Prospects for controlling reproductive processes in cattle, sheep, and swine from recent findings in reproduction. J. Dairy Sci. 62:651-665.

Britt, J. H., N. M. Cox and J. S. Stevenson. 1981. Advances in reproduction in dairy cattle. J. Dairy Sci. 64:1378-1402.

Foote, R. H. 1978. General principles and basic techniques involved in synchronization of estrus in cattle. Proc. 7th Tech. Conf. on Artif. Insem. and Reprod., Nat'l Assoc. Anim. Breeders, pp 74-86.

Hansel, W. and S. E. Echternkamp. 1972. Control of ovarian function in domestic animals. Amer. Zool. 12:225-243.

Jones, R. E. (Ed.) 1978. The Vertebrate Ovary. Comparative Biology and Evolution. Plenum Press, New York.

Lee, C. N., R. L. Ax, J. A. Pennington, W. F. Hoffman and M. D. Brown. 1981. Reproductive parameters of cows and heifers injected with GnRH. 76th Ann. Mtng. of the Amer. Dairy Sci. Assoc. Abstract P228.

Maurice, E., R. L. Ax and M. D. Brown. 1982 Gonadotropin releasing hormone leads to improved fertility in "repeat breeder" cows. 77th Ann. Mtng. of the Amer. Dairy Sci. Assoc. Abstract P233.

Nalbandov, A. V. 1976. Reproductive Physiology of Mammals and Birds. W. H. Freeman and Co., San Francisco.

20

ACCELERATED LAMBING: ITS ROLE IN THE SHEEP INDUSTRY

Clarence V. Hulet,
John N. Stellflug

INTRODUCTION

The cost of maintaining a ewe for a year is often quite high, and unless the returns from lamb and wool exceed these costs by a reasonable amount, a sheep enterprise will not be profitable. Thus, increasing the frequency of lambing or the number of lambs born per lambing, or both, without a comparable increase in maintenance and labor costs, should increase net return. Any increase in the number of lambings in relationship to the age of the ewe is referred to as accelerated lambing and includes: 1) starting lamb production at a younger age, and 2) breeding and lambing at intervals more frequent than 12 months.

With few exceptions, ewe lambs should be bred, selected, and managed to lamb first at 1 year of age. When an appropriate breed or carefully selected strain is used, the income realized from yearling production will far surpass the extra cost of adequate nutrition and management required to achieve production the first year. Ewes managed and selected for production at 1 year of age will be more productive throughout their life than are ewes managed to lamb first at 2 years of age (table 1).

Of the several accelerated lambing systems now being used, an 8-month-interval lambing appears to be generally the most popular. However, success appears to be limited to those using light control in the northern United States or breeds or breed crosses with long breeding seasons in the southern states.

Another system using Polypay sheep in Idaho is essentially twice-a-year (6-month-interval) lambing on a flock basis, and variable intervals of 10-7-7-months. This results in approximately three lambing opportunities every two years on a ewe basis--yet does not limit exceptional ewes. A ewe, in fact, does have the opportunity to lamb every 6 months. A few exceptional Polypay ewes have lambed at 6-month intervals over extended periods of time--this system also offers important advantages over the 8-month-interval system (detailed discussion of this system follows.)

TABLE 1. LAMB PRODUCION OF TARGHEE RANGE EWES BRED AND SELECTED ON PREGNANCY AS COMPARED TO EWES LAMBING FOR THE FIRST TIME AS 2-YEAR-OLDS

Management practice	Age of ewe	Ewes lambing	Lambs born	Lambs weaned[a]	Total lambs weaned
	Years	%	%	%	lb
Lambed 1st as yearlings	1	100[b]	111	83	56
	2	98	143	115	84
	3 or over	97	158	134	107
Lambed 1st as 2-year-olds	1	0	0	0	0
	2	88	102	82	58
	3 or over	89	141	115	87

[a] Percent lambs born or weaned of ewes bred.
[b] Only those ewe lambs diagnosed pregnant were saved.

Accelerated lambing systems should be used only after one has gained considerable experience and competence at once-per-year lambing. Other important considerations include: a good local source of economical feed or pasture available when the demands for feed are high such as during lactation and when finishing lambs for market; early puberty and high prolificacy to enhance the possibility of satisfactory economic returns; starting relatively small and testing the system.

Accelerated lambing will facilitate the distribution of quality lamb throughout the year and should be a positive factor in promoting the increased consumption of lamb.

Accelerated lambing can lead to the more efficient use of lambing facilities and can possibly justify, under certain conditions, semi-confinement or total-confinement, lamb-production systems.

The use of hormones to facilitate accelerated lambing during the nonbreeding season will not be discussed because it seems unlikely that these hormones will be approved by the Food and Drug Administration for use in food animals, the lack of commercial availability of the hormones, and the technical problems in using the hormones in commercial operations.

IMPORTANT FACTORS THAT AFFECT SUCCESS OF ACCELERATED LAMBING PROGRAMS

- Day length is the primary factor controlling breeding season in sheep.

- Ewes breed naturally only during their normal breeding season.
- Length of breeding season varies with breed and individuals within breeds.
- The sudden introduction of a ram during the transition period between the nonbreeding season and the breeding season stimulates earlier breeding and conception in the ewes.
- Lactation inhibits the occurrence of heat near the end of the breeding season but not during early or midseason.
- Breeds and individuals appear to vary in the interval of time required for them to return to breeding condition and to exhibit fertile heat following lambing.
- The length of the breeding season appears to be moderately heritable and, therefore, may be changed by selection.

ACCELERATED LAMBING SCHEMES

Eight-Month-Interval Lambing

In order for the fixed 8-month lambing interval to be successful, one must either (1) have a breed of sheep that will consistently breed naturally at the fixed breeding dates that are established, such as that used by Outhouse (1974) in Indiana (table 2), or (2) use light control to induce fertile estrus at the planned breeding periods. In most 8-month-interval lambing schemes, ewes are bred every 4 months. This permits those that fail to conceive at the 8-month interval an opportunity to breed 4 months later (12-month interval) and improve the overall efficiency of production.

TABLE 2. BREEDING AND LAMBING SCHEDULE FOR 8-MONTH INTERVAL ACCELERATED LAMBING

Bred	Lambed
December 1972	May 1973
August 1973	January 1974
April 1974	September 1974

Source: Outhouse, J. B. (1974).

Rambouillet, Dorset, and Rambouillet x Dorset breeds and crosses have been used most successfully in this program. Outhouse of Indiana and a few commercial Dorset flock owners claim satisfactory results using the 8-month accelerated program without light treatment (table 3) (Outhouse, 1968). However, some have attempted this program but have found that the occurrence of estrus and fertility have not been satisfactory during the spring breeding period. If the spring breeding fails, the entire program fails.

TABLE 3. PRODUCTIVITY OF RAMBOUILLET EWES FOR 8 YEARS AND 12 LACTATIONS, INDIANA AGRICULTURAL EXPERIMENT STATION, SEPTEMBER, 1964

Item	Winter (1966-72)	Fall (1964-70)	Spring (1965-71)	Total (1964-72)
No. ewes exposed	197	189	192	578
% ewes lambing	83.2	84.1	91.7	86.3
% born of ewes exposed	133.0	124.3	128.6	128.7
% born of ewes lambing	159.8	147.8	140.3	149.1
% raised of ewes exposed	105.1	101.1	109.4	105.2
% raised of ewes lambing	126.2	120.1	119.3	121.8
% lamb loss	21.0	18.7	15.0	18.3

Because of the difficulty many have experienced getting their ewes to breed out of season, some have established a light-control system to permit fertile estrus at the prescribed times. Leslie and David Chalmers of Choteau, Montana, (personal communication) are working with Polypay ewes in total confinement, using a system developed by the Rowett Institute, Buckland, Scotland, which seems to work well for them. Beginning 40 days before the start of breeding they change from 16 hours light and 8 hours dark to 8 hours light and 16 hours dark. This treatment is continued until breeding is completed; the treatment is then reversed again. During this past year, they reported that of 27 ewes exposed between June 10 and July 10 using the above light treatment, 22 lambed between November 15 and December 20. The Chalmers also practice early weaning (25 to 35 days of age), which may facilitate accelerated breeding. They report marketing a 341% lamb crop in 1980 using this system. In 1981 they plan to bred 250 ewes in this accelerated program.

Russ Beattie of Rexburg, Idaho, also working with Polypay ewes in a semi-confinement system, uses an accelerated lambing schedule, but has no light treatments to control breeding. Lamb production of this system is shown in table 4. Beattie early weans his lambs at 31 to 45 days of age. He then fasts his ewes for 3 days. This is followed by 10 days of average-level feeding while isolated from rams. He then introduces sterile rams and flushes the ewes with alfalfa hay and barley for 5 days. The sterilized rams are then removed and the fertile rams put in.

Beattie claims that farm operating costs, especially fuel and fertilizer costs, have been reduced dramatically and profits have increased greatly since he has started raising sheep on his farm.

TABLE 4. ACCELERATED LAMB PRODUCTION ON THE RUSS BEATTIE RANCH

Year	No. of ewes	No. ewes lambing	Lambs raised to market age	
			No.	%
1979	64			
Winter		64	112	175
Fall		43	70	163
Annual			182	284*
1980				
Winter	103	101	207	205
Fall		47	131	196
Annual			349	339*

* Percentage lambs raised to market weight per year of ewes in the flock.

Six-Month-Interval Lambing

The 6-month-interval lambing permits two breedings and lambings at the same time each year. Table 5 illustrates the system. This system can be adapted to range operations and fits in better with most farm operations. It also requires fewer breedings and lambing per year than most 8-month-lambing schemes while permitting an opportunity for shorter lambing intervals in exceptionally fertile ewes. This provides an opportunity to select for twice-a-year lambing on a ewe basis.

We have found that the average lambing interval in ewes lambing at two consecutive breedings to be about 200 days or slightly less than 7 months.

Ewes that lamb in the period from July to September breed readily while lactating (95% in 1981)--provided that they are fed well or are on good pasture. However, we have found that ewes lambing in the period from January to March must have their lambs weaned from them or they fail to conceive in significant numbers.

Table 6 clearly indicates that fertile estrus occurs much more readily following early lambing (January) rather than later lambing (February - March).

Thus it is clear that given a 200-day lambing interval and a declining breeding response with the advance of the breeding season, most ewes will not lamb at each lambing time for more than three consecutive times before they skip. When they skip, they will usually lamb early at the subsequent lambing period resulting in a lambing interval of

TABLE 5. SIX-MONTH INTERVAL BREEDING AND LAMBING SYSTEM

Bred	Lambed	Weaned
Aug. 10 - Oct. 25[a]	Jan. 5 - Mar. 30	31 days of age[b]
Feb. 10 - Apr. 10	July 5 - Sept. 14[a]	50-70 days of age

[a] Ewes lambing late in the fall and ewe lambs are permitted to stay in breeding until Nov. 20 to maximize winter lambing.

[b] Lambs are weaned at weekly intervals when they average 31 days of age.

TABLE 6. EFFECT OF 1980 WINTER LAMBING DATE ON PERCENTAGE OF EWES LAMBING THE FOLLOWING AUGUST-SEPTEMBER

Lambing day of year	No. of ewes	Percent ewes lambing the following fall
12-30	199	30.7
31-45	46	21.7
46-60	32	9.4
61-90	34	0.0

about 10 months. Thus good-performance ewes lamb two or three consecutive times at 7 month intervals and then skip one period for a 9 or 10 month lambing interval.

One important advantage of this system is that ewes are always being bred during the normal breeding season. This enhances good estrous response and increases the probability of a good lamb crop.

To facilitate early breeding and early lambing, which enhances rebreeding following winter lambing, we put sterilized rams with the ewes on about July 25 (15 to 18 days before desired breeding time) and put the fertile rams in about August 10. More ewes bred early following this procedure--72% vs 37% for controls.

We have elected to use Polypay ewes in this accelerated program after several years of testing Rambouillet, Targhee, Columbia, Finn x Rambouillet, Dorset x Targhee, and Polypay ewes in accelerated lambing. The Polypays consistently outperformed the other breeds. We believe that some of the factors accounting for their better adaptability, (supported by some data) include: short gestation, short postpartum interval, long breeding season, and high fertility. An additional advantage of the Polypay over most of the above breeds is early puberty and high prolificacy.

Light treatments also can be used if desired to enhance the late winter breeding performance of ewes on a twice-a-year lambing program. Conventional light treatments require light-tight barns. We have been experimenting with a new system that can probably be achieved with yard lights and a time switch.

When the ewes are taken out of fall breeding on about October 25, they are placed each night before dark under lights controlled by a timer. The lights go off for about 4 hours during the middle of the night. This treatment is continued until December 21 (the shortest day of the year). At this time ewes are again subject to the natural short day length. Ewes commence lambing in January. For night lambing low lighting (1 foot candle) is used without apparent detrimental effects. Starting February 10 when lambs are on average 31 days of age, they are weaned at weekly intervals onto their high-protein, high-energy creep diet. The ewes are put with fertile rams the next day. The ewes remain in breeding until about April 10. A higher percentage of treated ewes lambed when compared to the untreated group (44% vs 27%).

RAM FERTILITY

In any accelerated system, good ram fertility is critically important. If ewes are treated with light to enhance fertility, rams also should be treated with light. Precautions should be taken to ensure that rams are kept cool and healthy. They should not be overly fat nor too thin. Shearing can be helpful 3 weeks before breeding in hot or warm seasons.

Semen testing is highly recommended. If this is not possible, one should, as a minimum, carefully palpate testes and use only those rams with large, firm testes that are free from any lesions or abnormalities.

CONCLUSIONS

The role of accelerated lambing is two-fold: 1) to provide a more uniform supply of lamb throughout the year; 2) to increase lambs born and raised in relation to the number of ewes in the flock.

- Accelerated lambing can only be justified if it increases the net return over conventional once-a-year lambing systems.
- Low-cost local feeds available at the time of high feed demands are important to the success of an accelerated system.
- Successful accelerated lambing requires dedication and technical expertise above that required for once-a-year lambing. Time and experience usually improve the success of accelerate lambing systems.

- Very efficient, fertile, prolific sheep that have shown adaptability to accelerated lambing will greatly contribute to the success of an accelerated system.

REFERENCES

Outhouse, J.B. 1968. Four-year summary of accelerated lambing program. Purdue Univ. Ext. Mimeograph AS-383.

Outhouse, J.B. 1974. Ewe productivity on accelerated lambing programs. Purdue Univ. Agr. Exp. Sta. Bul. No. 49.

21

MANAGEMENT OF SHEEP TO MAXIMIZE LAMB PRODUCTION

Maurice Shelton

Other papers in this series or this proceedings will emphasize the importance of increasing lamb production and the potential it holds for the U. S. sheep industry. The U. S. differs from many other sheep producing countries in that we receive a reasonably good price for lamb and have the feed resources necessary to maximize lamb production. Many would argue with this statement about prices; however, we are discussing U. S. prices compared with those of other countries such as Australia.

A current major industry effort is that of improving the genetic potential for lamb production. However, except where breed substitution is possible, improvement requires a long time. Many producers are not willing to take this long-term view--and, even if this were not a limitation, the commercial industry must attempt to maximize current levels of production. In this context, we would be primarily concerned with management, which could well be the subject of a book rather than this short paper. The two important components of lamb production are the percentage of lambs raised and the growth rate of the lambs produced. This paper will concentrate primarily on reproduction.

REPRODUCTION

Since the males constitute only 3% to 5% of the sheep flock, they should simply not be allowed to be a limiting factor in production.

1. Use an adequate (or surplus) number of rams,
2. Make sure the rams are healthy, vigorous, and capable of service,
3. Take care to ensure that heat stress does not result in lowered breeding efficiency.

Ram Number

Defining an adequate number of rams is not easy. A standard rule of thumb is 3 rams per 100 ewes, although under favorable conditions fewer than this can be used with good results. However, in a commercial program, minor genetic differences within a breed seldom warrant over-use or over-dependence on a few presumed superior males. Under adverse conditions such as hot weather and large pastures, I would tend to use more rams than 3% of the flock. Under some conditions, rotational use of rams may be desirable, but this also would suggest a larger number of rams.

Ram Health

To ensure that rams are healthy, vigorous, and capable of service means 1) a good nutritional regime, 2) avoiding dependence on young or very old rams, and 3) preventing disease problems or discarding those showing evidence of disease. This paper deals only with conditions associated with reproduction. Epididymitis or orchitis is the primary concern, resulting primarily from infection. The primary organisms are *Actinobacillus seminis* and/or *Brucella Ovis*. Epididymitis technically means an infection of the epididymis, but in practice the term is commonly used for almost any abnormality of the testes. Orchitis is the term usually used for a feverish, swollen, or enlarged testicle or testicles. These conditions can be determined by palpation or physical examination of the testes. Thus, one of the first things a producer should do is to determine the normal condition of the scrotum and testicles of the ram. Each ram should be examined one or more times per year--and certainly in advance of the breeding season. Any ram that appears abnormal should be sold for slaughter or evaluated further. On inspection, the producer should look for small testicles (one or both), enlarged or feverish testicles, or knots or hard spots in the epididymis or connecting ducts. Aside from small testes, such symptoms are usually infections. They occur in greater frequency in blackface than fine wool rams, with the frequency of infection increasing with age. Infected rams may not be totally sterile, but as a group they will show reduced fertility and will tend to propagate or maintain the infection.

In some areas, it is possible to vaccinate against epididymitis/orchitis with some degree of effectiveness. (This vaccine is not available in Texas.) A more commonly recommended practice is to start with a clean group of young rams (based on either or both manual and laboratory tests), separate them from aged rams, and attempt to maintain them free of this problem. Dealing with this problem is not difficult, but in flocks that have not received attention, as many as one-third of the rams may be infected.

Ram Stress

A more difficult problem to deal with in many areas is the adverse effect of high temperature on the ram. In much of the United States, germ cell production or semen quality and fertility of rams is diminished in the summer, with the effect ranging from minimal to complete sterility. Young, highly fitted animals in heavy or full fleece of the less well-adapted breeds (Down breeds) are more seriously affected. Under winter conditions, the animal is less likely to be seriously affected, but these generalities do not provide adequate assurance of freedom from environmental stress.

The most useful method to test for heat-induced infertility or reduced breeding value is palpation of the testes. This can be done prior to use while checking for epididymitis or orchitis. Animals seriously affected by high temperatures will have smaller and softer testes as compared to those in good breeding condition. Unfortunately, this requires knowledge of their normal condition. Table 1 reports "normal" testes measurements for Rambouillet rams. A second approach to evaluating the effect of temperature, or overall semen quality, is that of semen evaluation. For most animals this can be done by electroejaculation, but good semen samples are seldom obtained by electroejaculation. Most rams will show some live sperm but that does not assure good breeding performance. If all rams are discarded that do not show good semen samples based on electroejaculation, a number of good rams could be discarded. The same may also apply to rams, particularly inexperienced rams, that fail to show a high level of mating efficiency based on short time (a few minutes) pen tests. Lack of libido is not a major problem, but males used in single-sire groups should be observed to be sure they are breeding.

TABLE 1. MEAN TESTES CIRCUMFERENCE OF YEARLING RAMBOUILLET RAMS AT DIFFERENT BODY WEIGHTS

Body weight (lbs)	Number of rams	Testes circumference (cm)
Less than 110	8	18.43
111 - 130	22	20.96
131 - 150	50	21.86
151 - 170	52	22.79
171 - 190	55	24.71
191 - 210	99	25.73
211 - 230	147	26.57
231 - 250	88	27.50
251 - 270	39	28.40
271 - 290	5	28.05
290 +	2	30.65

Ram problems can usually be overcome by simply obtaining another ram. However, the ewe constitutes a more serious problem.

EWE FERTILITY

Factors affecting ewe fertility are:

Age	Culling practices
Season (photoperiod)	Condition
Temperature	Nutritional level
Genetic potential	or flushing
Lamb death losses	

Age

Age has an important bearing on lamb production. Young ewes (lambs or two-year-olds) produce noticeably lower lamb crops. From age 3 years through 6 years, ewes generally produce at their maximum. If a producer is producing his own replacement stock, he has relatively few options with respect to age. The age at which ewes are routinely culled is a management decision. Generally, it is advisable to keep ewes as long as they can be expected to produce well. Aged ewes are only a small percentage of the total and affect production only slightly. The decision to sell or to keep older ewes should probably be influenced more by their market value than any presumed effect on the lamb crop.

Seasonal Breeding

The tendency for seasonal breeding has long been a major limitation or frustration for sheep producers. Seasonality can be shown to be greatly affected by latitude or day/night ratio. Genetic control over this variable can be inferred only by observed and by often minor differences among breeds or among populations that have evolved in certain climatic or management conditions. The tendency for seasonal breeding is under some degree of genetic control, but if other components of reproductive efficiency are taken as a guide, the heritability of this trait is likely to be low. In practice, producers should breed their sheep in periods of maximum fertility (late summer and fall) unless there is a good reason to want lambs at other seasons. Unfortunately, in warmer climates, fall is generally the preferred season since lambs do poorly during the hot summer season. It is a mistake to mate ewes for a controlled-mating period in March, April, and May. The months of February, June, July, and to a lesser extent August, are transitional months depending on breed and conditions. It is generally ill-advised to expose ewes to rams for a normal mating period during these dates and wait to see if they lamb. However, if there is an important economic advantage to mating at these periods, this can be successfully done if the ewes that failed to settle early in the season are exposed to the ram a second time later in the season. Such mating is widely practiced by producers who prefer fall or winter lambs, but essentially all of these rely on a clean-up mating period for spring lambs. It is also successfully used by producers practicing accelerated lambing, but this would be the case only if the ewes are re-exposed at the next breeding season of not more than four months later, and preferably at two-month intervals. Ewe flocks that are bred for early lambs will usually have many unsettled ewes. If mated only in spring or early summer, there will likely be some increase in the number of dry ewes even if re-exposed to rams in the fall. This increase occurs because some ewes will settle early in the season but lose the fetus during the hot weather of late summer or early fall and will not resettle for an early fall mating.

Temperature

Temperature also has an effect on the ewe. Under extremely hot conditions, ewes may fail to show estrus. (This is a minor problem with adapted types such as the Rambouillet ewe.) Those that fail to show estrus will do so when conditions moderate. Ewes that have been bred but that are unable to maintain their body temperature in the normal range will usually lose the fertilized ovum within 8 days after mating. The explanation for this is not known. However, the ewe will normally recycle and remate at the next estrus. This is one of the reasons for extended lambing from summer matings.

Ewes should not be worked or driven during hot weather in the mating season. High summer temperatures also effect reproductive efficiency of the ewe through fetal dwarfing. Ewes that go through the later stages of gestation during periods of heat stress (late summer and early fall in Texas) will invariably give birth to smaller lambs. The explanation is that in trying to dissipate the heat load, blood circulation is shunted to the lungs or superficial tissue and the reduced blood supply to the uterus causes undernutrition to the fetus and smaller lamb. This problem is more severe with unadapted sheep and those carrying twin lambs. To reduce heat stress, provide conditions that can contribute to the comfort of the ewe such as shade, shearing, and limited movement. When lambing is avoided until the cool season, the birth weight of lambs improved markedly especially for lambs born after October 15 in Texas.

Genetic Potential

The genetic potential of the ewe has an important bearing on lamb production, which is discussed in more detail in another paper in this book.

Culling Practices

Culling practices can have an important bearing on lamb production. Many people construe this as being a form of selection. In practice, culling practices for the ewes have relatively little effect on future generations but can have more marked effect on performance of the ewe flock in subsequent years. The distinction is the difference between heritability and repeatability. The heritability of reproductive traits is low, and selection differential based on culling ewes is small. For instance, if a ewe is culled because she is infertile, there can be no influence on future generations since she will have no offspring. Repeatability is a measure of the tendency of the ewe to repeat her performance in subsequent years. Thus, in this context, culling practices can have an important bearing on what happens the next year.

In commercial programs, it is generally ill-advised to cull ewes during their peak production years for minor presumed genetic differences. However, they may well be removed from flocks from which replacements (ewes or rams) are kept and put into flocks that are crossbred for market lamb production. Culling of breeding ewes should be based almost exclusively on predictions of her ability to raise lambs in subsequent years. This is largely based on soundness, condition of udder, and current lamb production. It would be obvious that ewes with damaged udders should be culled and it is a universal and sound recommendation that ewes be identified that are dry in a given year. Those that pass two opportunities to lamb should automatically be culled. The more important question is what to do with a ewe that is dry for one year. This decision should be based on other considerations as well: (1) age, (2) condition of the ewe, and (3) the number of dry ewes. Mature or aged ewes in good condition, but dry, should be culled because they have limited future potential. Young ewes, and especially those in poor condition, probably should not be culled. Under conditions of nutritional stress, they may gain in weight compared to those nursing lambs and actually be better producers in future years. Regardless of age, dry ewes in good condition or fat ewes should be culled. Their condition eliminates undernutrition as a cause of infertility and, in some cases, the overfat condition may influence the ability to breed in the future. However, in the case of fine wool ewes, a fat ewe is much more likely to be the result of infertility than a cause of infertility. If the number of dry ewes is small, consider culling these--but if the number of dry ewes is large, it is better to find and try to correct the cause. If a relatively small percentage (for example 1% to 2%) of the ewes is biologically infertile and 5% of the ewes in the flock are dry, these infertile ewes would make up 20% to 40% of the dry ewes. On the other hand, if there were 30% dry ewes, the infertile individuals would constitute 3% to 6% of this total, and the improvement in condition following one dry season would more than make up for the difference. The number of totally infertile ewes in the flock will increase with increasing age.

Condition

Condition (amount of fat or body weight), nutritional level, and/or flushing at breeding have a positive and beneficial effect on the lamb crop dropped. Extremely emaciated ewes may breed very poorly and will show a response to improved condition or level of nutrition. However, the same may not be true of ewes that are already in the normal range of condition and body weight for the breed involved. Condition and body weight are more important at the periphery of the breeding season than during the peak of the breeding season. Flushing under Texas range conditions is not widely practiced because it has not shown a consistent response in research. In some situations, ewes have been bred on lush, new-growth grass (such as small grain fields) with apparently good results. Under range conditions ideal situations are difficult to plan, either in research or in practice. The general conclusion is that producers should attempt to improve the nutritional level or condition of the ewe at breeding but should not pay too high a price to obtain it.

Lamb Death Loss

Lamb death loss is a major obstacle to production efficiency in sheep. A loss of 10% is near normal or minimal under production conditions. Many factors affect the death loss, especially multiple births. Death loss among ewes that twin is significantly higher than with single births. Very heavy losses can be explained by starvation, predation, or cold stress. Starvation of the lamb is generally explained by poor condition of the ewe (no milk), damaged udder, and disruption of the bonding between the ewe and the lamb. Essentially all lamb death loss due to starvation is correctable with culling or management. Death loss due to predation and cold stress defies simple, pragmatic solutions.

BREEDING OF EWE LAMBS AND PREGNANCY DIAGNOSIS

Breeding of ewe lambs and pregnancy diagnosis are practices that have the potential to improve reproductive efficiency. The breeding of ewe lambs offers great potential but only when the lambs are well developed at 7 to 8 months. This may occur under good feed conditions but is seldom practiced under range conditions. Cross-bred ewes, especially Finn crosses, are much better suited for lambing at one year of age.

Techniques for pregnancy diagnosis are available. The most practical of these is likely the use of ultrasound on purchased or trader sheep and on flocks mated in the spring. The utility of this practice is highly dependent on the percentage of the flock that is likely to be open. It has limited value with good flocks of fall-mated ewes if close to 100% of the ewes are expected to be pregnant.

22
HOW ENVIRONMENT AFFECTS REPRODUCTION IN SHEEP

Richard O. Parker

INTRODUCTION

These days, much is written and spoken about the environment of humans and animals. Recently, some groups have concerned themselves with the quality of animal environment provided by producers. It may surprise these groups, but the producer's concern for an animal's environment precedes theirs by many years. Sheepmen are no exception. A favorable environment combined with heredity improves performance. Improved performance means more dollars.

All too often environment is only thought of as that which can be measured, seen, or felt. This includes such things as temperature, humidity, light-dark, and poor nutrition. Sometimes extremes in these environmental conditions produce traumatic outward manifestations. Other times the effects of environment occur more subtly. To fully appreciate "environment" it must be understood from a different level. This understanding will aid in an awareness that allows sheepmen to make choices and adjustments in their programs to maximally "control"---or at least avoid---environmental conditions that limit production, primarily reproduction.

REPRODUCTION

Successful reproduction is the process by which sheep produce individuals like themselves that go on to market or to reproduce. Success of the reproductive process is the foundation of a successful operation.

Reproductive processes include: mating, conception, pregnancy, birth, and lactation. Failure or reduced performance at any point reduces efficiency, and any time efficiency is down, so are profits.

When viewing the complexity of the reproductive processes, it is a miracle that reproduction succeeds as often as it does. As with anything complex, numerous factors can influence the outcome. In a broad sense,these factors can be divided into those attributed to genetics (inheritance) and those attributed to the environment.

To measure the outcome of reproduction, a variety of "yardsticks" are available, but the following are considered: 1) multiple births, 2) birth weight, 3) weaning weight, and 4) number of lambs reared. Indirectly these traits measure fertility, embryo mortality, prenatal maternal influences, lactation, and postnatal survival (table 1).

TABLE 1. REPRODUCTIVE TRAITS IN SHEEP AFFECTED BY HEREDITY AND ENVIRONMENT

Trait	Source of variation	
	Inheritance	Environment
Multiple birth	15%	85%
Birth weight	30%	80%
Weaning weight	10%	90%
Number of lambs reared	13%	87%

HEREDITY

Within every cell of each sheep there are 54 chromosomes. They are composed of the substance deoxyribonucleic acid (DNA) which, because of its composition, contains the coded master plans for producing a sheep from the embryo to the adult. Each animal receives one-half of its DNA from each parent at the time of conception. In turn, breeding animals pass on one-half of their DNA again when they mature and breed.

The master plan contained within the DNA, in part, directs the formation and the function of sheep, including those traits relating to reproductive efficiency. Depending on the trait, heredity varies in its influence, leaving the action of the environment to "fill the gap" in influencing the formation and the function of sheep. For example, heredity largely determines the wool fineness but has only a small effect on the weaning weight.

ENVIRONMENT

Environment (defined broadly) is all of the surrounding things, conditions, and influences affecting the growth, development, and production of sheep. For a better understanding, environment can be considered at three levels: external, internal, and uterine.

External Environment

To the producer, the external environment includes the obvious and measurable conditions and influences, such as heat, cold, wet, dry, light, dark, wind, feed supply, and quality of that supply. Producers relate to these and are aware of their changes, but the external environment is capable of altering an environment at a more subtle level--the level of the cells of the body.

Internal Environment

The internal environment is the environment to which the cells are exposed--the level of all life processes. Despite the insults of the external environment, sheep strive to maintain a constant internal environment. This is accomplished through two cooperating systems: the nervous system and the endocrine system. Unfortunately, producers can only measure and relate to the external environment. There is no way to measure the internal environment before problems develop.

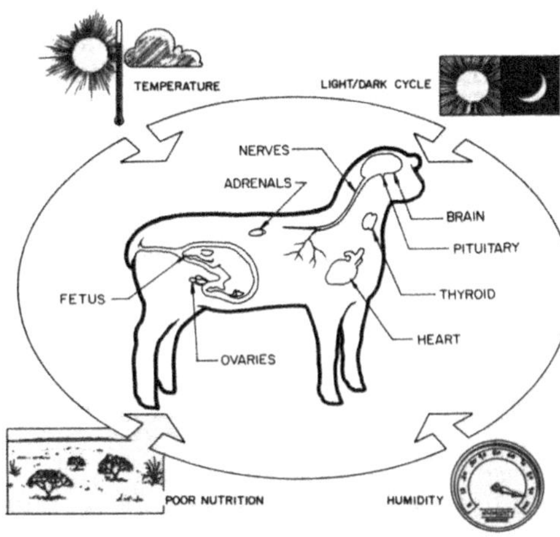

Figure 1. Two cooperating systems control the internal environment in response to the external environment: the nervous system and the endocrine system. The brain and nerves represent the nervous system, while the pituitary, thyroid, adrenals and ovaries are some important components of the endocrine system. The heart and blood vessels distribute the hormones from the endocrines.

Uterine Development

Uterine environment refers to that experienced by the embryo-fetus in the uterus. The conditions and influences affecting the developing lamb can have profound and possibly long-lasting effects, though this environment is temporary. During residency within the uterus, the embryo-fetus is totally dependent upon the ewe for an environment that is optimal for growth and development.

Heavier birth weights are associated with increased survival, and birth weight is largely determined by the uterine environment. Thus, through this environment, the maternal contribution to fetal size (birth weight) is greater than the paternal contribution.

One of the classical experiments demonstrating this effect of the uterine environment was reported by A. Walton and Sir John Hammond in 1938. They performed a reciprocal cross between the large Shire horse and the small Shetland pony. At birth, the crossbred foal from the Shire mare was three times larger than the crossbred foal from the Shetland mare. The environment provided by the Shetland mare limited the size of the foal. Even at 4 years of age, the crossbred from the Shire mare was one and a half times larger--demonstrating the possible long-lasting effects of uterine environment. Experiments in sheep using embryo transfer have demonstrated similar effects on embryo-fetal growth. A Welsh embryo (small breed) transferred to a Border Leicester ewe (large breed) produces lambs that are larger than Welsh lambs born to Welsh ewes.

Since the uterine environment is nested within the internal environment, it follows that changes in the internal environment affect the uterine environment.

GENETICS VS. ENVIRONMENT

Are genetics more important than environment in determining performance? This is an age-old question that will likely never be completely resolved or understood. When determining the effects of one or the other, the importance of individual variation becomes apparent. Often in individuals of similar breeding and environment there is a wide variation in their reproductive efficiency. Hence, most of the time stockmen and scientists talk in terms of averages.

Table 1 shows to what extent some reproductive traits in sheep are influenced by heredity (genetics) and environment. Regardless of the trait, environment (whatever it may be) exerts a greater influence on reproductive traits than does heredity. Hence, manipulation of the environment offers a tremendous potential for improving reproductive efficiency. Conversely, it provides a major avenue for inefficiency. No one should forget, however, that genetics and environment enjoy a complementary relationship.

INTERPRETATION OF THE ENVIRONMENT

Interpretation of the environment, or how the body of sheep reacts to the environment, is determined by the brain that continually receives and interprets information relative to the internal and external environment. Then through nervous signals, the brain alters the internal environment. While the whole brain is important, the hypothalamus area is most important to understanding how the external environment is capable of altering reproductive traits.

Environment-Brain-Hormone Connection

For years, stockmen and scientists were aware that the environment altered reproductive traits. For example, they observed that hot weather lowered conception rates, and day length caused sheep to be seasonal breeders. They also were aware of the nervous system and hormones controlling reproduction, but the connection between the two eluded scientists for years.

The brain is like a computer with many sensors, and it constantly receives information about temperature, humidity, amount of light, sounds, smells, pain, surroundings, etc. After interpreting this information, the brain decides what adjustments are necessary to ensure the well-being (survival) of the sheep. In some cases, appropriate signals for adjustments are sent via the nerves to the various tissues, organs, or limbs. In other cases, the signals must go via the endocrine system as a hormone. It is the hypothalamus of the brain the coordinates this activity. Small, but powerful, it responds to the various signals received by the brain and then acts like a "switchboard" by switching on or or switching off the hormone(s) needed to alter the internal environment. The hypothalamus is the bridge between the nervous and endocrine systems.

For descriptive purposes, the nervous system may be likened to a "wired" system, while the endocrine system may be likened to a "wireless" system.

Power of Hormones

Endocrine glands secrete hormones. The word hormone comes from the Greek word hormon, which means "to spur on, to set in motion, or to excite to action." These phrases are all very descriptive of a hormone. Hormones are "chemical messengers" that travel via the blood to specific organs or tissues and direct such processes as growth, reproduction, metabolism, and behavior. In the blood they exist in extremely small quantities--millionths, billionths, and trillionths of a gram. Yet, their effects upon the body are profound.

Just to gain an appreciation of the minute amounts of hormones, a comparison of one billionth is one second out of the life of a 32 year old individual and one trillionth is one second out of 320 centuries. Extraction of the blood from about one million animals may yield about a pound of many of the hormones.

Figure 2 illustrates the endocrine glands, their hormones, target organs and the relationship of the brain and hypothalamus.

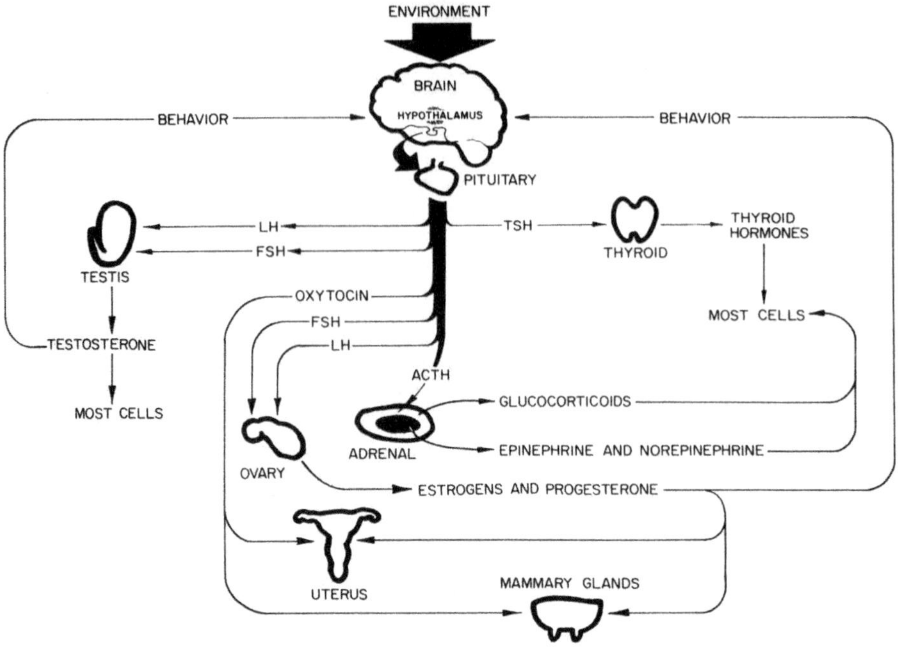

Figure 2. Interaction of the environment, brain, endocrine glands, hormones, and their target organs.

Environments and Hormone Release

There are numerous hormones but only those involved more directly in the reproductive processes are discussed here. Table 2 names the hormones, lists environmental conditions affecting their release and briefly describes their actions and the effects of a deficiency.

Secretion of the hormones should never be viewed as a single event but as a concert. As one hormone comes into play, another may fade out; or one hormone may cause the secretion of another; or the action of one may complement the action of another. The brain or nervous system acts as the conductor by signaling the proper time for increased or decreased secretion of a hormone.

TABLE 2

HORMONES AND THE ENVIRONMENT

Hormone	Environments Affecting Release	Hormone Action	Effects of a Deficiency[1/]
ACTH (Adrenocorticotropic hormone)	Anxiety Injury Light-dark Temperature Toxins	Release of glucocorticoids. Liberates free fatty acids from fat tissue into the blood. Necessary for normal growth and lactation.	Inability to cope with stressful situations. Depressed growth or lactation.
Epinephrine and norepinephrine	Fright Pain Restraint Shock Surprise Unfamiliar surroundings	Alters heart function. Raises blood pressure. Dilates or constricts blood vessels. Increases metabolic rate. Increases blood glucose. Increases alertness.	Unknown
Estrogens	Humidity Light-dark Nutrition Temperature	Female secondary sexual characteristics. Estrous cycles. Female behavior. Development of uterus and mammary glands (udder).	Failure to exhibit estrus. Failure of function of accessory sex organs.
FSH (Follicle stimulating hormone)	Humidity Light-dark Nutrition Temperature	Secretion of estrogens. Follicle growth in females. Sperm formation in the male.	Failure to show heat in female. Failure to develop eggs. Lack of secondary sex characteristics. Depressed sperm formation in males.
Glucocorticoids	Anxiety Injury Light-dark Nutrition Pain Restraint Temperature	Balanced metabolism of proteins, fats, and carbohydrates. Anti-inflammatory.	Inability to cope with stressful situations. Abnormal protein and carbohydrate metabolism.
Growth hormone	Anxiety Restraint Unfamiliar surroundings	Growth of all tissues. Mobilization of fat for energy. Role in lactation.	Improper growth. General depression of metabolism.
LH (Luteinizing hormone)	Light-dark Nutrition Temperature	Testosterone production in the male. Progesterone production in the female. Release of egg from follicle.	Lack of libido in the male. Depressed sperm production. Pregnancy failure. Ovulation failure.
Oxytocin	Auditory or visual cues Mating Stressful situations Suckling	Milk letdown. Uterine contractions aiding birth process or sperm transport.	Possible disruption in sperm transport. Lack of milk ejection.
Progesterone	Light-dark Nutrition Temperature	Implantation and pregnancy maintenance. Mammary and uterine gland development.	Pregnancy failure. Lactation failure.
Prolactin	Variety of nonspecific environmental changes.	Promotes lactation, though high levels may inhibit. May prevent estrous cycles.	Cessation of lactation, but other hormones are involved.
Testosterone	Light-dark Nutrition Temperature	Necessary for sperm production. Develops and maintains male sex organs and masculine characteristics.	Depressed sperm production. Lack of libido.
Thyroid hormones	High temperatures Low temperatures Poor nutrition Unfamiliar surroundings	Increases metabolic rate. Increases nervous system activity. Stimulates protein synthesis. Increases motility and secretions of digestive tract.	Limited growth. Depressed metabolism. Depression or cessation of lactation. Improper development.

[1/] Excesses may also be detrimental by carrying normal hormone action to an extreme or by preventing the action of another hormone.

ENVIRONMENTAL PROBLEMS FOR STOCKMEN

Table 2 indicates that hormones have a widespread effect on reproduction, whether directly or indirectly. Also, many environmental situations are capable of altering hormone levels in the body of sheep. Some of the common reproductive problems include: silent heats, repeat breeders, anestrus, fertilization failure, embryonic mortality, newborn mortality, lactation failure, lack of libido, dystocia, and testicular degeneration. Judged on the basis of table 2, many of these reproductive problems can be explained, or at least compounded, by environmental factors altering endocrine function. Increasingly, research suggests that environmentally induced alterations in the hormone levels may influence the physiological phenomena related to lowered reproductive performance in sheep.

PUTTING IT TOGETHER IN A SYSTEM

Genetics and environment complement each other. But genetics requires considerable time to change due to the generation interval of sheep and the identification of superior animals. An awareness of the profound effects of environment on reproductive traits and the potential for improvement offers stockmen an immediate method for improving reproductive efficiency. With a knowledge of the way sheep perceive the environment and the changes it causes, stockmen can continually evaluate their system for potential problems or improvements. After all, stockmen are the best "welfarists."

23

PRODUCING VIGOROUS, HEALTHY NEWBORN LAMBS

Richard O. Parker

INTRODUCTION

To ensure their survival, many animals produce vast numbers of offspring. Unfortunately, sheepmen rely on a species that produces one to two offspring and promotes their survival through maternal nurturing. Hence, sheepmen should take steps to ensure that lambs are vigorous and healthy from birth.

Production of vigorous, healthy lambs begins in the uterus of a vigorous, healthy ewe. At conception, the genetic material (DNA) from the ewe and the ram provides the plans for the formation of lambs. After conception, the work of producing lambs is automatic, proceeding quietly and guided by the genetic code. Ideally, at the end of gestation, the uterine environment should have provided for optimal growth and development to meet the adaptation of life outside the uterus. Along the way, however, conditions may arise that affect the developing embryo or fetus, thereby placing the newborn lamb at a disadvantage to adjust to life outside the uterus.

Factors governing embryonic and fetal growth are little understood, but certain factors are recognized as being capable of altering normal processes. Regardless of the factor, the importance of the placenta--the supply line--must be recognized. Nutrition, hormones, maternal influences, multiple births, placental size and function, and temperature may each influence the progress of embryonic and fetal growth and development--and ultimately the successful adaptation of the newborn. An awareness of the factors affecting the rapid and dramatic changes during development and growth of the embryo and fetus will aid sheepmen in identifying management practices that require change or improvement.

EMBRYONIC-FETAL DEVELOPMENT AND GROWTH

Development (differentiation) is the formation of major tissues, organs, systems, and the major external features.

Growth is defined as an increase in the number of cells and/or enlargement of existing cells.

Development and growth within the uterus of the ewe occur during two stages: the embryonic period and the fetal period. The embryonic period extends from conception to about day 34. This period is characterized by rapid growth and development. The fetal period extends from day 34 until birth. Primarily, this period is characterized by growth and some changes in external form.

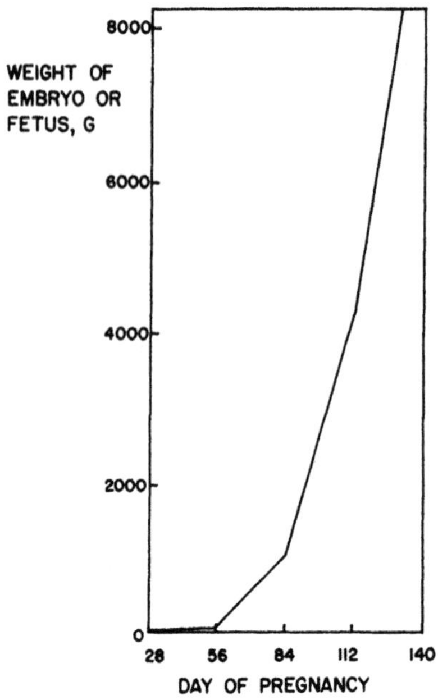

Figure 1. The absolute growth (the total increase in weight) for lambs from conception to birth

Absolute growth--the total increase in weight--increases exponentially, reaching a maximum during the last days of gestation (figure 1). Relative growth--the percentage increase in weight per unit of time--is most rapid in the earlier stages and declines as gestation advances. For example, on day 70, the fetal lamb weighs about 150 g and by day 90 it weighs about 1,250 g--an increase of 733%. Between day 100 and day 120, weight increases only about 190%, from 1,250 g to 3,630 g. Furthermore, organs develop at differing growth rates that seem to be related to functional necessity. For example, the central nervous system (brain and spinal cord), heart, and liver develop first, while muscle and fat develop last. Therefore, factors capable of altering fetal growth and development induce changes depending on the stage of development.

Besides actual weight and length measurements, scientists often employ other measurements to determine the progression of intrauterine development. Two common measures are 1) the estimation of cell numbers and 2) the estimation of cell size. The reasoning behind using these two measures is that a reduction in cell numbers or size may limit the ability of the lamb in achieving its genetic potential.

DNA (deoxyribonucleic acid) is the chemical that makes genes and is contained in the chromosomes within the cell nucleus. The DNA content per cell (except sperm and ova) within an animal's body is the same, i.e., it does not vary between organs. Thus, total DNA content in a tissue estimates the number of cells. In the same tissue, the amount of protein per unit of DNA estimates the size of the cells in the organs or tissues: the larger the number representing the ratio of protein/DNA, the larger the cells.

PLACENTA--THE SUPPLY LINE

The placenta is a versatile, important organ. It substitutes for the fetal digestive tract, lungs, kidneys, liver, and to some extent endocrine glands (hormone secreting). Also it separates the ewe from the fetal lamb(s), thus ensuring the separate development of the fetus(es). Most often there is a direct relationship between the size of the placenta and the size of the fetus. Placental growth occurs early in gestation, reaching its maximum size during the second trimester. Hence, the rapid fetal growth of the last trimester (figure 1) must be supported by a placenta that has virtually stopped growing.

While the blood of the fetus and the mother never come into direct contact, they are close enough for oxygen, nutrients, and waste products to transfer across the placenta. Transfer occurs via primarily two processes: simple diffusion and active transport--an energy-requiring process. Electrolytes (Na, K, Cl), water, and carbon dioxide are transferred by diffusion. Amino acids (building blocks of protein), sugars (energy), and most water-soluble vitamins, iron, calcium, and iodine are transferred by active transport systems. Fat-soluble vitamins are impeded by the placenta. In general, substances capable of traversing the placenta vary according to molecular size, charge, and placental type, which differs between species.

In combination, the fetus and placenta form the hormones progesterone and estrogen, which are important for maintaining later pregnancy and preparing for parturition.

FACTORS AFFECTING EMBRYONIC-FETAL GROWTH AND DEVELOPMENT

Much of the embryonic-fetal growth and development proceeds normally and automatically, without fanfare. But it is a complex process and, like any complex process, disruption is possible. Science has pinpointed some factors that will influence fetal growth and development, but growth and development within the uterus do not "play" by the same rules as those outside the uterus. For example, decapitated fetuses from various species will continue to grow, some quite normally.

Much of the information relative to embryonic-fetal growth and development is derived from laboratory animals rather than actual farm animals. Nevertheless, the following items can be discussed as factors that will affect embryonic-fetal growth and development: nutrition, hormones, maternal influences, multiple births, placental size and function, and temperature.

Nutrition

Of all the factors contributing to embryonic-fetal growth and development, nutrition is the most studied in the most species. A variety of studies have clearly shown that maternal feed restriction (protein and/or energy) limits fetal growth, thereby the birth weight of the newborn lamb. This is especially true when the restriction is imposed during the last trimester when fetal demands for protein and energy are the greatest (table 1), since fetal weight increases so rapidly. Abundant evidence indicates that smallness *per se* is a distinct disadvantage for newborn survival.

TABLE 1. DAILY GAIN OF ENERGY AND PROTEIN FOR LAMB FETUSES AT 100,120, AND 140 DAYS OF GESTATION

	Energy		Protein	
Days	Singles	Twins	Singles	Twins
	(kcal/day)		(g/day)	
100 - 105	64	136	7.6	18.4
120 - 125	131	222	17.8	28.4
140 - 145	220	221	34.7	33.0

Table 2 demonstrates the effects of a severe feed restriction in pregnant rats. Fetuses in this study were limited to 50% of normal weight, but more importantly the small fetuses possessed less DNA--indicating fewer cells--and a smaller protein/DNA ration,--which suggests the cells were smaller. Fewer cells and smaller cells may drastically limit the genetic potential of such individuals. Fortunately, most cases of undernutrition are not so severe.

TABLE 2. SEVERE FEED RESTRICTION DURING THE LAST HALF OF GESTATION AND ITS EFFECTS ON FETAL RAT DEVELOPMENT

Item	Treatment	
	Full-fed	10-day fast
Fetus weight	4.3 g	2.3 g
Total DNA	13.6 mg	9.9 mg
Protein/DNA	25	17

Table 3 provides some of the results from a study in which ewes were subjected to a 50% reduction in energy and protein intake compared to well-fed ewes. Underfed ewes produced smaller fetuses with smaller organs, with the exception of the brain. But none of the organs contain less DNA or had smaller cells. Furthermore, many studies have shown that the brain and possibly the skeletal system are protected from nutritional insults. The results in table 3 would support this.

TABLE 3. WHOLE BODY AND ORGAN WEIGHTS OF FETUSES FROM HIGH- AND LOW-FED EWES ON DAY 144 OF GESTATION[1]

Item	Ration	
	High	Low
Fetus weight (kg)	6.50	5.30
Brain		
Weight (g)	58.00	60.30
Percentage of body	.89	1.14
DNA (mg)	57.30	73.90
Protein/DNA	53.50	57.20
Liver		
Weight (g)	151.70	105.40
Percentage of body	2.30	2.00
DNA (mg)	1283.50	1092.00
Protein/DNA	12.60	10.3
Femur		
Length (cm)	9.70	8.90
Percentage of body	0.15	0.17

[1] Ewes were fed two levels of the same ration beginning on day 90 of gestation.

Most nutritional studies deal with undernutrition during the last half of gestation when the nutritional demands of the fetus are high. Some evidence, however, suggests that poor nutrition during the embryonic period can retard embryo development and possibly affect subsequent intrauterine growth. This is a possibility and is an area of research that should be pursued.

While considering nutrition in general, specific nutrient deficiencies should not be overlooked. Several minerals and vitamins affect fetal development. A deficiency of calcium leads to increased stillbirths. Vitamin D also may be involved. An iron deficiency of the mother may predispose the newborn to anemia. An iodine deficiency or ingestion of goitrogens causes fetal death, abortion, goiter, lowered birth weights, and debility of the newborn. Manganese and zinc deficiencies are associated with reduced birth weights, fetal resorption, and stillbirths. Copper deficiencies may cause the birth of weak ataxia offspring. In sheep this can be aggravated by excess dietary molybdenum. Vitamin A deficiencies can result in the production of lambs that, at birth, are weak, malformed, or dead.

Hormones

Fetal endocrine glands and the placenta secrete a variety of hormones during gestation. These hormones induce metabolic changes in the fetus and prepare the fetus for extrauterine adaptation. Any factor hindering or altering the release of the hormones shown in table 4 may affect the vigor of newborn lambs. Such factors could be nutrition of the ewe or some stress such as heat.

TABLE 4. HORMONES INFLUENCING FETAL AND NEONATAL GROWTH AND DEVELOPMENT

Hormone	General action
Epinephrine	Glucose mobilization
Estrogens	Behavior of newborn
Glucocorticoids	Activation of a variety of enzymes
Insulin	Glucose utilization
Placental lactogen	Nutrient supply in blood of ewe; tissue growth
Progesterone	Behavior of newborn; nutrient supply in blood of ewe
Thyroid hormones	Activation of a variety of enzymes

Maternal Influences

Faster embryonic and fetal growth is directly related to maternal size. Reciprocal crosses between large Border Leicester and small Welch breeds of sheep demonstrated that the crossbred lambs from the large mothers were heavier than those from the small mothers.

The age of the ewe also affects the size of the fetus. Older ewes have heavier lambs. Perhaps young ewes continue to grow through their first pregnancy and compete with the fetus for available nutrients. Also, the degree of vascularity (blood vessels) may increase with subsequent pregnancies.

Whatever the cause, individuals of the same breed vary widely in maternal influence. For example, in a group of ewes subjected to the same degree of undernutrition, some ewes will produce lambs of normal birth weight while others will produce lambs with extremely low birth weights.

Multiple Births

Twin and triplet births place an additional demand on the ewe for nutrients--multiple births also crowd the uterus and reduce placental size (table 5). As a result, lambs from multiple births are lighter.

TABLE 5. INDIVIDUAL FETAL AND PLACENTAL WEIGHT AND THE NUMBER OF FETUS OCCUPYING THE UTERUS OF THE EWE

No. of fetuses	Day of gestation							
	70		100		120		140	
	Fetus (kg)	Placenta (kg)	Fetus (kg)	Placenta (kg)	Fetus (kg)	Placenta (kg)	Fetus (kg)	Placenta (kg)
1	0.15	0.53	1.25	0.63	3.63	0.70	5.70	0.68
2	0.15	0.32	1.20	0.49	3.50	0.56	5.66	0.68
3	0.15	0.36	1.09	0.45	2.80	0.57	—	—

It is a general assumption that uterine capacity does not limit the ability of the ewe to carry twins if they are located in separate uterine horns. Some recent work using ewes in which uterine capacity was surgically reduced suggests that this may not be entirely correct.

Placental Size and Function

By the second trimester, placental growth is virtually complete. Thus, the capability of the placenta to provide for fetal growth the last trimester (figure 1) depends on the adequacy of early placental development. Poor nutrition, heat, or "stress" may hamper placental development.

Function of the placenta relies upon the blood delivered to it. Anything reducing placental blood flow is apt to limit placental transfer of nutrients. In laboratory animals, restricted feeding of the dam slows the transfer of amino acids across the placenta. It is unclear whether this is due to cardiovascular changes in the dam or the action of nutrition *per se* on the placenta.

Temperature

Heat stress (82° to 100°F) can limit fetal growth and result in smaller lambs at birth. This effect is similar, but independent of, the effects of undernutrition. Besides being smaller at birth, lambs from heat-stressed ewes are less viable at birth.

How heat stress exerts its effects on the growing fetus is uncertain. Two good possibilities seem to be reduced placental development or decreased blood flow to the uterus and placenta.

PUTTING IT TOGETHER IN A SYSTEM

Awareness is the key. Sheepmen need to be aware of the process of embryonic and fetal growth, realizing the many factors and interactions that ultimately produce vigorous, healthy lambs. Through an overall awareness sheepmen can view their total program, decide to produce vigorous, healthy lambs, and based upon awareness and experience, exercise that valuable commodity--judgment.

24

MANAGING FOR INCREASED REPRODUCTION EFFICIENCY IN INTENSIVE CROP-SHEEP PRODUCTION SYSTEMS

Hudson A. Glimp

INTRODUCTION

Integrated reproductive management is receiving major emphasis in extension and other educational programs for all livestock species. Optimizing reproductive efficiency includes utilization of appropriate genetic material, optimum nutrition throughout the reproductive cycle, control, or management of environmental factors (such as temperature) that may affect reproduction. No other factor affects the profitability of the sheep enterprise as much as does the percentage lamb crop (number of live lambs per ewe in the flock). It is also the factor that we as sheep producers can most rapidly improve by implementing existing technology and using good animal husbandry practices.

Two management systems for sheep production may be feasible in intensive crop-livestock production systems. The first of these includes confinement or semiconfinement production systems in which the ewes would graze only crop residues, noncropland grazing areas, and strategic grazing crops (on a limited basis). The lambs would not leave the confinement unit until marketed. This system may be more economically feasible during the projected near-term period of cheap feed-grain prices. The second production system involves intensive pasture management (as outlined in other papers presented by the author in this book) and only limited confinement during the lambing period, but the optimum reproductive rate may be somewhat less than the maximum.

GENETIC IMPROVEMENT OF REPRODUCTION

The first step in designing the sheep of the future for intensive crop-livestock production systems is to establish expected levels of performance and/or breeding-program objectives. For example, the basis objective of our breeding program is to develop the best breeds or breed crosses adapted to our environment that will use a maximum amount of forages and a minimum amount of concentrates in producing market lambs of choice grade. The ewe of the future should be capable of weaning two lambs per lambing, should breed anytime of the year, and must have good mothering and milking ability if she is to raise twins. Hardiness, adaptability to the environment and management system, and longevity have not been emphasized adequately in past breeding programs. Ewes need be of only moderate size (130 lb to 140 lb) to produce the currently desired 105 lb to 110 lb of market lamb.

The goal of two lambs per ewe is a maximum with the current systems that maximize the use of forage crops and minimize the use of concentrates. In fact, 1.5 to 1.7 lamb per ewe is a more realistic goal for such production systems. With confinement production systems requiring larger capital investments, two lambs per ewe per year is probably the minimum production level required to achieve the desired levels of profitability.

Sire breeds of the future must emphasize hardiness, ease of lambing, and rate and efficiency of gain to the desired market weight. Too many of these traits are being ignored in our current "show-ring" selection systems.

There are currently no breeds in America that combine all of the desired ewe breed traits. The low heritability of most reproductive traits further suggests that response can be maximized through crossbreeding. The ewe breeds in the U.S. that may be combined to provide these desirable traits are the Rambouillet, Dorset, and Finnsheep.

Research has clearly shown that those producers with intensive sheep-production systems are not planning for the future if they are not including Finnsheep breeding in their commercial ewe program. The commercial ewe of the future should have at least 25% Finnsheep breeding for increased prolificacy, excellent mothering ability, and lamb vigor. Our data suggest that 25% is about optimum for intensive pasture-management systems, while 50% Finnsheep blood may be desired for ewes in intensive confinement-management systems.

The Rambouillet is the hardiest breed, and has minimum lambing difficulties and the greatest longevity of any breed available in large numbers in the U.S. today. The Delaine Merino breed also has these traits but is available only in limited numbers. These breeds are good mothers, have excellent fine-wool clips, and are of adequate size and respectable growth potential and carcass merit. It is our opinion that the desirable commercial ewe should contain at least 50% Rambouillet or Merino breeding; however, it has become more and more difficult to find commercial Rambouillet ewes. The introduction of the Columbia to many commercial Rambouillet flocks is destroying the desirable aseasonal breeding characteristics of the Rambouillet.

The Dorset is a ewe breed with several desirable characteristics. Dorsets are excellent mothers and milkers and produce lambs with desirable carcass merit. They are one of the more prolific domestic breeds, and if one is careful to select pure, fall-born Dorsets, they are likely to breed out of season. I regret having to say this, but pure Dorsets are increasingly hard to find.

We are breeding 1/2 Finn x 1/2 Dorset rams to Rambouillet ewes to produce the 1/4 Finn x 1/4 Dorset x 1/2 Rambouillet replacement ewes that we consider optimum for our intensive pasture-management system. We would prefer to purchase these ewes rather than have to manage a separate breeding program, but we have not been able to find adequate number available from other sources.

It would be highly desirable to find all of the desired ewe breed traits in a single breed, thus greatly simplifying breeding and management programs. The Polypay is a breed developed at the U.S. Sheep Experiment Station that includes one-fourth of each of the Rambouillet, Targhee, Finnsheep, and Dorset breeds. The Polypay breed does not currently exist in adequate numbers for widespread inclusion in commercial sheep-production programs and has not been adequately tested in intensive production systems.

From around the world, the breed of greatest potential in our intensive pasture-production system appears to be the New Zealand Coopworth. Unfortunately, the economic constraints imposed by U.S. quarantine requirements have prevented importation of this breed into the U.S. in adequate numbers for testing under our production system. Lamb crops weaned in excess of 170% with zero assistance under pasture-management systems have been recorded in the Coopworth breed. This is impressive and clearly of interest to anyone seriously interested in intensive sheep production systems.

INCREASING REPRODUCTIVE EFFICIENCY THROUGH NUTRITION

Research has indicated that relatively wide fluctuations in ewe body weight are possible without significantly affecting the lamb and subsequent ewe performance. Relatively large weight changes are in fact expected throughout the production cycle. Undernutrition may be a serious problem affecting reproductive efficiency in the arid regions of the U.S., but is generally not a problem in intensive crop-livestock production system. In fact, controlling obesity under improved pasture management systems is more of a problem than undernutrition. Emaciation under these management conditions is almost always caused by severe internal parasitism or other health-related problems.

In general, a ewe nursing twins should be fed to expect a 10% to 15% decrease in body weight from parturition to weaning at 70 to 90 days. This weight loss should be recovered during the postweaning-prebreeding period. Flushing, or making certain that ewes are healthy and gaining weight during the last 30 days prior to the breeding season, is important to assure high ovulation rates. Ewes carrying excessive fat will not respond to flushing and, in fact, can have problems at breeding time. Ewes carrying twins can be expected to gain approximately 20% of their normal body weight during gestation. In general, no more than one-third of this gain should be during the first two-thirds of the gestation period, and at least two-thirds of the gain should be during the last one-third of gestation.

Rams are generally kept too fat for proper breeding condition in our region of the U.S. Our philosophy is to keep our rams "lean and mean" throughout the year. Our mature rams do not receive concentrates at any time during the year and are expected to forage for themselves throughout the winter.

EFFECTS OF TEMPERATURE ON REPRODUCTION

The effects of temperature on reproductive efficiency are well-documented and are somewhat related to the previous discussion on body condition. Heat stress caused by ambient temperatures in excess of 90° F, or stressful exercise at somewhat lower temperatures, may result in lower ovulation rates, reduced fertilization rates, and increased embryonic mortality in ewes and temporary sterility in rams. Heat stress can be worse in ewes or rams carrying excessive body fat.

These problems can be avoided by maintaining proper body condition, using pastures with adequate natural shade, and avoiding moving or handling the sheep for any reason when the temperature is in excess of 75° F. We also shear all rams and ewes approximately 50 to 70 days prior to the onset of the breeding season. Wool clips longer than one inch or shorter than 1/4 inch are, in our opinion, undesirable during the early breeding season. One other minor practice we consider important is not to initiate breeding until the weather forecast includes 2 to 4 days of cooler weather. Rams will be very active and aggressive when first turned with the ewes, so we try to wait until weather projections provide at least 2 or 3 days in succession with a maximum temperature of 80° F. In some areas of the U.S. where extremely high temperatures are a problem, good results have been achieved by turning the rams with the ewes only at night.

EFFECTS OF ANIMAL HEALTH ON REPRODUCTION

Maintaining proper animal health cannot be over-emphasized. Those producers who have not developed an integrated flock or herd health management program (with their veterinarians) should go home and immediately develop this plan.

The major reproductive problems such as ram epididymitis and vibriosis in ewes are not endemic to the midwestern and eastern sections of the U.S. These can be introduced, however, by importing breeding stock from infected areas or flocks. Caution in purchasing breeding stock cannot be over-emphasized.

Most other disease problems affecting reproductive efficiency are sanitation or management related. Strategic dosing of anthelmintics and pasture rotation systems will control internal parasites. Cleanliness and sanitation in lambing barns and other sheep working areas also are critical to disease control.

Part 5

ENVIRONMENT, BUILDINGS, AND EQUIPMENT

25

MEASURING AN ANIMAL'S ENVIRONMENT

Stanley E. Curtis

ASSESSING ANIMAL ENVIRONMENTS

Animal environments are characterized according to problems suspected of being associated with environmental stress on the animals and the effectiveness of control measures. Most of the elements of animal environments can be measured but, interpreting the results in terms of animal well-being, facility operation, and production economics often remains a dilemma because of the interaction of environmental factors. The effect of one stressful factor on an animal quite often depends on the nature of the rest of the environment.

ANIMAL ENVIRONMENT PROBLEMS

Troubleshooters often must engage in trial-and-error to identify animal-environment problems, but there are several points to be kept in mind:

- The environment results from all external conditions that the animal experiences, so all elements must be considered. Those which cannot be measured or controlled readily might influence animal health and performance nonetheless, so even they must be considered so far as is possible.
- The environmental complex acts as a whole on the animal, so interactions must be kept in mind. The combined effects of two or more environmental components may be difficult to evaluate, but they must be considered.
- Time and space affect the environmental factors. Environmental variables should be measured where the animal experiences them--in its micro--environment--taking into account the lack of spatial uniformity that occurs in all facets of the surroundings. Most important are vertical stratification of various parameters of the thermal environment and horizontal and vertical

variation in airflow due to design or mode of operation of the ventilation system.

- Environmental elements change with time at a given place. Because weather and facility occupancy vary with time, control requirements do also. Thus, environmental assessments and control schemes must take daily and seasonal environmental cycles into account. Most animals adjust to environmental cycles readily as long as extremes of the excursions are not unduly stressful.
- The rate of environmental change is critical. Abrupt environmental changes tend to be more stressful than those occurring over a longer period. For example, preconditioning young animals to a cool environment before moving them from a warm, closed house to a cool, open one during cold weather reduces the stress. It is sometimes difficult to identify a single index of environmental stress or even an adequate multiple index or combination of indices. For example, daily temperature range per se may be of little consequence as long as the day's maximum and minimum values do not exceed or fall below respective trigger levels. Likewise, a certain rate of temperature change might be stressful if it occurs at extremes of temperature, but not within more moderate ranges. With modern statistical techniques, it is possible to develop many environmental indices, with relative ease but it is a very difficult job to determine their respective significances in terms of the environment's impingements on the animal.
- Animals modify their own environments by giving off heat, water vapor, urine and feces, disease causing microbes, and others. The animals' own processes help determine the nature of their microenvironment. Changes in age or number of animals in a facility alter these impacts and, therefore, the control measures required.
- Anthropomorphism is a common pitfall in the assessment of animal environments. A comfortable environment for a human is not necessary for an animal. Animals send signals of discomfort or uneasiness to alert caretakers. These behavioral indications are always useful signs that the environment could be improved. In some cases, the way animals behave is the only clue that stress is present.

MEASURING ENVIRONMENTAL FACTORS

A well-planned and organized approach to environmental-measurement programs is important. The means employed will depend on the amount of accuracy, the extent of detail required and the effort devoted to the actual work of measurement. Insights into sampling theory, as well as descriptions of some of the instruments and techniques that have proved especially applicable to measuring outdoor and indoor animal environments, are provided in the following sections.

Sampling Theory

Environmental assessments are based on interpretations of measurements of pertinent variables. Hence, the observations must represent the situation faithfully. To ensure this, environmental sampling programs must be planned carefully. The main reasons for this have been alluded to already: most environmental elements vary over time and space. Some of these variations are regular and predictable, others are not. The times and the places the environment is measured determine whether the resultant information reflects the character of the environment well enough to be of use. Of course, it is possible to gather more data than necessary, too.

Time considerations. To estimate the average impingement of an environmental factor over a period of hours or days, or to learn about excursions of these values over time, a continuous sample is needed. In short, continuous sampling from more or less permanent instrument stations, usually coupled with recording equipment, provides the data needed for detailed analyses of animal environments.

The most important consideration in regard to time is the length of the observation period. It should be a multiple of a well-established environmental cycle. In studying an animal facility in a temperate climate, for instance, seasonal periodicities must be accounted for to appreciate the facility's nature all year long. On the other hand, a particular problem may be limited to one season, in which case the observation period would be a multiple of the day, to account for diurnal cycles in meteorologic phenomena.

Of course, there is considerable variation among years and even among days within seasons. For example the number of cycles--the number of years or days--that should be observed to give meaningful results depends partly on the nature of this variation already known to occur in the facility's locale. Furthermore, interpretation of any results should include an historical perspective. For example, if observed values for air temperature are at the lower end of the acceptable range during a winter known to be relatively mild by local standards, the problem of a too-cold animal micro-environment might well be encountered during a more nearly normal winter.

The frequency of observations needed within a sampling period is another important decision. In general, measurements should be made as frequently as feasible, especially if automatic recording equipment is being used. Then, after observations have been completed, key periods of environmental extremes or change can be evaluated in detail, while less interesting periods can be ignored. Also, when observations are made too infrequently, errors can be made in characterizing both the ranges and the average values of the variables.

Time-averaging can obscure extremes of environmental factors that may trigger animal responses, hence it must be done only when warranted. For example, effective environmental temperature in an outdoor environment might range from -10° to 20°C one day, from 0° to 10°C another. Average temperature might be around 5°C on both days but the nature of the animals' thermoregulatory reactions would be different on these days, and thus averaging the environmental data over a day could lead to misleading impressions.

Nonlinearity in interactions among environmental factors causes additional difficulties so far as the time-averaging of environmental measurements is concerned. In other words, effects of combined factors must be calculated carefully, even when their relations for steady-state conditions are well-known. Take as an example the case of the wind-chill index (table 1). The average of the wind-chill indices for the three sets of conditions in the table is 1.6×10^3 Kcal m^{-2} hr^{-1}, while the wind-chill index for the average of the three conditions (namely, temperature -34°C and wind speed 9 m sec^{-1}) is 2.1×10^3 kcal m^{-2} hr^{-1}.

TABLE 1. AN EXAMPLE OF NONLINEAR INTERACTION AMONG ENVIRONMENTAL VARIABLES: WIND-CHILL INDEX

	Condition		
	A	B	C
Temperature (°C)	-18	-34	-51
Wind speed (m sec^{-1})	18	9	0
Wind-chill index (kcal m^{-2} hr^{-1})	1.8 x 10	2.1 x 10	.8 x 10

Time lags between environmental occurrences and animal responses also must be recognized--taking them into account could improve the probability of defining a connection between animal and environment.

Another approach is discontinuous sampling, sometimes called spot- or grab-sampling. This usually involves more portable equipment and is aimed at gaining information over short periods, such as hours. Discontinuous sampling is best suited to determining extremes in an environmental element when the basic nature of the variation of the factor is known. For example, air temperature might be measured only in the early afternoon on relatively hot days to gain information on upper values of air temperature, or concentrations of air pollutants might be measured in closed animal houses on relatively cold days when ventilation rate is relatively low.

Discontinuous sampling has several advantages. it requires less time and often less equipment. Further, fewer data are generated, so data-processing equipment requirements are less than when continuous sampling is practiced. Finally, equipment portability often facilitates economical use of a single instrument at many locations in a facility to learn more about environmental variation over space than might be feasible otherwise.

Space considerations. Two prime considerations should determine where environmental measurements are to be taken. In the first place, environmental factors generally should be measured in the immediate surroundings of the animals. For example, most elements of the environment vary consistently with distance above the floor, but most animals reside at discrete heights within a facility. Thus, the height at which the environment is measured is crucial if the measured values are to reflect the conditions to which the animals are being exposed.

In animal facilities it is sometimes tempting to sample the environment in an alleyway or some other place where the instruments will be relatively safe from animal damage. For the most part, these temptations should be resisted. The animals affect their own surroundings so greatly that most variables differ even from animal microenvironments to nearby areas where the animals are not permitted.

Second, the environmental and animal features known to affect the variable under study should be clearly in mind when the measurement sites are chosen. Major items include heat sources: air inlets and outlets; orientation of the facility to winds, the sun, and other structures; location of mechanical services, such as feed-delivery systems; and animal size and population density.

There also is the substantial problem of the effect of the measurement instrument itself, and its protective hardware, on the environment. Some equipment stations obstruct airflow, for example, and the air samplers may be drawing air from different heights and extracting components from the air around the sampler, thereby modifying it.

Instrument Choice

Dozens of instruments for measuring environmental factors are available on the commercial market today. The choice of instrument or set of instruments to be used in a given environmental-measurement program is based on several considerations: (1) the kind of information needed, (2) the relative efficiencies of the various instruments and their reliabilities under field conditions, (3) ease of use, cost, and availability, and (4) personal choice--often based on past experiences--is an important point.

These instruments have accompanying instructions. If the manufacturer's written advice proves inadequate, get in touch with a technical representative of the manufacturing firm directly; most will assist customers by mail or telephone with problems in specific applications of their product.

In the sections that follow, some instruments commonly used in measuring environmental factors in animal facilities are described in brief detail.

MEASURING AIR TEMPERATURE

Liquid-In-Glass Thermometers

A variety of commercially available mercury- or alcohol-in-glass thermometers are used widely to measure air temperature. As temperature rises, the liquid expands to occupy more of the capillary tube in which it is held. Of course, as in all thermometry, the temperature registered is actually that of the thermometer, not necessarily that of the environment, so factors apart from the temperature of the surrounding air that affect thermometer temperature lead to errors. Chief among these are solar and thermal radiations and air movement. Precautions must be taken to shield the thermometer against them. Furthermore, the thermometer may be placed in a spot where the air is stagnant or otherwise unrepresentative of the general area.

Lag time. When environmental temperature changes, the value required for a thermometer to reduce the difference between registered temperature and actual air temperature to 36.8% of the original difference is called the lag time of the thermometer. Lag time for most mercury-in-glass thermometers is around 1 min, while that for alcohol-in-glass is roughly 1.5 min. Some electronic thermometers have much shorter lag times. Especially in discontinuous sampling, lag time is an important consideration because a measurement can be made more quickly with the shorter lag time.

Maximum-minimum thermometer. A thermometer designed to register the maximum and minimum temperatures experienced during the measurement period often has been used in animal facilities. It consists of a U-tube with bulbs at both ends. One side of the U serves as the scale for maximum temperature, the other for minimum. The bulb on the maximum side serves as a safety reservoir and is partially filled with a liquid such as creosote solution. That on the other side is completely filled with the liquid. Between these two portions of liquid, in the bottom of the U-tube, is mercury. As the thermometer becomes warmer, the liquid in the filled bulb expands, pushing the mercury up on the maximum-scale side (the opposite side). Atop the mercury on both sides is an iron index (a sliver of iron), and as the mercury moves up the maximum side it pushes the index ahead of it. As the thermometer cools, the mercury retracts, but the index remains at its highest point due to friction with the inside of the U-tube that can be overcome by the mercury, but not by the other liquid. Of course, as the mercury column retracts upon cooling, it pushes the minimum-temperature side's index ahead of it, so this one registers the lowest temperature of the measurement period on the maximum-temperature scale, which is upside-down. When maximum and minimum temperature have been observed, the thermometer is reset by replacing the indices atop respective mercury meniscuses by means of a magnet.

Although maximum and minimum temperature may be all the information needed in some situations, and despite the fact it is relatively inexpensive and straightforward in design, this instrument has serious drawbacks. Chief among them are the tendencies for the mercury column to become separated or broken and for the indices to become permanently lodged in the U-tube.

Bimetallic-strip thermometer. The sensor of a bimetallic-trip thermometer comprises a sheet of each of two metals having dissimilar coefficients of linear thermal expansion, which are joined along their faces. When such a strip's temperature changes, it bends. Lag time is relatively short--around 10 seconds.

Thermograph. The bending movement noted above for the bimetallic strip is usually magnified by shaping the strip appropriately--and in a thermograph the movement is recorded by affixing a pen to its free end and applying this pen to a piece of graph paper on a drum rotated by a clockworks. By this relatively inexpensive means, a permanent record of temperature variation is made. Further, it requires no electrical supply. Of course, a thermograph is so large that it affects the microenvironment it is used in and, because of its construction, is prone to error due to radiation and stagnant air pockets.

Calibration of a thermograph is a critical matter. The recording element is very sensitive to physical shock, which often occurs when the instrument is moved from place to place, thus, after the instrument is moved, it is absolutely necessary to calibrate a thermograph by adjusting the recording pen several days in a row, preferably near the times of the daily high and low temperatures. An artificially ventilated psychrometer is commonly used as the standard instrument for calibration of thermographs.

Electric thermometers

There are two general kinds of electric thermometers. One kind depends on the principle that as the temperature of a substance changes, so does its electrical resistance. The other kind--thermocouple thermometry--depends on the principle that when wires of two specific metals are joined at both ends, and when these two junctions are kept at two temperatures, an electromotive force is generated in that circuit. Voltage in such a circuit, when measured with a potentiometer, is directly proportional to the temperature difference between two junctions.

Thermistor. Certain semiconducting materials have negative temperature coefficients of electrical resistance: as temperature rises, resistance decreases. Resistance in the circuit to which a current has been applied is measured by an indicating unit. These sensing elements are called thermistors--parts of the electric thermometers most applicable to air thermometry in animal environments.

Thermistor probes, indicating units, and recording units are commercially available in a wide range of models. At one time, thermistor systems were notoriously unstable, but nowadays stable, calibrated probes are on the market and in recent years they increasingly have become the sensors of choice for routine assessment of animal environments.

Thermocouples. A variety of combinations of dissimilar wires are used for the two sides of the circuit in thermocouple thermometry. Copper and constantan are frequently chosen. One of the thermocouple junctions (the reference junction) is held at a constant or known temperature so changes in electromotive force measured reflect changes in the temperature of the measurement junction, which is placed in the environment to be monitored. The voltage generated can be used to drive a millivolt recorder or registered on a millivoltmeter.

In general, use of thermocouples for air thermometry in animal facilities has some disadvantages. The needed equipment, especially the constant-temperature bath for the reference junction, is relatively cumbersome, the physical integrity of the measurement and reference junctions critical and sometimes difficult to maintain, and careful calibration of the thermocouples very important.

Metal-resistance thermometer. In metals, as temperature rises, so does electrical resistance. Small-diameter platinum wire wound on a support having a small thermal expansion coefficient is a frequent choice. A small current is introduced into the wire, and resistance of the whole circuit is measured using a Wheatstone bridge circuit. The most sensitive resistance thermometers tend to be fragile, and for this reason alone they are of limited use in animal environments. In addition, the measuring equipment is relatively expensive.

MEASURING AIR MOISTURE

The measurement of water vapor in air is called psychrometry or hygrometry. Several principles have been used to measure air moisture, and three have been applied widely in quantifying animal environments.

Hair Hygrometer

Hair is hygroscopic, and its length is related directly with the amount of water it contains. Further, there is a nonlinear, direct correspondence between length of hairs and the relative humidity of the air surrounding them. Hair hygrometry is most accurate when relative humidity ranges between 20% and 80%. This principle has been employed extensively in hygrometry in animal facilities, partly because hair hygrometers require no electrical supply and they are affected little by other factors.

Hygrograph. Elongation and shortening of hair bundles can be magnified by an appropriate level system and transformed into movement of an arm holding a pen that is applied to graph paper on a rotating drum, giving a record of changes in relative humidity. Just as for the thermograph, hygrographs must be calibrated carefully over a period of several days after they have been moved to a new location before reliable measurements can be made.

Psychrometers

Psychrometers are a class of instrument by means of which the air's moisture content or relative humidity can be estimated indirectly. They use both a wet-bulb thermometer and a dry-bulb, and their measurement principle is based on the thermodynamic relation between the air's moisture content and wet-bulb temperature. (Wet-bulb temperature is affected by dry-bulb temperature and air pressure.) Once dry-bulb and wet-bulb temperatures of the air are known, the air's moisture content and relative humidity can be estimated from a psychrometric chart.

The dry-bulb thermometer can be of any type, while the wet-bulb thermometer is a similar instrument having a water--saturated wick closely surrounding the bulb or sensing the element. As the psychrometer is operated, evaporation from the wick occurs, and the temperature of the wet bulb is depressed. Of course, the drier the air, the greater the evaporation, and the greater the wet-bulb depression.

To give an accurate estimate of wet-bulb temperature, the thermometers must be ventilated adequately; maximum cooling of the wet bulb does not occur in still air. Also, temperature readings must not be taken until equilibrium has occurred. Other sources of error when using any psychrometer are heat conduction down the wet-bulb thermometer (the wick is ordinarily extended up the stem), receipt of solar and thermal radiation (the latter even from the operator of the instrument), and wicks that are too thick or dirty (one way to minimize mineral crust is to wet the wick with distilled water only).

Wet-bulb temperature should always be read before that of the dry-bulb, as it will begin to rise as soon as ventilation ceases. It is also good practice to repeat the measurement several times to make sure the lowest wet-bulb temperature has been attained.

Sling psychrometer. The sling psychrometer, once the standard instrument for spot-sampling psychrometry, consists of dry- and wet-bulb liquid-in-glass thermometers in a frame that can be revolved around a handle. The thermometers are whirled--usually for a minute or more--to permit wet-bulb depression to occur. Larger sling psychrometers must be revolved at least two times per second to provide sufficient ventilation, and smaller ones five times. The movement of the operator mixes the air in the region.

Artificially ventilated psychrometer. Various models of psychrometer are now available in which air is drawn artificially past the temperature sensors. One popular model employs dry-cell batteries that supply a small fan, which pulls air at speeds up to 5 m sec^{-1} past the bulbs. Of course, this kind of instrument may draw air from as far away as 1 foot, hence it is not applicable to some microenvironmental measurements. Still, it is reliable, rugged, and portable.

Electrical Conductivity of Hygroscopic Materials

Salts such as lithium chloride are hygroscopic, and their electrical conductivity increases--thus, their resistance decreases--with increasing water content.

Electric hygrometer. Commercially available sensing units usually involve a film of lithium chloride on a nonconducting frame through which an electrical current is passed for the purpose of measuring electrical resistance changes. The logarithms of the resistance and the atmospheric humidity parameters are inversely related.

Dew cell. When a film of lithium chloride is applied to a heating-element frame, and the temperature is so high the salt is dry, the salt is also highly resistant to conducting electricity. The electrical circuitry of a dew-cell apparatus is designed so that when the salt is a conductor, the element is being heated, but the heating stops when the salt becomes warm enough to become dry. Thus, the lithium-chloride film is kept more or less at the same temperature and dry at all times. This equilibrium salt temperature is related directly with the air's dew-point temperature.

Electric hygrometers and dew cells make it possible to monitor air humidity continuously, but in animal environments they often become dirty, and this can lead to errors. Further, they remain operational for periods of only a few months under the best of conditions.

MEASURING AIR MOVEMENT

Drafts, stagnant spaces, and inadequate removal of moisture or noxious gases in animal houses are among the common symptoms of improper design or operation of a ventilation system. Air speed and distribution throughout an animal facility must be known if ventilation problems are to be remedied.

Anemometers

Several kinds of instruments to measure air speed are available. Each is best suited to a particular application in animal-environment measurement.

Pitot-tube measurement. When air moves into or across the mouth of an open tube, the air pressure in tne tube changes; it increases in the former case, decreases in the latter. Such pressure changes are proportional to air speed and serve as the basis of pitot-tube anemometry.

The most common pitot-tube anemometer used today is the Velometer, a rugged and portable instrument that gives a direct reading of air speed and comes supplied with a variety of probes for different velocity ranges. This instrument is most adaptable to measuring air velocity at inlets, in areas of strong drafts, and in air ducts.

Because of its very nature, the pitot-tube is extremely directional; it is sensitive to air movement in one direction. Large errors can result when a probe is not properly oriented.

Hot-wire anemometer. Several modes of hot-wire anemometer are on the market. Some are directional and therefore applicable to measuring air speed in ducts and at inlets; others are more nearly omnidirectional. They operate on the principle that as air passes across a fine platinum or nickel wire that has been heated electrically, the wire tends to cool by an amount proportional to air speed. This instrument is designed so current flowing through the wire automatically changes so as to keep wire temperature nearly constant. The amount of current required to achieve this is thus related directly with air velocity.

Hot-wire anemometers are relatively sensitive and thus especially applicable to situations--such as animal microenvironments--where velocity can be as low as .5 cm sec^{-1}. Another advantage is that some designs are very portable and fairy rugged, and so small they do not interfere much with the environment. They also respond very quickly to changes in air speed. However, the sensing wire can be affected over time by atmospheric pollutants, and for this and other reasons frequent calibration is necessary. Further, the sensing wire itself is exceedingly fragile and subject to damage. The hot-wire anemometer cannot be used outdoors during rainy periods or in any environments where water can reach the sensing wire.

Rotating-vane anemometer. There are various kinds of anemometers employing a lightweight propeller with eight or more blades. All are highly directional, and for applications where air direction changes greatly they are commonly attached to a vane device that keeps them aimed into the main flow of the air.

A variety of physical and electronic metering devices are used in conjunction with these anemometers. They are most useful at high air speeds, and thus are most applicable in outdoor settings. Indoor models are relatively fragile and subject to damage by corrosive gasses.

Cup anemometer. The device used to measure air velocity outdoors consists of several cups, each connected to a rod radiating from a rotor. A revolution of the assembly is directly related to air velocity. Cup anemometers are not suitable for measuring low air speeds. They are most applicable to estimating the average speed of the wind during periods of at least several hours.

Kata thermometer. A simple instrument was developed over a century ago to estimate the cooling power of the air. The kata thermometer is capable of measuring very low air velocities. It is simply an alcohol-in-glass thermometer with the stem marked at the 37.5° and 35°C levels.

The thermometer is warmed in a water bath to around 40°C, removed from the water and wiped dry, and placed in the environment to be measured. The time required for temperature to fall from 37.5° to 35°C is determined using a stopwatch, and this value substituted into a formula (which also includes air temperature and an individual instrument-calibration factor supplied by the manufacturer) for calculating air velocity.

Airflow-pattern measurement

Certain visible particles suspended in the air can be used as an aid to tracing how the air is distributed in an animal house. The source of the particulate matter is simply placed at the air inlet or in the microenvironment to be studied, and the course of the pollutant followed visually through the space of interest. Much cigar and pipe smoke has been used quite effectively for this purpose in the past. More reliable sources of larger amounts of visible tracers are now on the market.

In the absence of an anemometer, low and moderate air speeds can be estimated to a first approximation by briefly interrupting the tracer's flow and monitoring the movement of the turbulence so induced with the aid of a measuring tape and a stopwatch.

"Smoke"vials. When a solution of titanium tetrachloride is exposed to air, it gives off copious whitish fumes. Small glass vials of this solution areideal for use in studying airflow patterns in microenvironments or parts of a room. A vial is broken open and simply can be held in the location to be observed. An open vial is commonly placed in a cup, which may be attached to a pole in order to reach certain parts of the room of interest. Several devices are also available to increase the evolution of fumes by passing air over the solution. Titanium tetrachloride fumes in the concentrations used are not toxic to animals.

Talcum aerosolizer. Fine talcum powder is also used as a tracer in small-scale air-distribution studies in animal houses. Special talcum-powder aerosolizers are available commercially. Talcum tends to precipitate faster than does titanium tetrachloride, and it is generally more difficult to generate an adequate aerosol of the powder. For these reasons, it is inferior to titanium tetrachloride for this purpose.

"Smoke" pellets and candles. When long-term or large-scale observations are desired, more tracer might be needed than can be supplied by titanium tetrachloride or talcum aerosols. The commercial market has a wide range of pellets and candles that, when lighted, produce very large amounts of fumes. These are generally set in the way of the incoming air. The fumes of some of these devices are toxic and therefore must not be used in occupied houses.

Static-Pressure Measurement

In negative-pressure or exhaust-ventilation systems, adequate negative-static pressure must be maintained by the fans. Static pressure inside an animal house is usually measured by means of a manometer sloped 1:10 to increase accuracy. One end of the manometer is open to the outside atmosphere, the other to the inside, and the pressure difference is registered by the manometer.

Every animal house employing mechanical ventilation should be equipped with its own static-pressure manometer.

MEASURING SURFACE TEMPERATURE

The temperature of an animal's surface is an important determinant of convective and radiant exchanges of heat.

The temperatures of environmental surfaces likewise play a central role in determining the magnitude and direction of thermal-radiant flux. Accurate measurement of surface temperature is difficult, but there are two ways this can be approached.

Portable radiation thermometer. A variety of battery-powered instruments now available provide rapid measurement of surface temperature by a technique that doesnot involve contact with the surface. The portable radiation thermometer is simply aimed at the surface to be measured, and the surface's temperature is registered on the meter. Models providing different fields of view are available.

This kind of thermometer must be calibrated before every observation by means of a Leslie cube or some other temperature-calibration device. Solar radiation reflected by a surface leads to overestimation of that surface's temperature when measured with this instrument. Also, surfaces in the surroundings substantially cooler than that being measured tend to bias these estimates in the opposite direction. The instrument is reliable, reasonably rugged, and is an excellent means for spot-sampling animal and environmental surface temperatures.

Contact thermometers. Thermometers in a variety of styles involving contact with animal and environmental surfaces have been used sometimes to measure surface temperature. These include thermistor probes in hypodermic needles, contact discs (banjo probes), as well as thermocouples taped or glued to surfaces.

There are serious drawbacks to contact thermometers. First, the contact must be flawless--otherwise insulative air pockets, even very small ones, between thermometer and surface will introduce measurement errors--and this is rarely achieved on either animal or environmental surfaces.

There are other problems of introducing artifacts with animal surfaces in particular. It is a practical impossibility to achieve the necessary contact with covered areas of an animal's surface without disrupting that cover and altering surface temperature. Even on nude areas, affixing a thermometer to the skin in such a way as to ensure adequate contact alters cutaneous blood flow and surface temperature.

MEASURING SOLAR AND THERMAL RADIATION

Solar Radiation

The total direct and diffuse (sky) solar radiation received by a horizontal surface is measured by an instrument called a pyranometer. Several designs are available on the market. The standard instrument in the United States is the Eppley pyranometer.

Eppley 180° pyranometer. This instrument consists of horizontal concentric silver rings, one black and one white. The glass hemisphere that covers the rings transmits only radiation with wavelengths less than 3.5 μm (solar radiation), for which the black and white rings have different absorptivities. The temperature difference between the two rings generates an electromotive force in a thermopile (a series of thermocouple junctions), alternate junctions of which are in thermal contact with respective rings.

The Eppley 180° pyranometer measures direct and diffuse solar radiation received by a horizontal flat surface from the upper hemisphere. When direct sunshine is blocked by a shade, only the sky radiation is measured.

This instrument is fragile and must be sited carefully. Its glass bulb must be cleaned daily.

Thermal Radiation

The rate of incoming thermal radiation is usually estimated as the difference between the rate of total incoming radiation having wavelengths between .1 and 100 μm, and the rate of total incoming solar radiation as estimated by an unshaded pyranometer. The rate of total incoming radiation can be measured by any of several kinds of total radiometer.

Beckman and Whitley total radiometer. This instrument is ventilated to minimize errors due to variable convective heat loss and measures the rate of incoming radiation at wavelengths between .1 and 100 μm. It essentially consists of three layers of plastic: the top one is painted black and exposed to radiation from the upper hemisphere; the middle one contains a thermopile; and the lower one is shielded to minimize its receipt of radiation. The temperature difference between top and bottom of the middle plate is related directly to the upper plate's rate of radiation absorption.

MEASURING LIGHT INTENSITY

Photovoltaic Meters

Light meter. A selenium photovoltaic cell comprises the basis of most light-intensity meters used by photographers. Illumination of the sensitive layer of selenium by visible radiation (wavelengths from .3 to 7 μm set up a flow of current in an appropriate circuit. Photographic light meters are portable and very useful in measuring light intensity in animal environments.

In most commercially available light meters, sensitivity over the visible spectrum is trimmed to resemble that of the human eye; that is, it peaks at a wavelength of about .55 μm instead of the .63 μm wavelength radiation that drives photoperiodisms in animals.

Illuminometer. A recording hemispherical photometer called the Illuminometer is also on the market. It is particularly suited to stationary installations where light intensity is known to vary over time, such as it does outdoors.

MEASURING SOUND LEVEL

Sound-level meter. Several battery-powered models of sound-level meter are available commercially. All provide a simple, portable, and reliable means of measuring sound level in decibels. The microphones on such instruments are relatively nondirectional, and the operation of sound-level meters is straightforward. Necessary precautions include recognizing the possible presence of obstacles to sound waves; locating the microphone at the observer's side, not between observer and sound source; shielding the microphone from any moving air; and making sure interfering electromagnetic fields from other electrical equipment are accounted for.

MEASURING AIR PRESSURE

Aneroid barometer. Measurement of air pressure in conjunction with animal production is ordinarily accomplished by means of an aneroid barometer--an instrument in which the walls of an evacuated cell move as air pressure changes. The movements are transmitted to a pointer, which indicates air pressure. While not as accurate as a mercury barometer, the aneroid version is nonetheless quite useful in animal work. If moved, it must be recalibrated against a mercury barometer.

Barograph. Movements of an aneroid barometer cell's wall can be recorded on graph paper affixed to a rotating drum when a pen is linked to that wall.

MEASURING AIR POLLUTANTS

Measuring Aerial Gases and Vapors

Colorimetric indicator tubes. Several systems of the same general type are available commercially for the convenient and, when properly used, reasonable accurate measurement of aerial gases and vapors. These consist of an indicator tube and a precision piston or bellows pump operated manually to draw air. The detector tube contains a specific chemical that reacts with the gas or vapor being measured. These small detector tubes are available for all the major gases and some of the vaporous compounds commonly present in animal-house air.

When air is pulled through an indicator tube, the pollutant for which the tube's indicating gel is specific reacts with the chemical, resulting in discoloration. The extent of this change is related to the concentration of the pollutant in the air. One problem with such a measurement system is that gases associated with dust particles--for example, some of the ammonia in dusty air--are filtered out of the air before it reaches the colorimetric indicator. This tend to cause underestimation of the pollutant's concentration in the air.

The pump for this kind of system must be kept leakproof and must be calibrated. Also, the detector tubes must be handled carefully and their predicted shelf lives observed.

Measuring Aerial Dust

High-volume sampler. Several models of dust samplers in wide use draw through a filter made of cellulose paper, glass or plastic fibers, or organic membrane. Dust particles too large to pass the filter are collected on it.

In practice, a filter is dried and tared before sample collection, and the particle-laden filter is dried and weighed again at the end of a sampling period. The difference between the two is an estimate of the mass of the particles collected during the sampling period. This usually is divided by the product of sampling period and average airflow rate to give the concentration of the pollutant. Filters are commonly handled with tweezers and transported outside the laboratory in large, covered petri dishes.

Another critical factor is calibration of airflow rate. most high-volume air samplers are equipped with some sort of airflow meter, but these instruments should be calibrated frequently because errors in this estimate are perpetuated as errors in all concentration estimates.

Particle counting and sizing. Several dozen models of instruments to count airborne particles are on the market. Some are primarily collectors--based on the principle of impaction on a solid surface, impingement in a liquid medium, centrifugation, or settling--and used in conjunction with subsequent visual observation. Others are direct-reading instruments employing optics and electronics.

Impactors are most commonly used for discontinuous sampling in animal environments. Two popular instruments are the Anderson six-stage, stacked-sieve, nonviable sampler, and the four-stage cascade impactor. Both feature impaction of dust particles on pieces of glass, which are then inspected microscopically for counting. As for sizing of the particles, both instruments have several stages designed so that the polluted air is drawn through a series of jets with progressively smaller cross-sections. The result is that relatively large particles are impacted in early stages, smaller ones at later stages. The size ranges monitored by these instruments are pertinent to the site of deposition of the particles within an animal's respiratory tract.

MEASURING AERIAL MICROBES

Qualitative studies of airborne microbes in animal environments have long involved opening a petri dish of culture medium, permitting viable particles to settle out of the air onto the medium's surface. Of course, special media can be used when there is interest in particular kinds of microbes. When the aerial concentration of microbes or microbe-carrying particles must be determined, another method must be used.

<u>All-glass impinger</u>. One method of quantifying airborne microbes is to impinge them in an isotonic solution that can then be diluted appropriately, combined with nutrients, and cultured in preparation for counting. This method is well-adapted to situations where aerial microbic level is high, but has the disadvantage of disintegrating airborne particles containing more than one microbe so that, for instance, an airborne particle that would give rise to one colony in an animal's respiratory tract might give rise to hundreds to be counted in the culture dish.

<u>Andersen six-stage viable sampler</u>. The viable version of the Andersen stacked-sieve sampler holds special culture-medium plates instead of flat pieces of glass as in the nonviable model. Particles are impacted onto the solid medium's surface where colonies can grow and be counted.

The Andersen viable sampler is a very useful instrument for both counting and sizing airborne microbic particles in animal environments. Special media can be used when desirable, and both the size and the number results can be interpreted in terms of the challenge the aerial microbic particles present to the animals' respiratory tracts. On the other hand, because of its relatively high air-sampling rate, the Andersen sampler is less well-adapted to air environments in which microbic populations are very high, as in some closed animal houses during cold weather. In such case, the sampling period may have to be short as 15 seconds, and thus special care must be exercised to ensure accuracy in estimating the volume of air drawn through the instrument during the sampling period.

<u>Andersen disposable two-stage viable sampler</u>. A less expensive device is a disposable-plastic, two-stage sampler fashioned after the original Andersen six-stage model. Commercial petri dishes available in hospital microbiology laboratories and a variety of vacuum sources can be used with this system.

This instrument has a critical orifice providing an air-sampling rate of 1 ft^3 per min when a vacuum of at least 10 in. of mercury is maintained. When operated in this way, the colonies that grow on the upper stage have arisen from particles having an aerodynamic diameter greater than 7 μm, and hence they would not have deposited in the lungs of an animal. The particles that are impacted on the lower stage are between 1 and 7 μm in diameter, and many of these could have reached the animal's lungs.

Like the Andersen viable sampler, the disposable sampler sometimes must be operated for a short sampling period in commercial animal houses.

Also, the collection efficiency of the Andersen disposable sampler seems to be less than that of the standard version. Despite these drawbacks, the disposable model is an inexpensive, relatively accurate means of estimating the concentration of microbe-bearing particles in the air, and whether they are of such a size as to directly threaten pulmonary health.

REFERENCES

Anonymous. 1972. Air Sampling Instruments for Evaluation of Atmospheric Contaminants. Fourth Ed. Am. Conf. Gov. Indust. Hygienists. Cincinnati.

Curtis, S.E. 1981. Environmental Management in Animal Agriculture. Animal Environment Services, Mahomet, Illinois.

Gates, D.M. 1968. Sensing biological environments with a portable radiation thermometer. Appl. Optics 7:1803.

Hosey, A.D. and C.H. Powell (Eds.). 1967. Industrial noise--a guide to its evaluation and control. Pub. Health Serv. pub. 1572. U.S. Gov. Printing Off., Washington.

Johnstone, M.W. and P.F. Scholes. 1976. Measuring the environment. In: Control of the Animal House Environment. Vol. 7, Laboratory Animal Handbooks. Laboratory Animals, Ltd., London.

Kelly, C.F. and T.E. Bond. 1971. Bioclimatic factors and their measurement. In: A Guide to Environmental Research on Animals. Nat. Acad. Sci., Washington.

Munn, R.E. 1970. Biometerological Methods. Academic Press, New York.

Platt, R.B. and J.F. Griffiths. 1972. Environmental Measurement and Interpretation. Krieger, Huntington, NY.

Powell, C.H. and A.D. Hosey (Eds.). 1965. The industrial environment--its evaluation and control. Pub. Health Serv. Pub. 614, U.S. Gov. Printing Off., Washington.

Schuman, M.M., et al. 1970. Industrial Ventilation. Eleventh Ed. Am. Conf. Gov. Indust. Hygienists, Cincinnati.

Spencer-Gregory, H., and E. Rourke. 1957. Hygrometry. Crosby Lockwood, London.

Stern, A.C. (Ed.). 1976. Measuring, Monitoring, and Surveillance of Air Pollution. Vol. III, Air Pollution. Third Ed. Academic Press, New York.

Tanner, C.B. 1963. Basic instrumentation and measurements for plant environment and micrometerology. Soils Bull. 6, Univ. of Wisconsin, Madison.

Wolfe, H.W., et al. 1959. Sampling Microbiological Aerosols. Pub. Health Serv. Pub. 686. U.S. Gov. Printing Off., Washington.

26

MEASURING ENVIRONMENTAL STRESS IN FARM ANIMALS

Stanley E. Curtis

Relations between agricultural animals and their surroundings always have been important. Those species recruited for domestication generally differ from their wild cousins in that the domesticated animals are adaptable to a wider range of environments than are their cousins (Hale, 1969). Hence, these animals we keep are more amenable to being confined and managed by the humans they serve (Bowman, 1977).

Ecology always has been at the heart of animal production. The shelter aspect of environmental management has been applied for a long time. shepherds kept their flocks in folds at night thousands of years ago. Only with the advent of widespread spacewise and time wise intensiveness in animal agriculture have animal-environmental relations become so important relative to other factors of production. And only with this intensiveness has major environmental modification been possible not to mention economically feasible. Now in addition to increasing the fit of the animals to the environment, we are coming closer to meeting the animals' needs by modifying their environments.

A DIGRESSION

"Measuring Environmental Stress in Farm Animals," calls to mind that which we should keep in mind. Let us examine the last five words of the title first and the first word last.

Environmental Stress in Farm Animals

Stress is of the environment, not of the animals (Fraser, et al., 1975; Curtis, 1981). Nevertheless, we measure stress in the animals, not in the environment. An animal is under stress when it is required to make extreme functional, structural, or behavioral adjustments in order to cope with adverse aspects of its environment. Thus, an environmental complex is stressful only if it makes extreme demands on the animal.

In other words, an environment is not stressful in and of itself; it is stressful only if it puts an animal under stress. And because animals differ in the ways they perceive and respond to the environmental impingements, the very same environment can be stressful to one animal and not to another.

An environmental factor that contributes to the stressful nature of an environment is called a stressor. When we "measure stress in an animal" we really measure the effects of the stress: the changes the stressor causes in the animal (such as the rise in body temperature when the animal is experiencing a net gain of heat from the environment) or the responses the animal invokes in an effort to establish a normal internal state in the face of a stressor (such as the rise in breathing rate when the animal needs to increase its heat-loss rate to bring body temperature back down to the desired point).

Measuring Environmental Stress

Scientists in a wide range of disciplines have been "measuring stress in animals" with increasing frequency over the past century and a half. Almost twenty years ago, The American Physiological Society published an epic tome of some 1056 pages called Adaptation to the Environment (Dill, 1964). The means of measurement have continued to develop as the scientific inquiry in animal ecology has blossomed profusely in the intervening two decades.

Yet the measurement of stress in animals is but the first step in applying ecological knowledge to animal production. The second step is the interpretation of the values. And of the two, the second step is by far the more difficult. In particular, it is necessary to determine where stress leaves off and distress (excessive or unpleasnt stress) begins.

Interpretation of stress parameters and indices is thus the real challenge as we continue to generate more knowledge and endeavor to use more completely what is already known for the purpose of increasing the fit between agricultural animals and their environments. And so it is this interpretation step on which we shall dwell.

STRESS RESPONSES: TRADITIONAL CONCEPTS

It is the unusual moment when an animal--in the wild or on a farm--is not responding to several stressors at once. Stress is the rule, not the exception. And nature has endowed the animals with a marvelous array of reactions to these impingements.

External environment comprises all of the thousands of physical, chemical, and biological factors that surround an animal's body. Each environmental factor varies over space and time. The animal's environment is therefore exceedingly complex.

The animal must maintain a steady state in its internal environment despite fluctuating external conditions. Claude Bernard (1957) said: "All vital mechanisms, however varied they may be, have only one object, that of preserving constant the conditions of life in the internal environment." This is the concept based on negative-feedback control loops that Walter Cannon later called homeostasis. More recently it has been called homeokinesis to emphasize its dynamic, yet consistent, nature.

All sorts of external environmental elements tend to modify corresponding internal environmental elements in an animal. Ultimately, if no homeokinetic mechanism acted, the internal environment would resemble the external, and life would cease.

Homeokinetic Control Loops

The homeokinetic animal attempts to control all aspects of its internal environment via adaptive responses similar in principle to a house's temperature-control system. Neural mechanisms participate in input reception and analysis, decision-making, and effector activation. Neuroendocrine mechanisms link neural and endocrine elements and activate effectors. Endocrine mechanisms take part in neural-endocrine and endocrine-endocrine linkages, as well as effector activation, and in some cases even effector action. These processes occur in specific configurations in the animal's many specific control loops. Muscles and glands are the body's chief effectors. Effector action is usually specific for the particular remedial reaction required.

Nonspecific stress response. In addition to specific stimulus/effector activation loops, Hans Selye (1952) has developed the concept of a nonspecific initial reaction to diverse stimuli. According to this facet of Selye's general adaptation syndrome, the rate of adrenal glucocorticoid secretion increases abruptly following any insult to the body. The teleological reason for this is that glucocorticoids promote mobilization of proteins from tissues. The amino acids liberated in this way can be used either as fuel or for synthesis of other proteins, such as immunoglobulins or scar tissue, that might be crucial at the moment.

This nonspecific reaction no doubt occurs, but specific impingements sooner or later require specific counteractions. Further, for domestic animals, the nonspecific alarm reaction seems to be superfluous, if not counterproductive, whereas insult or injury might interfere with a wild animal's getting food, and thus crucial amino acids, food-getting is ordinarily not a problem for domestic animals. Finally, all productive processes involve protein synthesis, so a high glucocorticoid secretion rate can be detrimental to food-animal performance at least in the short term.

Adaptation: Stress and Strain

An environmental adaptation refers to any functional, structural, or behavioral trait that favors an animal's survival or reproduction in a given environment, especially an extreme or adverse surrounding. Rates of life processes are the criteria used most often to assess adaptation. Adaptation can involve either an increase or a decrease in the rate of a given process.

A strain is any functional, structural, or behavioral reaction to an environmental stimulus. Strains can be adaptive or nonadaptive. Many enhance the chances of survival, but others are seemingly of little consequence.

A stress is any environmental situation--and a stressor any environmental factor--that provokes an adaptive response. A stress might be chronic (gradual and sustained) or acute (abrupt and often profound). Thus, by definition, environmental stress provokes animal strain, or in other words environmental stress provokes a stress response.

Environmental stress occurs when a given animal's environment changes so as to stimulate strain (as when environmental temperature falls below the crucial level) or when the animal itself changes in relation to a given environment (as when shearing reduces a sheep's cold tolerance).

Kinds of adaptation. There are several categories of environmental adaptation. A given animal represents one stage in a continuum of evolutionary development. An animal's heredity determines the limits of its environmental adaptability. Hence, there are genetic adaptations to environment. Genotypic changes occur naturally due to genetic mutations. Environmental stress theoretically permits mutations having adaptive utility to be realized and ultimately to become fixed in animal populations. Artificial selection pressures for productive traits reduce such natural selection pressures. But individuals selected on the basis of productive performance are at least adequately adapted to the production environment; otherwise, they would neither perform at relatively high levels nor reproduce.

There are also induced adaptations. A given stressful environmental complex provokes various responses depending on the individual animal's current adaptation status, which is determined by heredity and by its life history, as well.

Acclimation is one kind of induced adaptation. It refers to an animal's compensatory alterations due to a single stressor acting alone, usually in an experimental or artificial situation, over days or weeks. A hen in a layer house might acclimate to altered day length, for example.

Acclimatization, on the other hand, refers to reactions over days or weeks to environments where many environmental factors vary at the same time. A ewe at pasture acclimatizes to seasonal variations in day-length in conjunction with variations in other environmental factors.

Finally, an animal may become habituated to certain stimuli when they occur again and again. Sensations and effector responses associated with particular environmental stimuli tend to diminish when these stimuli occur repeatedly. A pig raised near an airport becomes habituated to the roars of jet airplanes, for example.

Level of Adaptation. An animal's environmental adaptation can be analyzed at several levels of organization. At one end of the spectrum, adaptations can take the form of enzyme inductions or of changes in other modifiers of catlyzed biochemical reactions. In the middle are changes associated with adaptive responses to environmental stress in sensory, integrative, and coordinative neural functions, in neuroendocrine and endocrine functions, and in effectors' outputs. At the other end of the range, the animal's behavior often changes in response to environmental stimuli. Malcolm Gordon (1972) said: "There is certainly no logical basis for any claim that understanding the nature of life at one level of organization is more fundamental to overall understanding than comprehension at any other level."

STRESS RESPONSES: PSYCOLOGICAL COMPONENTS

It is now generally recognized that the amount of stress an animal is under depends not only on the intensity and duration of the noxious agent (the traditional concept), but on the animal's ability to modify the effects of the stressor as well (Mason, 1975; Archer, 1979).

Lack of Control

A recent study of stress effects on tumor rejection demonstrated psycological components of stress responses (Visintainer, et al., 1982). Stressors such as mild electrical shock depress an animal's ability to reject certain tumors in experimental settings. In this particular experiment, individually held rats were inoculated with a standard dose of tumor-causing cells and assigned to three treatments: control (no shock), mild shock that could be stopped by pressing a switch (escapable shock), and mild shock that stopped anytime the escapable-shock rate in the trial pressed its switch but over which this rat itself had no control (inescapable shock).

In other words, the amount of physical impingement received by animals in the two shock treatments was the same, but those in one group (escapable shock) could control the duration, while those in the other group (inescapable shock) could not.

Fifty-four percent of the control rates rejected their tumors. Inescapable shock caused so much stress that tumor rejection occurred in only 27% of the rats, while 63% of those subjected to escapable shock rejected their tumors. The conclusion: the low rate of tumor rejection was due not to the shock itself, but to the animal's inability to control this stressor.

Alliesthesia Modification

Central perception ("alliesthesia") of stress intensity depends on the context within which it occurs. Alliesthesia in the form of comfort rating or pleasure rating is affected by the animal's internal state and, hence, by its external surroundings as well.

For example, a thermal stimulus can feel pleasant or unpleasant depending on the body's thermal status. Hypothermic humans find cold stimuli very unpleasant and hot very pleasant, while hyperthermic humans have the opposite perceptions (Cabanca, 1971). Similarly, gastric loading with glucose decreased the human subjects' pleasure rating of the sweet taste of sucrose in a thermoneutral environment (26°C), but this negative alliesthesia due to glucose loading was eliminated when ambient temperature was reduced to 4°C (Russek, et al., 1979).

These findings remind us that "variety is the spice of life" and suggest that "taking the bitter with the sweet" is pleasurable in the long run. Extrapolating the concept of alliesthesia modification to agricultural animals' lives, it would appear that stress of one sort often primes the animal to receive pleasure from some other aspects of its environment.

In any case, the fact that an animal's psychological state can modify its perception of stress makes it all the more difficult to interpret how a specific stressor is affecting a specific animal.

MEASURING AND INTERPRETING STRESS RESPONSES

The scientific literature stores report of hundreds of experiments purported to measure stress in food animals (Hafez, 1968; Hafez, 1975; Johnson, 1976a,b; Stephens, 1980; Craig, 1981; Curtis, 1981). It is a relatively simple task to subject experimental animals to a controlled stressor and measure a resultant change in some physiological, anatomical, or behavioral parameter. Hormonal, cardio-respiratory, and heat-production parameters have been studied most in the past. Behavioral and anatomical changes are being characterized more lately. But an objective index of stress in terms of animal health, performance, and well-being has been elusive.

As Graham Perry (1973) said: "Even marked physiological changes may indicate only that an animal is successfully adapting to its environment--not necessarily that it is succumbing to adversity." And, similarly, Ian Duncan (1981) said: ". . . it should be of no surprise that chickens behave differently in different environments. This may simply demonstrate how adaptable they are." Again: at what point does stress become dis-stressful.

As for methodology, it is very difficult to study the effects of specific supposed stressors on an animal without introducing artifacts due to the stressfulness of the investigative techniques themselves (Adler, 1976). This is especially so in real or simulated production situations. What is the baseline adrenal-glucocorticoid secretion rate of an animal? Will it ever be certain beyond a reasonable doubt that the experimental manipulation necessary to obtain the needed samples or observations is not itself so stressful as to compromise the results?

Also, interpretation of the results of this kind of research is hampered by the fact that, by and large, there is not yet consensus as to the meaning of data on specific behavioral and hormonal changes in responses to stressors. What does it mean when an animal increases breathing rate by 250% in one environment compared with another? Does a 65% increase in plasma glucocorticoid concentration indicate the animal is under stress? If so, is the stress mild or severe? Scientists still do not understand how findings such as these relate to an animal's well-being, its health, and its productivity; consequently, we cannot rely on physiological or behavioral traits as valid indicators of the amount of stress an animal actually perceives, let alone how these might be related to the animal's health and productivity.

STRESS AND PRODUCTIVITY

Environmental stress generally alters animal performance (Curtis, 1981). The stress provokes the animal to react, and this reaction can influence the partition of resources among maintenance, reproductive, and productive functions in one or more of five ways:

1. The reaction may alter internal functions. Many bodily functions participate in productive processes as well as in reactions to stress. Survival responses may thus unintentionally affect productive preformance. For example, increased adrenal glucocorticoid secretion in response to stress can impair growth.

2. The reaction may divert nutrients. When an animal resonds to stress, it in effect diverts nutrients to use in higher-priority maintenance processes. Adaptive reactions are implemented even at the expense of productivity.
3. The reaction may reduce productivity directly. The animal's response sometimes partly comprises intentional reductions in productive processes. This generally frees some nutrients for maintenance uses. For example, an animal might reduce its productive rate in a hot environment in an attempt to re-establish heat balance with its surroundings.
4. The reaction may increase variability. Individual animals within a species differ from each other in functions, behaviors, and structures by what have been called "individuality differentials." Individual animals therefore differ in their responses to the same environmental stressor. In other words, two animals in the same group might successfully cope with the same stressful situation by calling different mechanisms into play. Then, if the complements of mechanisms used by the two individuals differ in the energy expenditure required to achieve them, the amount of energy diverted from productive processes will be different for the two animals. The result of this is that the amount of variation in individual performance in a group of animals tends to be related directly with the environmental adversities to which the animals are subjected.
5. The reaction may impair disease resistance. Because the animal's reaction to stress can impair disease resistance, that reaction influences the frequency and severity of disease. Of course, infection itself is a stress, so once established it in turn can influence the animal's productive performance. The mechanisms involved in the relations between environmental stress and resistance against infectious disease are just now being elucidated.

Kelley (1980) identified eight stressors: heat, cold, crowding, regrouping, weaning, limit-feeding, noise, and movement restraint. He documented the fact that all of these have been accorded a central role in stress-induced alterations of resistance against infection.

Having developed a framework for analyzing relations between adverse environments and animal productivity, it would be unrealistic to leave the impression that the link between stresses and productive processes are clear and simple. Consider two examples.

Lactating dairy cows held in a natural subtropical summer environment and provided no shelter are obviously under severe stress at mid-afternoon. They have markedly higher body temperatures and respiratory rates than their herdmates under the shade. Yet there might be no significant difference in fat-corrected milk yield between the two groups of cows (Johnson et al., 1966).

Socially and physically deprived animals often grow faster than do their counterparts in more enriched environments (Fiala et al.) So there is a risk in assuming that an animal stressed by a specific environmental complex is necessarily unfit for productive use in that environment. While one often might be justified in presuming that strain against stress reduces animal productivity, the animal can still be putting out an acceptable amount of product per unit of resource input.

Robert McDowell (1972) refers to "physiological adaptability" to environment (measured by physiological traits such as breathing rate) that is associated with survival responses, and to "performance adaptability" to environment (measured by productive-performance traits such as growth rate). These two often bear little positive relation to each other.

Thus, it is not sufficient for an animal producer to be concerned only with physiological and behavioral indices of environmental adaptability. Producers are more interested in the size of decrement, if any, in production associated with an animal's living in a particular environment. And to learn the quantitative effects of a given environment on animal performance, the productive traits themselves must be measured. After all, knowing a hen's breathing rate tells one little or nothing about her rate of lay.

There has been unfortunate ambiguity on this point among researchers and producers alike. An animal exhibiting marked strain has generally been assumed to be having markedly depressed performance. This is not necessarily so. Indeed, visible strain signifies that the animal is attempting to compensate for an environmental impingement. These attempts might succeed, and they might interfere with production only slightly or not at all.

The marvelous homeokinetic phenomena they possess make for resilient beasts and birds on our farms and permit profitable performance in a wide range of circumstances. The response flexibility that animals demontrate in the face of myriad stressors seem more remarkable than those instances when defensive reactions are inadequate and the environmental complex drastically reduces health or performance.

REFERENCES

Adler, H. C. 1976. Ethology in animal production. Livestock Prod. Sci. 3:303.

Archer, J. 1979. Animals Under Stress. Edward Arnold, London.

Bernard, C. 1957. An Introduction to the Study of Experimental Medicine. Dover, New York.

Bowman, J. C. 1977. Animals for Man. Edward Arnold, London.

Cabanac, M. 1971. Physiological role of pleasure. Science 173:1103.

Cannon, W. B. 1932. The Wisdom of the Body. Norton, New York.

Craig, J. V. 1981. Domestic Animal Behavior. Prentice-Hall, Englewood cliffs.

Curtis, S. E. 1981. Environmental Management in Animal Agriculture. Animal Environment Services, Mahomet, Illinois.

Dill, D. B. (Ed.) 1964. Handbook of Physiology. Section 4: Adaptation to the Environment. American Physiological Society, Washington.

Duncan, I. J. H. 1981. Animal rights-animal welfare: a scientist's assessment. Poul. Sci. 60:489.

Fiala, B., F. M. Snow, and W. T. Greenough. 1977. "Impoverished" rates weigh more than "enriched" rats because they eat more. Devel. Phychobiol. 10:537.

Fraser, D., J. S. D. Ritchie, and A. F. Fraser. 1975. The term "stress" in a veterinary context. Brit. Vet. J. 131:653.

Gordon, M. S. 1972. Animal Physiology: Principles and Adaptations. (Second ed.) Macmillan, New York.

Hafez, E. S. E. (Ed.) 1968. Adaptation of Domestic Animals. Lea and Febiger, Philadelphia.

Hafez, E. S. E. (Ed.) 1975. The Behavior of Domestic Animals (Third ed.) Williams and Wilkins, Baltimore.

Hale, E. B. 1969. Domestication and the evolution of behavior. In: E. S. E. Hafez (Ed.). The Behavior of Domestic Animals. William and Wilkins, Baltimore.

Johnson, H. D. (Ed.) 1976a. Progress in Animal Biometerology, Volume 1, Part I. Swets and Zeitlinger, Amsterdam.

Johnson, H. D. (Ed.) 1976b. Progress in Animal Biometerology, Volume 1, Part II. Swets and Zeitlinger, Amsterdam.

Johnston, J. E., J. Rainey, C. Breidenstein, and A. J. Gidry. 1966. Effects of ration fiber level on feed intake and milk production of dairy cattle under hot conditions. Proc. Fourth Int. Biometerological Cong., New Brunswick.

Kelley, K. W. 1980. Stress and immune function: A bibliographic review. Ann. Vet. Res. 11:445.

Mason, J. W. 1975. Emotion as reflected in patterns of endocrine integration. In: L. Levi (Ed.) Emotions--Their Parameters and Measurement. Raven, new York.

McDowell, R. E. 1972. Improvement of Livestock Production in Warm Climates. Freeman, San Francisco.

Perry, G. 1973. Can the physiologist measure stress? New Scientist 60 (18 October):175.

Russek, M. M. Fantino, and M. Cabanac. 1979. Effect of environmental temperature on pleasure ratings of odor and testes. Physiol. Behav. 22:251.

Selye, H. 1952. The Story of the Adaptation Syndrome. Acata, Montreal.

Stephens, D. B. 1980. Stress and its measurement in domestic animals: a review of behavioral and physiological studies under field and laboratory situations. Adv. Vet. Sci. Comp. Med. 24:179.

Visintainer, M. A., J. R. Volpicelli, and M. E. P. Seligman. 1982. Tumor rejection in rats after inescapable or escapable shock. Science 216:437.

27

LIVESTOCK PSYCHOLOGY AND HANDLING-FACILITY DESIGN

Temple Grandin

Handling your cattle and sheep will be much easier if you learn a little livestock psychology. Many people do not realize that cattle and sheep have panoramic vision and they can see all around themselves without turning their heads (Prince, 1977; McFarlane, 1976). Sheep with heavy fleeces would have a more restricted visual field depending on the amount of wool on their head and neck. Both cattle and sheep depend heavily on their vision and are easily motivated by fear (Kilgour, 1971). Livestock are sensitive to harsh contrasts of light and dark around loading chutes, scales, and work areas. "Illumination should be even and there should be no sudden discontinuity in the floor level or texture" (Lynch and Alexander, 1973).

Solid shades should be used over the working, loading, and scale areas (Grandin, 1981). Slatted shades are fine for areas where the animals live and feel familiar. However, when the animals come into the handling areas they are often nervous. The zebra stripe pattern cast by the slatted shades constructed from snow fence or corrugated sheets suspended on cables will cause balking. The pattern of alternating light and dark has the same effect as building a cattle guard in the middle of the facility. Contrasts of light and dark have such a deterrent effect on cattle that in Oregon lines are painted across the highway to take the place of expensive steel cattle guards.

Shadowy stripes will cause balking problems with sheep. A single-file chute for sorting sheep should be oriented so that the sun does not form a shadow down the middle of the chute. The worst possible situation for sheep is to have half the floor of the chute in the shade and the other half in the sunlight. In shearing sheds and sheep holding areas, the wooden slats on the floor should face so that the sheep walk across the slats instead of in the same direction as the slats (Hutson, 1981). If you get down on your hands and knees and look at the floor, the floor appears more solid if you move across the slats. The floor should also be constructed to prevent sunlight from shining up through the slats.

A single shadow that falls across a scale or loading chute can disrupt handling. The lead animal will often balk and refuse to cross the shadow. If you are having problems with animals balking at one place, a shadow is a likely cause. Balking can also be caused by a small bright spot formed by the sun's rays coming through a hole in a roof. Patching the hole will often solve the problem. Handlers themselves should be cautious about causing shadows. Figure 1 illustrates a shadow that was formed when the handler waved at the cattle. The animals refused to approach the shadow of the waving handler cast at the entrance to the single-file chute.

Figure 1. The handler's shadow cast on the entrance to the single-file chute caused the cattle to balk. This is just one of the many kinds of shadows which can cause balking problems in your cattle handling facility.

APPROACH LIGHT

Both cattle and sheep have a tendency to move towards the light. If you ever have to load livestock at night, it is strongly recommended that frosted lamps that do not glare in the animals face be positioned inside of the truck (Grandin, 1979). However, loading chutes and squeeze chutes should face either north or south; livestock will balk if they have to look directly into the sun.

Sometimes it is difficult to persuade cattle or sheep to enter a roofed working area. Persuading the animals to enter a dark, single-file chute from an outdoor crowding pen in bright sunlight is often difficult. Cattle are more easily driven into a shaded area from an outdoor pen if they are first lined up in single file.

Many people make the mistake of placing the single-file chute and squeeze chute entirely inside a building and the crowding pen outside. Balking will be reduced if the single file chute is extended 10 to 15 feet outside the building. The animals will enter more easily if they are lined up single file before they enter the dark building. The wall of the building should NEVER be placed at the junction between the single file chute and the crowding pen. Either cover up the entire squeeze chute and crowding pen area or extend the single file chute beyond the building. If you have just a shade over your working area, make sure that the shadow of the shade does not fall on the junction between the single file chute and the crowding pen.

PREVENT BALKING

Drain grates in the middle of the floor will make both sheep and cattle balk because the animals will often refuse to walk over them. A good drainage design is to slope the concrete floor in the squeeze chute area toward an open drainage ditch located outside the fences. The open drainage ditch outside the fences needs no cover and so it is easier to clean.

Animals will also balk if they see a moving or flapping object. A coat flung over a chute fence or the shiny reflection off a car bumper will cause balking. You should walk through your chutes and view them from a cow's eye level before moving or loading animals. You will be surprised at the things you may see. When cattle and sheep are being worked, the handlers should stand back away from the headgate so that approaching animals cannot see them with their wide angle vision. The installation of shields for people to hide behind can facilitate the movement of livestock (Kilgour, 1971; Freeman, 1975).

Problems with balking tend to come in bunches; when one animal balks, the tendency to balk seems to spread to the next animals in line (Grandin, 1980). When an animal is being moved through a single-file chute, the animal must never be prodded until it has a place to go. Once it has balked, it will continue balking. The handler should wait until the tailgate on the squeeze chute is open before prodding the next animal (Grandin, 1976). A plastic garbage bag attached to a broom handle is a good tool for moving cattle in pens. The cattle move away from the rustling plastic. When livestock are being moved, well-trained dogs are recommended for open areas and large pens. Once the animals are confined in the crowding pen and single-file chute, dogs should not be allowed near the fences where they still can bite at the cattle or sheep.

SOLID CHUTE SIDES

For both cattle and sheep the sides of the single-file chute, loading chute, and crowding pen should be solid. Solid sides prevent the animals from seeing people, cars, and other distractions outside the chute. A study with sheep showed that they moved more rapidly through a single-file chute that had solid sides (Hutson and Hitchcock, 1978). The principle of using solid sides is like putting blinkers on the harness horse. The blinkers prevent the horse from seeing distractions with his wide-angle vision. Cattle and sheep in a handling facility should be able to see only one pathway of escape--this is extremely important. They should be able to see other animals moving in front of them down the chute, when sheep are being sorted, the approaching animals should be able to see the previously sorted sheep through the end of the sorting chute.

Livestock will balk if a chute appears to be a dead end (Brockway, 1975; Hutson, 1980). Sliding and one-way gates in the single-file chute must be constructed so that your animals can see through them, otherwise the animals will balk (figure 2). The sides of the single-file chute and the crowding pen should be solid. The crowding-pen gate also should be solid so that animals cannot see through and will head for the entrance to the single-file chute (Rider, 1974). Mirrors could be used to attract sheep into pens and other areas that appear to be a dead end. The sheep are attracted to the image of sheep in the mirror (Franklin & Hutson, 1982).

HERD BEHAVIOR

All species of livestock will follow the leader and this instinct is strong in both cattle and sheep (Ewbank, 1961). Many people make the mistake of building the single-file chute to the squeeze too short. The chute should be long enough to take advantage of the animal's tendency to follow the leader. The minimum length for the single-file chute is 20 ft. In larger facilities 30 to 50 lineal ft is recommended.

Cattle and sheep are herd animals and, if isolated, can become agitated and stressed. This is especially a problem with Brahman-type cattle. An animal left alone in the crowding pen after the other animals have entered the single file chute, may attempt to jump the fence to rejoin its herdmates. A lone steer or cow may become agitated and charge the handler. A large portion of the serious handler injuries occur when a steer or cow, separated from its herd-mates, refuses to walk up the single file chute. When a lone animal refuses to move, the handler should release it from the crowding pen and bring it back with another group of cattle.

Figure 2. The single-file chute to the squeeze should have solid sides to prevent the cattle from seeing distractions outside the fence. Sliding gates in the single-file chute must be constructed from bars so that the cattle can see through them. Solid sliding or one-way gates will cause balking.

EFFECTS OF SLOPE AND WIND

To prevent livestock from piling up against the back gate in the crowding pen, the floor of the pen must be level. A 10° slope in the crowding pen will cause the animals to pile and fall down against the crowding gate. A small 1/4 in. to 1/8 in. slope per foot for drainage will not cause a handling problem. Livestock move more easily uphill than down, but they move most easily on a flat surface (Hitchcock and Hutson, 1979).

Research by Hutson and Mourik (1982) indicates that sheep will move more easily when they are heading into the wind. Heading into the wind can stimulate sheep to start moving along a chute.

WHY A CURVED CHUTE WORKS

A curved chute works better than a straight chute for two reasons. First it prevents the animal from seeing the truck, the squeeze chute, or people until it is almost in the truck or squeeze chute. A curved chute also takes advantage of the animal's natural tendency to circle around the handler (Grandin, 1979). When you enter a pen of cattle or sheep you have probably noticed that the animals will turn and face you, but maintain a safe distance (figure 3). As you move through the pen, the animals will keep looking at you and circle around you as you move. A curved chute takes advantage of this natural circling behavior.

Figure 3. When you walk through a pasture the cows will turn and look at you. They will circle around you as you move about the pasture. Curved chutes take advantage of the cow's circling behavior.

Cattle can be driven most efficiently if the handler is situated at a 45° to 60° angle perpendicular to the animal's shoulder (Williams, 1978) (figure 4). A well-designed, curved single-file chute has a catwalk for the handler to use along the inner radius. The handler should always work along the inner radius. The curved chute forces the handler to stand at the best angle and lets the animals circle around him. The solid sides block out visual distractions except for the handler on the catwalk.

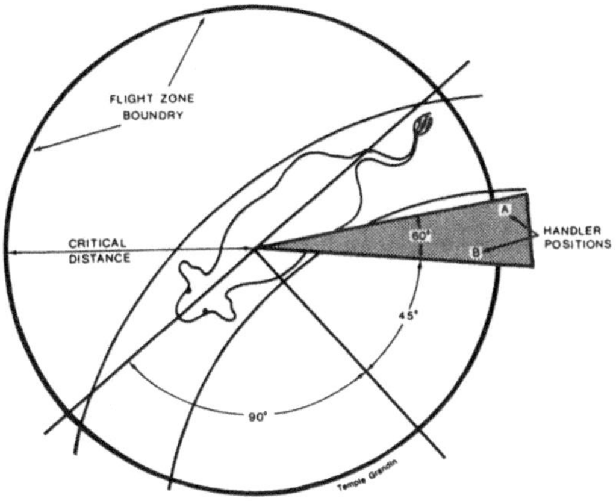

Figure 4. **The shaded area shows the best position for moving an animal. To make the cow move forward the handler moves into Position B which is just inside the boundary of the flight zone. The handler should retreat to Position A if he wants the animal to stop. The solid curved lines indicate the location of the curved single-file chute.**

The catwalk should run alongside of the chute and NEVER be placed overhead. The distance from the catwalk platform to the top of the chute fence should be 42 inches. This brings the top of the fence to belt-buckle height on the average person.

Figures 5 and 6 illustrate curved facilities for handling cattle and sheep. Curved designs are recommended by both Grandin (1980) and Barber (1977).

FLIGHT DISTANCE

When a person penetrates an animal's flight zone, the animal will move away. If the handler penetrates the flight zone too deeply, the animal will either turn back and run past him or break and run away. Kilgour (1971) found that when the flight zone of bulls was invaded by a mechanical trolley, the bulls would move away and keep a constant distance between themselves and the trolley. When the trolley got too close the bulls bolted past it. The best place for the handler to work is on the edge of the flight zone. This will cause the animals to move away in an orderly manner. The animals will stop moving when the handler retreats from the flight zone.

The size of the flight zone varies depending on the tameness or wildness of the animal. The flight zone of range cows may be as much as 300 ft whereas the flight zone

Figure 5. Cattle handling facility utilizing a curved single-file chute, round crowding pen, and wide curved lane. Up to 600 cattle per hour can be moved through the dip vat with only three people. The handlers work along the inner radius of the single-file chute and the wide curved lane (designed by Temple Grandin).

Figure 6. Sheep handling and sorting facility with a curved bugle crowding pen. The inner radius is solid to prevent the sheep from seeing the handler standing at the sorting gates (designed by Adrian Barber, Australia).

of feedlot cattle may be only 5 to 25 ft (Grandin, 1978). Extremely tame cattle or sheep are often difficult to drive because they no longer have a flight zone.

Many people make the mistake of getting too close to the cattle when they are driving them down an alley or putting them in a crowding pen. Getting too close makes cattle feel cornered. If the cattle attempt to turn back, the handler should back up and retreat to remove himself from the animal's flight zone instead of moving in closer.

Cattle will often rear up and get excited while waiting in the single-file chute. The most common cause of this problem is the handler leaning over the single file chute and deeply penetrating the animal's flight zone. The cattle will usually settle down if the handler backs up.

When sheep are being handled in a confined area, pile-ups can occur if their flight zone is deeply penetrated. This is why dogs should not be used in the crowding pen or the single-file chute, because a dog, in a confined area, deeply penetrates the flight zone and the sheep have no place for escape. Dogs are recommended only for open areas and larger pens where there is room for the sheep to move away. During handling, minimize yelling and screaming so as to avoid enlarging the size of the animal's flight zone.

BREED DIFFERENCES

The breed of the cattle or sheep can affect the way it reacts to handling. Cattle with Brahman blood are more excitable and may be harder to handle than the English breeds. When Brahman or Brahman-cross cattle are being handled, it is important to keep them as calm as possible and to limit use of electric prods. Brahman and Brahman-cross cattle can become excited; they are difficult to block at gates (Tulloh, 1961) and prone to ram into fences. With this type of cattle it is especially important to use substantial fencing. If thin rods are used for fencing, a wide belly rail should be installed to present a visual barrier. Angus cattle tend to be more nervous than Herefords (Tulloh, 1961). Holstein cattle tend to move slowly (Grandin, 1980). Brahman cattle tend to stay together in a more cohesive mob than English cattle.

Brahman and Brahman-cross cattle can become so disturbed that they will lie down and become immobile, especially if they have been prodded repeatedly with an electric prod (Fraser, 1960). When a Brahman or Brahman-cross animal lies down, it must be left alone for about five minutes or it may go into shock and die. This problem rarely occurs in English cattle or European cattle such as Charolais.

There are distinct differences in the way various breeds of sheep react during handling (Shupe, 1978; Whately et al., 1974). Rambouillet sheep tend to bunch tightly together and remain in a group; crossbred Finn sheep tend to turn, face the handler, and maintain visual contact. If the handler penetrates the collective flight zone of a group of Finn sheep, they will turn and run past the handler.

Cheviots and Perendales are the easiest to drive into a crowding pen; the Romney, Merino-Romney cross, and the Dorset-Romney are the most difficult. The Romney tends to follow the leader but it is easily led into blind corners. Cheviots have a strong instinct to maintain visual contact with the handler and to display more independent movements than other breeds.

DARK BOX AI CHUTE

For improved conception rates, cows should be handled gently for AI and not allowed to become agitated or overheated. The chute used for AI should not be the same chute used for branding, dehorning, or injections. The cow should not associate the AI chute with pain. Cows can be easily restrained for AI or pregnancy testing in a dark box chute that has no headgate or squeeze (Parsons & Helphinstine, 1969; Swan, 1975). Even the wildest cow can be restrained with a minimum of excitement. The dark box chute can be easily constructed from plywood or steel. It has solid sides, top, and front. When the cow is inside the box, she is inside a quiet, snug, dark enclosure. A chain is latched behind her rump to keep her in. After insemination the cow is released through a gate in either the front or the side of the dark box. If wild cows are being handled, an extra long dark box can be constructed. A tame cow that is not in

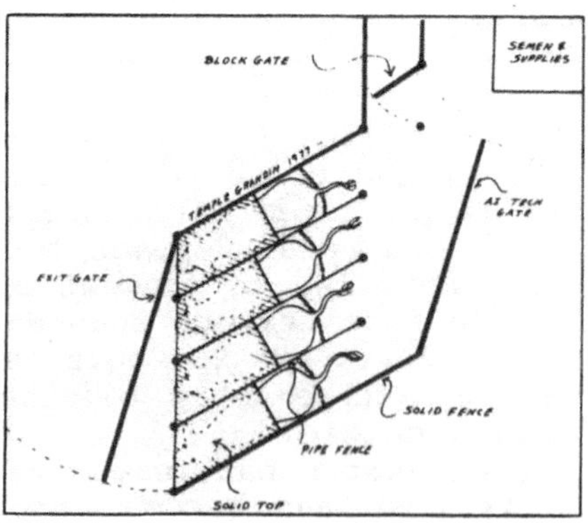

Figure 7. Chutes for A.I. can be laid out in a herringbone design. The two outer fences and the grates should be solid. The inner partition in between the cows should be constructed from bars. Cows will stay calmer if they know they have company.

heat is used as a pacifier and is placed in the chute in front of the cow to be bred. Even a wild cow will stand quietly and place her head on the pacifier cow's rump. After breeding, the cow is allowed to exit through a side gate, while the pacifier cow remains in the chute.

If a large number of cows have to be pregnancy checked or inseminated, two to six AI chutes can be laid out in a herringbone pattern (figure 7). This design is recommended by McFarlane (1976) from South Africa. The chutes are set on a 60° angle. They are built like regular dark box AI chutes except that the partitions inbetween the cows are constructed from open bars so the cows can see each other. The cows will stand more quietly if they have company. The two outer fences should be solid. If the cows are reluctant to enter the dark box, a small 6 in. by 12 in. window can be cut in the solid front gate in front of each cow.

LOADING CHUTE DESIGN

Loading chutes should be equipped with telescoping side panels and a self-aligning dock bumper. These devices will help prevent foot and leg injuries caused by an animal stepping down between the truck and the chute. The side panels will prevent animals from jumping out the gap between the chute and the truck.

A well-designed loading ramp has a level landing at the top. This provides the animals with a level surface to walk on when they first get off the truck. The landing should be at least 5 feet wide for cattle. Many animals are injured on ramps that are too steep. The slope of a permanently installed cattle ramp should not exceed 20°. The slope of a portable or adjustable chute should not exceed 25° (Grandin, 1979). Steeper ramps may be used for loading sheep but they are NOT recommended for unloading. Sheep will move up a steep ramp readily.

If you build your ramp out of concrete, stairsteps are strongly recommended. For cattle the steps should have a 3.5 to 4 in. rise and a 12 in. tread width. The surface of the steps should be rough to provide good footing. For sheep the steps should have a 2 in. rise and a 10 in. tread width.

On adjustable or wooden ramps, the cleats should be spaced 8 in. apart from the edge of one cleat to the edge of the next cleat (Mayes, 1978). The cleats should be 1 1/2 to 2 in. high for cattle and 1 in. by 1 in. for sheep.

Chutes for both loading and unloading cattle should have solid sides and a gradual curve (figure 8). If the curve is too sharp, the chute will look like a dead end when the animals are being unloaded. A curved single-file chute is most efficient for forcing cattle to enter a truck or a squeeze chute. A chute used for loading and unloading cattle should have an inside radius of 12 ft to 17 ft, the bigger radius is the best. A loading chute for cattle should be 30 in. wide and no wider. The largest bulls will fit through a 30 in. wide chute. If the chute is going to be used exclusively for calves, it should be 20 to 24 in. wide.

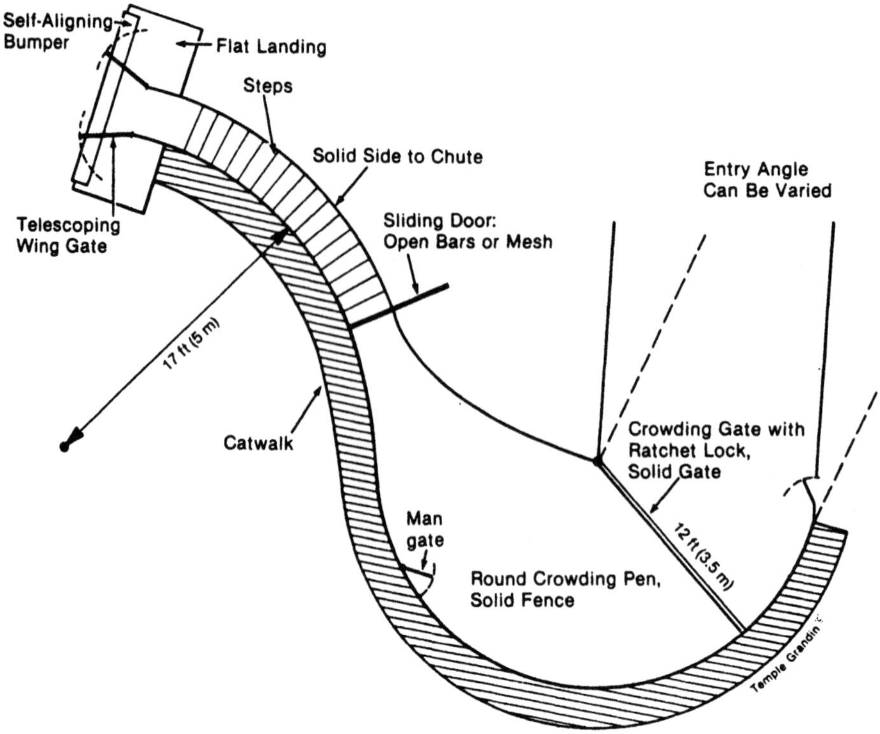

Figure 8. Curved loading chute with a round crowding pen. The sides of the chute are solid.

In auctions and meat packing plants where a chute is used to unload only, a wide straight chute should be used. This provides the animals with a clear path to freedom. These chutes can be 6 to 10 ft wide. A wide, straight chute should not be used for loading cattle.

SHEEP LOADING

Since most trucks have a 30 in. wide door, a good chute design for sheep is a 30 in. wide ramp that enables two sheep to walk up side by side. If a single-file chute is used, it should be 17 in. to 18 in. wide. The chute should be designed so that the animals walk up either in a single file or two abreast. Don't build a chute that is one and one-half animals wide--this creates jamming problems.

A wide ramp is recommended for loading sheep into shearing sheds or onto trucks that can be opened up the full width of the vehicle. In Australia sheep moved easily up ramps 8 to 10 ft wide when loading sheep onto ships for shipment to the Middle East (figure 9). The entrance ramp into a raised shearing shed should be 8 ft to 10 ft wide (Simpson, 1979).

Figure 9. A wide ramp is used to load sheep onto a ship in Australia. Once the flow of sheep was started the animals moved easily up the ramp.

REFERENCES

Barber, A. 1977. Bugle sheep yards. Fact Sheet, Dept. of Agriculture and Fisheries South Australia, Adelaide, Australia.

Brockway, B. 1975. Planning a sheep handling unit. Farm Buildings Center, Nat. Agr. Center, Kenilworth, Warickshire, England.

Ewbank, R. 1961. The behavior of cattle in crises. Vet. Rec. 73:853.

Franklin, J. R., G. D. Hutson. 1982. Experiments on attracting sheep to move along a laneway. III Visual Stimuli, Appl. Animal Ethology 8:457.

Fraser, A. F. 1960. Spontaneously occurring forms of "tonic immobility" in farm animals. Canad. J. Comp. Med 24:330.

Freeman, R. B. 1975. Functional planning of a shearing shed. Pastoral Review 85:9.

Grandin, T. 1981. Innovative cattle handling facilities. In: M.E. Ensminger (Ed.). Beef Cattle Science Handbook, 18:117. Agriservices Foundation, Clovis, Calif.

Grandin, T. 1980. Livestock behavior as related to handling facilities design. Int. J. Stud. Animal Problems 1:33 etc.

Grandin, T. 1979. Understanding animal psychology facilitates handling livestock. Vet. Med. and Small Animal Clinician 74:697.

Grandin, T. 1978. Observations of the spatial relationships between people and cattle during handling. Proc. Western Sec. Amer. Soc. of Animal Sci. 29:76.

Grandin, T. 1976. Practical pointers on handling cattle in squeeze chutes, alleys, and crowding pens. In: M.E. Ensminger (Ed.). Beef Cattle Science Handbook 13:228.

Hitchcock, D. K., G. D. Hutson. 1979. The movement of sheep on inclines. Australian J. Exp. Agr. and Animal Husbandry 19:176.

Hutson, G. D., S. C. van Mourik. 1982. Effect of artificial wind on sheep movement along indoor races. Australian J. Exp. Agr. and Animal Husbandry 22:163.

Hutson, G. D. 1981. Sheep movement on slatted floors. Australian J. Exp. Agr. and Animal Husbandry 21:474.

Hutson, G. D. 1980. The effect of previous experience on sheep movement through yards. Appl. Animal Ethology 6:233.

Hutson, G. D. and D. K Hitchock. 1978. The movement of sheep around corners. Appl. Animal Ethology 4:349.

Kilgour, R. 1971. Animal handling in works, pertinent behavior studies. 13th Meat Industry, Res. Conf. Hamilton, New Zealand. pp 9-12.

Lynch, J. J. and G. Alexander. 1973. The Pastoral Industries of Australia. pp 371. Sydney University Press, Sydney, Australia.

Mayes, H. F. 1978. Design criteria for livestock loading chutes. Technical Paper No. 78-6014, Amer. Soc. Agr. Eng. St. Joseph, Michigan.

McFarlane, I. 1976. Rationale in the design of housing and handling facilities. In: M. E. Ensminger (Ed.). Beef Cattle Science Handbook 13:223.

Parsons, R. A. and W. N. Helphinstine. 1969. Rambo AI breeding chute for beef cattle. One-Sheet-Answers, University of California Agricultural Extension Service, Davis, California.

Rider, A., A. F. Butchbaker and S. Harp. 1974. Beef working, sorting, and loading facilities. Technical Paper No. 74-4523, Amer. Soc. Agr. Eng. St. Joseph, Michigan.

Shupe, W. L. 1978. Transporting sheep to pastures and markets. Technical Paper No. 78-6008, Amer. Soc. Agr. Eng. St. Joseph, Michigan.

Simpson, I. 1979. Building a modern shearing shed. Division of Animal Industry Bulletin A3.7.1. New South Wales Dept. of Agr., Australia.

Swan, R. 1975. About AI facilities. New Mexico Stockman. Feb., pp 24-25.

Tulloh, N. M. 1961. Behavior of cattle in yards: II. A study of temperament. Animal Behavior 9:25.

Whately, J., R. Kilgour and D. C. Dalton. 1974. Behavior of hill country sheep breeds during farming routines. New Zealand Soc. Animal Production 34:28.

Williams, C. 1978. Livestock consultant, personal communication.

28

DESIGN OF CORRALS, SQUEEZE CHUTES, AND DIP VATS

Temple Grandin

CORRALS

A corral constructed with round holding pens, diagonal sorting pens, and curved drive lanes will enable you to handle cattle more efficiently because there is a minimum of square corners for the cattle to bunch up in. The principle of the corral layout in figure 1 is that the animals are

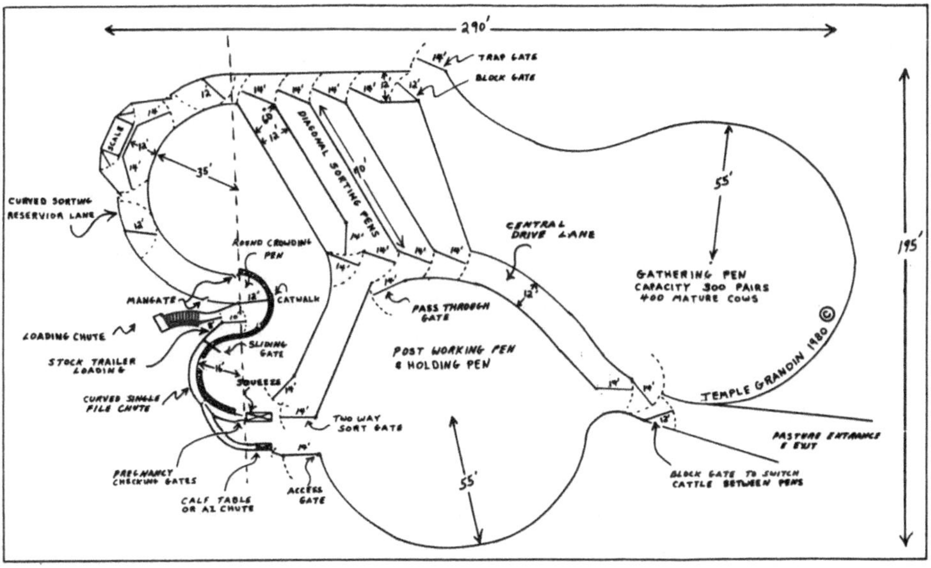

Figure 1. General purpose corral system for shipping, branding, sorting, and AI. It can handle 300 cow and calf pairs or 400 mature cows. Capacity can be increased by adding more diagonal pens and holding pen space (Grandin, 1981).

gathered into the big round pen and then directed to the curved sorting reservoir lane for sorting and handling. The curved sorting reservoir lane serves two functions: It holds cattle back into the diagonal pens, that are being sorted. It also holds cattle waiting to go to the squeeze chute, AI chute, or calf table.

Large Corral

The corral shown in figure 1 is a general-purpose system for shipping calves, working calves, sorting, pregnancy checking, and AI. It can handle 300 cow-calf pairs or 400 mature cows. It is equipped with a two-way sorting gate in front of the squeeze chute for separating the cows that are pregnant from cows that are open. Depending upon your needs, you can position either the squeeze chute, AI chute, or calf table at the sorting gate. If the cattle are watered in the large gathering pen, they will become accustomed to coming in and out of the trap gate. When you need to catch an animal, you merely shut the trap gate and direct her up the curved reservoir lane to the chutes. This is an especially handy feature for AI.

The curved sorting reservoir terminates in a round crowding pen and curved single-file chute. The crowding gate has a ratchet latch that locks automatically as the gate is advanced behind the cattle. To load low stock trailers, open a 8 ft gate that is alongside the regular loading chute. This provides you with the advantage of the round crowding pen for stock trailers. All fences in the curved single-file chute and the round crowding pen are solid. The ratchet crowd gate also should be covered with sheet metal or plywood.

Figure 1 can also be adapted for use with a prefabricated steel circle crowding pen and curved single-file chute. Since the prefabricated units have a 12 ft radius instead of the 16 ft radius shown in the drawing, you will have to move the sorting gate. If you plan to build the entire setup yourself, out of either wood or steel, keep the 16 ft radius, especially if you have large cows.

Diagonal Sorting Pens

When cows and calves are being separated, the calves are held in the diagonal pens and the central drive lane, and the cows are allowed to pass through one of the diagonal pens into the large post working pen. The diagonal pens and the central drive lane in figure 1 can hold 300 weaned calves overnight or 500 weaned calves crowded together. Each 70 by 12 ft diagonal pen holds 60 weaned calves overnight or 85 weaned calves crowded together. If the mother cows are put in the diagonal pens, each pen holds 40 cows overnight or 50 cows crowded. These capacities may vary depending on the size of your cattle.

To expand the corral system to handle more cattle, you can add more diagonal pens. Do NOT increase the length of the diagonal pens! If they are too long, the cattle will bunch up. You can increase the diagonal pen capacity to 1000 calves. It is NOT recommended to increase the size of the round gathering pen beyond the 55 ft radius shown. If the round gathering pen is too large, you may have difficulty getting the cattle into the curved reservoir lane. (Grandin, 1980a).

In order to increase the gathering area, you can build an additional round gathering pen at the pasture entrance. After the first 300 pairs are worked or sorted you can bring in 300 more pairs. The post working pen can be enlarged to hold cows after sorting or handling in the squeeze or AI chute.

Small Corral

The layout in figure 2 is designed for smaller ranches as a main working corral or a pasture corral on larger

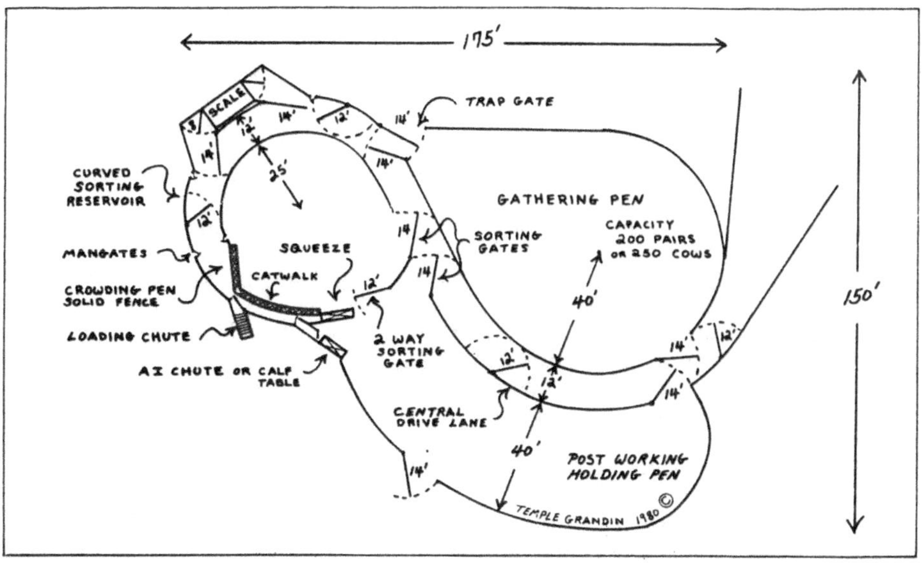

Figure 2. This is an economical corral system for a smaller operation or a pasture corral on a large ranch. It can handle 200 cow and calf pairs or 250 mature cows. It can be expanded to handle 300 cow and calf pairs (Grandin, 1980a).

operations. It is economical to build but still retains many of the features of the larger corral. It can handle

200 cow-calf pairs or 250 mature cows. By increasing the radius of the gathering pen to 55 ft and lengthening the central drive lane, it can be expanded to 300 pairs or 400 cows.

In figure 2 you can sort two ways out of the squeeze chute and three ways from the curved reservoir lane. Groups of cattle held in the curved reservoir lane can be sorted back into the post working pen, the central drive lane, or the round pen that is formed by the inner radius of the curved reservoir lane. When calves are being separated from the cows, the cows can be sorted into the post working pen and the calves into the central drive lane. For additional photos of corrals and a diagram of a minicorral, see Grandin, 1981, Beef Cattle Science Handbook 18:117.

Corral Construction Tips

Five foot high fences are usually sufficient for cattle such as Hereford and Angus. For Brahman cross and exotics a 5 1/2 ft to 6 ft fence is recommended. Solid fencing should be used in the crowding pen, single-file chute, and loading chute. If your budget permits, solid fencing should be used in the curved reservoir lane. If solid fencing is too expensive, then a wide belly rail should be installed. This is especially important if the corral is constructed from sucker rod (figure 3).

A V shaped chute can be built that will accommodate both cows and calves. It should be 16 to 18 in. wide at the bottom and 32 in. wide at the top. The 32 in. measurement is taken at the 5 ft level. If the single-file chute has straight sides it should be 26 in. wide for the cows and 18 to 20 in. wide for calves.

When a funnel-type crowding pen is built, make one side straight and the other side on a 30° angle. This design will prevent bunching and jamming. The crowding pen should be 10 to 12 ft wide (figure 4).

To prevent animals from slipping in areas paved with concrete, the concrete should be scored with deep grooves. The grooves should be 1 in. to 1 1/2 in. deep in an 8 in. diamond pattern. A diamond pattern should be used because it is easier to wash. If cattle are falling down when they exit from the squeeze chute in an existing facility, a grid constructed from bars will prevent falls. Construct the grid from 1 in. steel rods in 12 in. squares. Each intersection must be welded and the grid securely fastened to the concrete floor.

In areas with solid fence, small man-gates must be installed so that people can get away from charging cattle. The best type of man-gate is an 18 in. wide, spring-loaded steel flap. The gate opens inward towards the cattle held shut by a spring. A person can quickly escape because there is no latch to fool with. The man-gates can be constructed from 10 gauge steel with a rim of 1/2 in. rod.

Figure 3. Curved corral system shown in figure 1 constructed from steel with sucker rod fences. A wide 24 in. belly rail has been installed in the curved reservoir lane to provide a visually substantial fence. The round crowding pen and loading chute are covered with 10 ga. steel sheets.

Figure 4. A funnel crowd pen should have one straight side, and the other side on a 30 degree angle. Jamming will occur if both sides are angled.

Many people have asked questions about how the corrals should be laid out. It is really very simple. In figure 1 the curved single-file chute, round crowding pen, and curved reservoir lane are laid out along the dotted line. The first step is to place a string on the site in the position of the dotted line. The radius points of the curved single-file chute, round crowding pen, and curved reservoir lane are all located along the string.

Layout steps for figure 1 (To be done in order.)

- Make a 16 ft 180° half circle for the single-file chute.
- Make a 12 ft 180° half circle for the round crowding pen.
- Make a 35 ft 180° half circle for the curved reservoir lane.
- Layout the diagonal pens on a 60° angle by placing a transit on the string.
- Layout gathering pen with a 55 ft radius. The radius point for the gathering pen is located 95 ft from the strike post of the last gate in the row of diagonal pens. The row of diagonal pen gates and the gathering pen radius point should be at a 90° angle relative to your string (dotted line).
- Layout the post working pen. The 55 ft radius point is found by measuring 55 ft from the hinge of the 14 ft sorting gate in front of the squeeze chute. The exact location of the hinge may vary depending on the length of the squeeze chute. Leave a 3 ft to 4 ft space between the end of the sorting gate and the headgate on the squeeze. This provides enough room so you can swing the sorting gate in front of the headgate without hitting the cow's head.
- After laying out the basics, finish laying out both sides of the lanes. The pasture entrance gates and the central drive lane will have to be laid out by eye.

Lay out everything in lime before building anything. This will prevent mistakes. Walk through the layout. If it looks like the drawing, then you have got it right. If there is a hill next to the site, look at the lime layout from there. If an aircraft is available, use it to check your layout.

SQUEEZE CHUTES

Herd health care is virtually impossible without a headgate or a squeeze chute for restraining animals. There are many headgates on the market and each type is especially suited for certain handling procedures. There are four basic types of headgates: scissors stanchion, full-opening stanchion, positive control, and self-catcher (Grandin, 1980b).

Scissors-stanchion headgates consist of two biparting halves that pivot at the bottom (figure 5). The full-opening stanchion consists of two biparting halves that work

Figure 5. Scissors stanchion headgate with curved neck bars. The gate opens like a pair of scissors and has pivots at the bottom. The curved neck bars provide a good combination of head control and protection against choking.

like a pair of sliding doors. A positive-control headgate locks firmly around the animal's neck. This type of gate completely restricts up and down movement. The self-catcher headgate can be set like a trap. When the animal enters, its forward movement will close the gate automatically around its neck. The advantages and disadvantages of the four types of headgates are summarized in table 1.

Self catching, scissors stanchion, and full-opening stanchion are available with either straight or curved stanchion bars. A straight-bar stanchion headgate is extremely safe and will rarely choke an animal. The disadvantage of a straight bar stanchion is that an animal can slide its head up and down unless a nose bar or other restraint is used. The straight bar stanchion is recommended if the headgate is going to be used primarily for AI or pregnancy testing.

TABLE 1—Types of Manually-Operated Headgates Compared

	Self-Catcher	Scissors Stanchion	Positive	Full-Opening Stanchion
Recommended for	Hornless cattle, gentle cattle, one-man AI	General purpose, big feedlots, wild cattle, minimum maintenance, cattle of mixed sizes adjustment)	Dehorning, wild cattle, horned cattle, good head control, big feedlots. Requires less strength to operate than stanchion gates.	General purpose, vet clinics, mixed cattle sizes (because the gate seldom needs adjustment). Large bulls can exit easily.
Not recommended for	Wild cattle, big feedlots, horned cattle, groups of mixed-size cattle (because the gate has to be readjusted to catch animals of different sizes)	Very large bulls (because they may have trouble exiting due to the narrow space between the two bottom pivots)	Vet clinics where the animal is held in the headgate for a prolonged time. When AI and pregnancy testing are the primary uses of the headgate.	Big wild cattle, big feedlots (because many full-opening stanchion headgates are not sturdy enough to withstand constant heavy usage)
Warnings	Mechanism requires careful maintenance. Head and shoulder injuries may result if the animals are allowed to slam into the gate.	Be careful not to catch the animal's legs or knees between the two halves of the gate or the animal may be injured.	More likely to choke than a self-catcher, scissors, or full-opening stanchion.	Mechanism requires careful maintenance to prevent jamming. Animal may trip over the lower gate track if it becomes excited.

Self-catcher, scissors-stanchion, and full-opening stanchion headgates are available in models with either a straight or curved stanchion. Refer to the text for discussion on choking hazard *versus* head control.

Source: T. Grandin (1980).

A curved-bar stanchion is a good compromise between control of the animal's head and protection from choking. It is more likely to choke than a straight-bar stanchion, but it is safer than a positive type gate. A nose bar is not needed for ear implanting or tagging, if the animal is backed up in a stanchion headgate.

The problem of choking in a curved-bar stanchion or positive type headgate can be reduced by adjusting the squeeze sides of the chute. The V shape of the chute should support the animal. The proper spacing at the bottom of the squeeze sides is 6 in. for 250 to 400 lb calves, 8 in. for 600 to 800 lb animals, and 12 in. for cows and most fed steers. The space should be 14 to 16 in. for large bulls. The measurements are taken on the inside of the chute at the floor level. The best type of chutes have two squeeze sides that fold in evenly when the squeeze is applied.

Operator Skill Important

Results of a survey conducted by the author indicated that the main cause of handling accidents in hydraulic squeeze chutes was a careless operator handling cattle too fast. The skill of the operator affected the incidence of choking and escape from the squeeze chute (tables 2 and 3). Problems such as balking and falling while exiting from the squeeze chute were largely determined by conditions such as slick floors or shadows in the handling facility. You should be able to do a better job of operating your squeeze chute than that indicated by the percentages on tables 2 and 3.

TABLE 2—Effect of Cattle Breed and Operator's Skill on the Frequency of Handling Accidents in Hydraulic Squeeze Chutes

	Average for All 22 Groups (2150 Head)	Average for 12 Brahman Groups (1210 Head)	Average for 10 Non-Brahman Groups (940 Head)	Number of Groups with a Perfect Score	Single Worst Group Score	Observed Cause of the Worst Score
Mild choke	0.40%	0.17%	0.77%	17	3.00%	Rushing and carelessness*
Severe choke	0.30%	0.17%	0.40%	18	2.00%	Inexperienced operator
Entry balk	11.25%	10.30%	12.40%	0**	30.00%	Electric pole in front of the chute
Exit balk	15.20%	17.90%	12.00%	0	25.00%	Brahman cattle backed up after release
Partial escape	2.30%	3.90%	0.55%	10	11.00%	Long horns
Total escape	0.74%	1.00%	0.20%	14	5.00%	Carelessness
Falling	4.90%	7.30%	2.00%	5	15.00%	Slick, smooth concrete floor in front of chute
Headgate leg	2.50%	2.20%	3.00%	7	12.00%	Rushing

*All occurred in the same feedlot.
**Best balking score: entry balk 1.00%, exit balk 3.00%.

TABLE 3—Frequency of Handling Accidents in Hydraulic-Stanchion Headgate Squeeze Chutes*

	Ear Implant Only** (1230 Head Weighing More Than 600 lb)***	Full Processing† (920 Head Weighing 250 to 600 lb)*** 45% Castrated
Mild Choke	0.00%	0.88%
Severe Choke	0.11%	0.30%
Entry Balk	10.95%	12.70%
Exit Balk	13.90%	15.65%
Partial Escape	2.20%	1.39%
Total Escape	1.18%	0.07%
Falling	4.68%	3.15%
Headgate Leg	0.88%	3.77%

*Bowman and Trojan hydraulic squeeze chutes.
**Cattle received an implant of growth promotant in the ear. No other treatment was given.
***The sample consisted of both Brahman-cross and English-type cattle.
†In places where no Brahman or Brahman-cross cattle are handled the percentages should be lower. Full processing consisted of a minimum of two brands, two injections, an ear implant, and at least one other treatment such as deworming, or pour-on insecticide.

Source: T. Grandin (1980).

Different breeds of cattle react differently to handling (Ewbank, 1968; Tulloh, 1961). More handling accidents occurred when Brahman-cross cattle were being handled. The Brahman-cross cattle had more total escapes and partial escapes. A partial ecape was recorded when the animal was caught around the middle by the headgate.

Choking in the headgate occurs when the headgate applies excessive pressure to the carotid arteries or the wind pipe (White, 1961; Fowler, 1978). Excessive squeeze pressure in a hydraulic squeeze chute can also cause choking. A hydraulic chute is safe if the pressure relief valve is set correctly. In fact it may be safer for both man and animal because the dangerous levers are eliminated. At several feedlots, some cattle died several days after going through the hydraulic squeeze chute. The animals weighed over 600 lbs and appeared to have pneumonia. An autopsy revealed that the cattle had been ruptured internally due to excessive squeeze pressure.

Cattle can also be injured if a fast-moving animal is stopped suddenly by clamping the headgate around its neck. Examination of beef carcasses revealed old healed injuries to the back and neck. Even if the animal appears normal it can sustain a spinal injury if it slams into the headgate. A skillful squeeze chute operator can slow the animal down in the squeeze before it reaches the headgate. Rubber strips can be placed on the headgate to absorb shock and help reduce injury. Old, split motorcycle tires will work well.

Too many people try to set the world speed record for working cattle and they end up injuring a lot of animals. The survey indicated that a skilled crew can actually handle more cattle per hour by handling them gently and skillfully. A four-person crew using a hydraulic stanchion-type chute in a well-designed circular cattle working facility could catch an animal and place an ear implant every 15 seconds. The crew could also brand, vaccinate (up to four injections), and ear implant an animal every 45 seconds. In 60 seconds the crew could castrate, brand, vaccinate, ear implant, and give at least one other treatment such as pour on insecticide or clip a tail (Grandin, 1980b). When the crowding pen was filled with cattle, these timed procedures could be achieved without rushing. If a crew goes faster than these times, they will be doing a sloppy job and may be injuring the cattle.

DIPPING VAT DESIGN

Building a slide to make the cattle slide into the dip vat on their rear ends is wrong. The animal should be provided with good footing as it enters the water (Grandin, 1980c).

A 9 ft downward sloping, hold-down rack (figure 6) prevents the animals from leaping to the center of the vat.

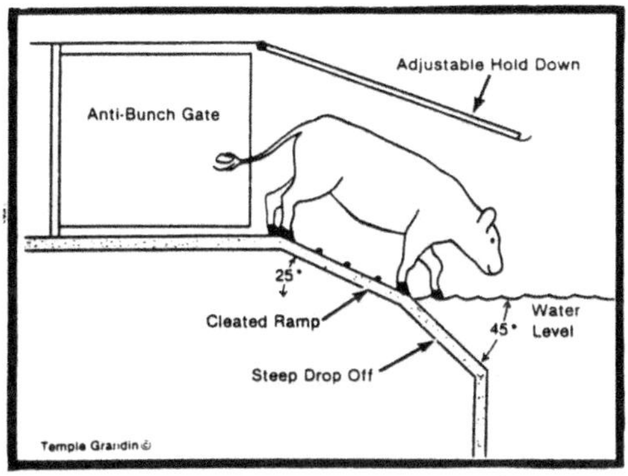

Figure 6. A grooved or cleated ramp with a nonslip surface enables the animal to enter the vat without hesitation. The steep drop off is hidden under the water. A 9 ft. adjustable hold-down rack makes the animal dive in and immerse its head. The distance between the hold-down rack pivot and the floor is 5 ft. The cleated portion of the ramp is 6 ft. long and on a 20 to 25 degree angle. The 45 degree angle portion is 4 ft. long (Grandin, 1980c).

The hold-down rack forces the animals to immerse their heads instead of being pushed under with a forked stick. It also helps prevent the chemicals from splashing out. Splashing can be further reduced by installing a 3 in. pipe along the inside and top edges of the dip vat. The pipe should be 3 to 4 ft above the surface of the water.

Each animal enters the vat by walking down a 6 ft gradual declining ramp that is on a 20° to 25° angle. This ramp has deep grooves in the concrete to provide the animal with good footing. The grooves should be 2 in. deep and 8 in. apart. The purpose of the ramp is to orient the animal's center of gravity towards the water. The steep drop-off is hidden under the water, but the ramp appears to continue on into the water (figures 6 and 7). When the animal steps out over the water it falls in. The ramp must have a **nonskid** surface or the animal may become scared and attempt to back out.

If both large and small cattle are going to be dipped, the entrance should be equipped with antibunch gates. These 7 ft gates allow only one animal to enter the vat at a time. The antibunch gates work on the same principle as the trigger trap one-way gates that are used to trap wild cows at the water hole. The opening between the ends of the two gates is adjusted to equal the width of one animal. The gates act as a valve to slow down incoming cattle. One of the antibunch gates can be spring loaded. The entrance should also be equippped with a semicircular block gate or a sliding gate to shut off the flow of cattle. A stanchion headgate powered by hydraulics also works well for a shut-off gate.

Figure 7. Cattle enter the vat and immerse their heads. They do not need to be pushed under with a forked stick. The hold down rack is easily constructed from 10 ga. steel and pipe.

Stairsteps are the best type of ramp for exiting from a dip vat. A steeper ramp can be used in a dip vat than for loading because the water supports the animal. Dip vat exit steps can have a 6 to 7 in. rise and a 12 in. tread width. The steps must be grooved to prevent slipping.

Drip Pen Design

A divided drip pen is recommended for dripping. When the cattle are drying on one side, the other side can be filled (figure 8). Figure 9 illustrates a dipping system layout with curved lanes and angled drip pens. Each side of the drip pen should be 30 to 40 ft long and 16 ft wide. The drip pens should be sloped 1/4 of an inch every foot towards the vat. The drip pen should be curbed with an 8 in. high

Figure 8. Divided drip pens with remote controlled exit gates. The pipes go to cylinders to operate the gates. The advantage of remote controlled gates is they are labor saving and they allow the handler to open the gates without entering the flight zone of the cattle.

curb. To prevent the cattle from slipping, the drip pen floor should be scored in an 8 in. diamond pattern. When the wet concrete is scored, the last pass should be made towards the vat so that the water will drain more easily.

Since often either hydraulic or air pressure is available at the site, placing cylinders on the exit gates is recommended. This will let the cattle out by remote control. The exit gates can be opened without the handler entering the flight zone of the cattle. In conventional systems, the cattle often become agitated when the handler walks up to open the gates manually. In some instances, the cattle will ram the fences and attempt to jump back into the vat when the handler approaches.

The drip pen exit gates and the dividing fence in between the two drip pens should be solid. This prevents the cattle that are confined on one side from pushing on the gate and attempting to follow the cattle that have just been released. If Brahman-cross cattle are being dipped, a one-way gate should be installed to prevent them from reentering the vat.

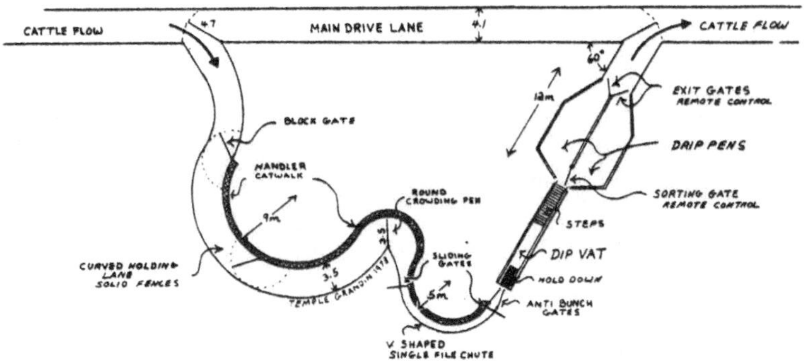

Figure 9. Dipping vat system with curved lanes. A squeeze chute can easily be incorporated into this layout. A 14 ft gate positioned in front of the squeeze chute can be used to divert animals you do not want to dip. English measurements for metric measurements are: inner radius of curved holding lane 9m (30 ft); width of holding lane and length of crowd gate 3.5m (12 ft); and inner radius of the single file chute 5m (16 ft) (Grandin, 1980a).

To help keep the vat clean, the single-file chute and the crowding pen should be installed to prevent hair and manure from reentering the vat from the dip pens. The sump should have a valve to divert rain water that falls on the drip pen away from the vat. It is also wise to install a 2 ft wide curbed concrete apron alongside the vat.

To help keep the vat clean, the single-file chute and the crowding pen should have a concrete floor to prevent the animals from tracking dirt into the vat. This floor should be washed down after each dipping session.

Agitation and Aeration

Regular agitation and skimming off of floating debris will keep a dip vat cleaner. A vat that is left standing will quickly become stagnant and foul smelling. An easy way to agitate a vat is to install an airline in the bottom. The line should be a 1 in. pipe with 1/16 in. holes drilled at 5 in. intervals. The pipe is mounted 1 to 2 in. off the vat bottom. Two rows of holes are drilled on a 45° angle facing downward on both sides of the pipe (Saulmon, 1972). Connect the pipe to an air compressor capable of delivering 0.5 cu ft per minute at 40 psi per foot of pipe.

Frequent agitation of the water with the air compressor wil aerate the water and prevent it from becoming septic and foul. Good results have been obtained by connecting a timer and a solenoid to control the release of air. A 30 sec blast of air every 30 min almost completely eliminates bad odors.

Filtering and Cleaning

The installation of a filtering system can double the life of the chemicals in the vat. Use of a Hydrasieve sloping screen to remove solids from the vat has reduced chemical disposal requirements by 50% and pesticide usage by 30% (Sweeten, 1976; Miller, 1975). Almost twice as many cattle can be dipped before the vat requires recharging.

A cleaning system with a Hydrasieve is very simple. It consists of a 28 in. wide Bauer Hydrasieve screen with a .2 in. screen spacing. A 3 in. centrifugal trash pump brings the water to the top of the screen. Don't use a smaller diameter pump, it will clog up. The water runs back through the screen and the solids fall off the screen into a container. The water returns to the vat by gravity. When this system is built, the suction line should be exposed. **Don't** bury the suction line under the slab. If you get an air leak, you will not be able to fix it. The water-return lines and other pipes may be buried.

Another type of cleaning system is a sluice box or settling basin. This system can also double the life of the vat. The advantage of the sluice box or settling basin is that it is inexpensive. The use of this system and a Hydrasieve together may enable you to quadruple the life of the chemicals in the vat (Sweeten, 1982).

The Hydrasieve removes the bigger solids, and the sluice box or settling basin removes the fine dirt. Settling systems can vary in capacity from 300 to 100 ga. The bigger systems will remove more solids but they are harder to clean. An easy sluice box system to construct consists of a metal or concrete box 10 to 18 ft long and 2 ft deep. The box contains a number of removable baffles. The highest baffles should be located at the end where the dirty water enters. The box should be narrower where the dirty water enters and becomes wider as the discharge point is reached. Two ft wide at the entrance and 4 ft wide at the discharge point will work well. The idea is to make the water move very slowly at the discharge point so the solids will be left in the bottom of the box. The water should flow through a 300 gal box at a rate of 25 to 66 gal per min (Sweeten, 1982). The sluice box or sedimentation tank must be run at least 4 hours for every 1000 cattle dipped. It is of the utmost importance to keep the sluice box clean. If it is not cleaned out regularly, it will put dirt back into the vat. After each use, the baffles should be pulled out and the box cleaned out with a shovel.

A good way to design a cleaning system is to connect the Hydrasieve and the sluice box in a tee circuit. By using two valves, the larger portion of the pump output is directed to the Hydrasieve and the smaller portion of the output is directed to the sluice box. If the flow rate is too great the sluice box will not work. Sluice boxes or settling basins must be used with care when using wettable powders because they will remove the chemical. Check with the pesticide manufacturer.

Chemical Disposal

Haphazard dumping of used chemicals around a feedlot or ranch is not recommended. Used dip can be evaporated in a shallow concrete basin (Fairbank et al, 1980). For large feedlots in the Southwest the evaporation basin should hold three times the vat volume and the water level should be less than 2 ft deep. In high rainfall areas, an evaporation basin should have cover. Check with a local engineer to determine evaporation rates for your area. Each state has its own regulations on pesticide disposal. Check with your own state.

REFERENCES

Ewbank, R. 1968. The behavior of animals in restraint, In: M. W. Fox (Ed.) Abnormal Behavior in Animals. W. B. Saunders, Philadelphia, PA.

Fairbank, W. C., T. Grandin, D. Addis, and E. Loomis. 1980. Dip vat design and management leaflet. 21190 Division of Agricultural Sciences, University of California, Davis, Calif.

Fowler, M. E. 1978. Restraint and handling of wild and domestic animals. Iowa State University Press, Ames, Iowa.

Grandin, T. 1980[a]. Efficient curved corrals. Angus Journal, October 1980, pp 95-97.

Grandin, T. 1980[b]. Good cattle restraining equipment is essential. Vet. Med. and Small Animal Clinician. 75:1291.

Grandin, T. 1980[c]. Safe design and management of cattle dipping vats. Technical Paper No. 80-5518. Amer. Soc. Agr. Eng. St. Joseph, Michigan.

Grandin, T. 1981. Innovative cattle handling facilities. In: M. E. Ensminger (Ed.) Beef Cattle Science Handbook.18:117. Agriservices Foundation, Clovis, Calf.

Miller, D. 1975. Evaluation of Bauer separation equipment as a dip vat filtration system. (Unpublished). USDA, Beltsville, Maryland.

Saulmon, E. E. 1972. Ticks and scabies mites...dipping vat management and treatment procedures. Veterinary Services Memorandum 556.1, USDA/APHIS, Washington, D.C.

Sweeten, J. M. 1976. Results of Hydrasieve cattle dip recycling study. Field Day on Dip Vat Management Systems for Cattle Feedyards. Texas A & M University.

Sweeten, J. M. 1982. Dipping vat sedimentation tanks compared. Beef. March, 1982, p. 104.

Tulloh, N. M. 1961. Behavior of cattle in yards: II A study of temperament. Animal Behavior 9:25.

White J. B. 1961. Letter to the editor. Vet. Record 73:935.

29

REDUCING TRANSPORTATION STRESSES

Temple Grandin

Improved transportation and handling methods can save money. Seven to ten percent of all feedlot steers are bruised during handling, loading, transporting, and weighing. The discounting on the sale price for bruised (damaged) meat costs the producer an average of $57 per 100 head of fat cattle marketed (Rosse, 1974). Thirty percent of all the bruises occur in the valuable loin area. Livestock Conservation Institute estimates that the beef industry is losing $22 million annually from bruises that cause damage to the marketed beef. A feedlot with rough handling has twice as many bruises as a feedlot with gentle handling. A survey I conducted indicated that cattle sold live weight had 14% discountable bruises and the carcass cattle had only 8%. The producers were more motivated to handle cattle carefully when they were sold on a carcass basis because bruises showed as the carcass hung in the cooler and the producer received a penalty price. A study by Marshall (1977) indicated that people take better care of livestock that belong to them. Cattle hauled by contract truckers had more bruises than cattle hauled by a packer's own truckers. A few bad truckers often account for many of the bruises (Rickenbacker, 1958).

Horns will greatly increase the amount of bruising. Loads of horned cattle had twice as much bruise trim compared to loads of polled cattle (Meischke, 1974). Ramsey (1976) found that tipping the horns will NOT reduce bruising. The answer to the bruise problem is to dehorn your baby calves or to breed polled cattle. Thin cows have to be handled gently because they bruise more easily than steers (Wythes et al., 1979). Interior surfaces in livestock handling facilities must be smooth. Structural members such as posts should be on the outside of the fences to prevent bruises.

SHRINK

Range cattle that have been placed in unfamiliar pens will shrink more than cattle held in familiar pens

(Brownson, 1979). You can reduce shrink losses by handling livestock quietly and with little excitement. During hot weather it is advisable to ship during the night or in the early morning. Psychological stresses can contribute to shrink losses. Calves that have been preconditioned prior to shipping will shrink less than calves that have been weaned at shipping time (Woods et al., 1973). Calves that have become accustomed to transport will shrink less and animals that have become accustomed to handling procedures will be less stressed. Calves lost less weight the second time they were transported (Ried and Mills, 1962; Hails, 1978).

Transporting your livestock directly to the packer will help reduce shrink losses. Mayes, et al. (1980) found that feedlot cattle hauled for 52 mi had higher carcass yields than cattle hauled for 373 mi. Shrink can be reduced by providing water up until the time of transport. It is important to have water continuously available so that the animals do not engorge themselves shortly before loading. Tissue shrink can start soon after loading, especially if the animals get excited. Asplund (1982), found that in many instances loss of gut fill was a minor portion of the overall shrink. A portion of the shrink occurs early in the journey when excited animals urinate, sweat, pant, and defecate.

A continuous water supply prevents the animal's tissues from losing water. Tissue shrink starts quickly when the animal is being transported, but it takes time to regain the tissue weight loss after the animal is able to drink on arrival. After unloading, long-haul cattle should be fed and allowed to eat before they are allowed to drink. This will help prevent engorgement.

The type of feed the animals have been on also will affect the amount of shrink. Cattle that have been fed concentrates will shrink less than cattle that have been on green feed. Typical shrink figures for feeder calves shipped during the fall for 600 miles is 7.2% to 9.1% (Self and Gay, 1972).

WIND CHILL

Many people do not realize that the wind whistling through a truck can chill the animals. If a truck is traveling at 40 miles per hour on a 32° F. day the wind chill factor will be a chilly -20°F. (figure 1) for cattle with summer coats. During cold weather the nose vents in trucks should be closed. If the truck is being hit by a stong cross wind it may be advisable to cover one side.

The important thing is to keep the animals DRY. If the hair becomes wet it loses its ability to insulate the animal from the cold. Wetting a calf has the same effect as lowering the outside temperature 40° to 50°F. Dry cold weather is usually less dangerous than wet weather around 32°F.

Even during very cold dry weather the coat retains its ability to insulate. Freezing rain is very hazardous and many animals have been lost due to wind chill under these conditions.

Ames Wind Chill Indexes

For Cattle with Summer Coats and Shorn Sheep (Dry Animals)

Wind speed mph	Actual Temperature (Fahrenheit)						
	−10	0	10	20	30	40	50
10	−20	−10	0	9	19	29	39
20	−37	−27	−17	−7	2	12	22
30	−53	−43	−33	−23	−13	−3	6
40	−60	−50	−40	−30	−20	−10	0

For Sheep with Full Fleece (Dry Animals)

Wind Speed mph	Actual Temperature (Fahrenheit)						
	−10	0	10	20	30	40	50
10	−10	0	9	19	29	39	49
20	−13	−3	6	16	26	36	46
30	−20	−10	0	9	19	29	39
40	−31	−21	−11	−1	8	18	28

Figure 1

There is no such thing as a single ideal temperature for an animal. The ideal temperature or thermal neutral zone, in which the animal feels neither hot nor cold is based on many factors, including wind speed, hair coat length, wetness, condition, and the level of nutrition. The amount of wool or hair will also affect the ability of the animal to withstand cold (figure 1) (Ames, 1974).

TRUCK TRAILER DESIGN

The design of the trailer can have an effect on the amount of bruising. In the West, side-loading trailers are used that unload through the side instead of through the rear. More bruises occur in this type of trailer because the cattle have to turn a 90° corner when they leave the trailer. If the animal becomes excited or rushed by the handler, it can easily bruise its loin on the door frame. The width of the door can have a significant effect on the amount of bruising. Replacing the standard 30 in. door with a door which was 42 in. wide at the top and tapered at the bottom reduced bruises. Tapering the door forced the animal to walk through the center and kept its hips from catching on the door frame (Grandin, 1980[a]).

There are still many unanswered questions about ventilation and the effects of exhaust fumes on livestock. Coleman and Sheldon (1974) reported that when the top of the exhaust stack was even with the trailer roof line, the calves on the top deck subsequently gained weight faster than the calves on the bottom deck. Calves loaded into trucks with low stacks 12 to 24 in. below the roof line gained faster if they rode on the bottom deck. Bottom-deck calves on trucks with high stacks may have been subjected to fumes sucked up in the vacuum that forms behind the trailer. Fortunately most trucks have higher stacks. Very few vehicles now have stacks 24 in. below the roof line. A recent study by Camp et al. (1981) indicated that there were no consistent differences in the average daily gains of calves transported on the bottom or top decks of a trailer. All the trailers surveyed were pulled by tractors with stack heights of 1 to 11.8 in. below the roof line. The elimination of the very low stacks could account for this result. Air movement through trailers should be researched. Muirhead (1982) is conducting studies on trailer streamlining to improve fuel efficiency and wind tunnel tests to determine air flow patterns inside the trailer.

Excessive vibration in a livestock trailer can be stressful to animals. Inflating the tires to 90 to 120 psi (to above mfg. specs.) prolongs the life of the tires but it greatly increases the vibration levels imposed on the livestock. Over-inflated tires also reduce the life of the trailer. In an aluminum trailer, the greatest amount of vibration occurs over the rear axles (Stevens and Camp, 1979).

SPACE REQUIREMENTS

Loading a truck correctly is essential for safe transport. Loading too few or too many animals can result in injuries. Livestock Conservation Institute has published recommendations for space requirements for different types of livestock (figures 2, 3, 4). Avoid overloading a truck with horned cattle. Too many animals can double the amount of bruising. If tiny baby ("bob") calves are being hauled, allow 2.9 sq ft per 100 lb calf and 3.5 sq ft for a 150 lb calf (Seubert, 1982).

TRUCKING TIPS

1. Use partitions to separate species transported on the same truck.
2. Clean trucks after each load to prevent the spread of disease.
3. Check truck load regularly enroute.
4. Load and unload animals quietly; limit use of electric prods. However, it is better to give a cow a small jolt than to break her

tail by twisting it. Abusive handling during loading and unloading is still a major problem.

Truck Space Requirements for Cattle

(Cows, range animals or feedlot animals with horns or tipped horns; for feedlot steers and heifers without horns, increase by 5 percent)

Av. Weight	Number Cattle per running foot of truck floor (92-in. truck width)
600 lbs.	.9
800	.7
1,000	.6
1,200	.5
1,400	.4

Examples (1,000 lb. cattle):

44 ft. single deck trailer—44 x 0.6 = 26 head horned, 27 head polled.

44 ft. possum belly (four compartments, 10 ft. front compartment; two middle double decks, 25 ft. each; 9 ft. rear compartment, total of 69 ft. of floor space)—69 x 0.6 = 41 head of horned cattle and 43 head of polled cattle.

Measure the total lineal footage of floor space in YOUR truck.

Figure 2

Truck Space Requirements for Calves

(Applies to all animals in 200 to 450 lb. weight range)

Av. Weight	Number Calves per running foot of truck floor (92-in. truck width)
200 lbs.	2.2
250	1.8
300	1.6
350	1.4
400	1.2
450	1.1

Examples (450 lb. calves):

44 ft. single deck trailer—44 x 1.1 = 48 head.

44 ft. double deck trailer—88 x 1.1 = 97 head.

Figure 3

Truck Space Requirements for Sheep

(Use for slaughter sheep, load 5 percent fewer if sheep have heavy or wet fleeces.)

Av. Weight	Number Sheep per running foot of truck floor (92-in. truck width)
60 lbs.	3.6
80	3.0
100	2.7
120	2.4

Example (120 lb. sheep):

44 ft. triple deck trailer—44 x 3 x 2.4 = 317 shorn sheep, 302 wooly sheep.

Figure 4

Source: Livestock Conservation Institute (1981).

5. Accelerate vehicle smoothly and avoid sudden stops.
6. Don't ship livestock that are full of green feed. Let them stand and empty out.

STRESS AND MEAT QUALITY

Stress can have a negative effect on meat quality in both cattle and sheep. Stress from fatigue, fighting, cold weather, or lack of food can use up the animal's store of glycogen (muscle energy). When the glycogen store is used up, the meat will be darker and drier than normal. This "dark cutting" meat is less palatable and has a shorter shelf life. Cattle that are "dark cutters" will receive USDA downgrading and discounting for the carcass.

The basic principle is that a long-term stress for 12 to 48 hours will use up the glycogen and make the meat darker than normal, but a short-term stress, such as excitement, immediately prior to slaughter will tend to make the meat tough. Stress-related meat quality problems are more likely to occur in the fall and spring when the days are hot and the nights are cool. When the weather turns more uniformly cold or warm the animal's body adapts and they become less prone to dark cutting.

To reduce stress, avoid mixing strange animals prior to loading or after they reach the meat packing plant. Everytime strange cattle are mixed they will fight to establish a new social order. Fighting is very stressful because large amounts of adrenalin are secreted. The rapid release of adrenalin that occurs during fighting uses up the animal's store of glycogen in the muscles. When this occurs, the pH of the meat rises and it becomes darker and drier.

When steers fight to establish a new social order during the 24 to 48 hours prior to slaughter, they increase the incidence of dark cutting meat (Grandin, 1980[b]) (figure 5). Fighting among bulls when unfamiliar animals are mixed will cause dark cutters within a few hours. When unfamiliar fed bulls were mixed overnight, prior to slaughter the resultant fighting caused 73% of the animals to become dark cutters (Tennessen and Price, 1980). Bulls fed for slaughter should travel to the packing plant with their penmates. Young bulls kept in penmate groups have a very low incidence of dark-cutters.

The shape of the pen that animals are held in can influence their behavior. Kilgour (1978) found that bulls tend to use the space more efficiently in a rectangular pen than in a square pen. A long narrow pen maximizes the length of fenceline space in relation to the floor space. Studies by Stricklin et al. (1979) indicated that cattle prefer to lie along the fenceline. A long narrow pen provides the animals with more fenceline space than does a square pen with the same area. The long narrow pen works on the same principle as that used by people when finding seats

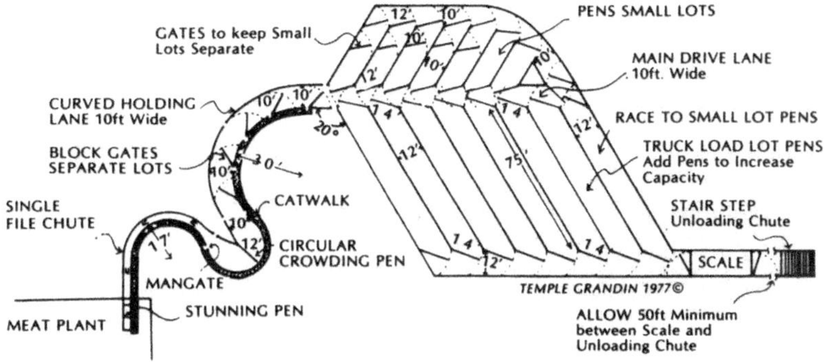

Figure 5. Stockyard for a slaughter plant with diagonal pens and curved chutes. One-way traffic flow and the elimination of sharp corners helps to keep animals calm (designed by Temple Grandin).

in a restaurant. People tend to prefer booths along the wall instead of center tables. The use of long narrow pens to hold livestock at the packing plant may help reduce stress. Figure 6 illustrates a layout for a packing plant with long narrow pens on a 60° angle. Each pen holds one truck load of cattle, and the livestock traffic flow is one-way. Pens on a 60° angle have been used for years in feed-lots for shipping cattle. McFarlane (1976) was one of the first people to recognize the benefits of pens in a diagonal layout.

ROUGH HANDLING

Rough handling and the excessive use of electric prods is very stressful to livestock. An animal that becomes excited and agitated immediately prior to slaughter is more likely to produce tough meat. Poorly designed handling facilities and rough abusive handling can greatly increase the amount of stress. Efficient, quiet sorting of feeder calves resulted in an increase of heart rate of only 7 beats per minute. Rough handling in poor facilities resulted in an increase of 48 beats per minute (Stermer et al., 1981). A noisy environment, yelling, and dogs biting at livestock will increase stress (Kilgour and deLangden 1970; Pearson et al., 1977). Being bitten by a dog was more stressful to sheep than being chased by a dog. Tiny baby calves should not be shipped until they are 5 to 7 days old to avoid stress and death loss problems (Seubert, 1982).

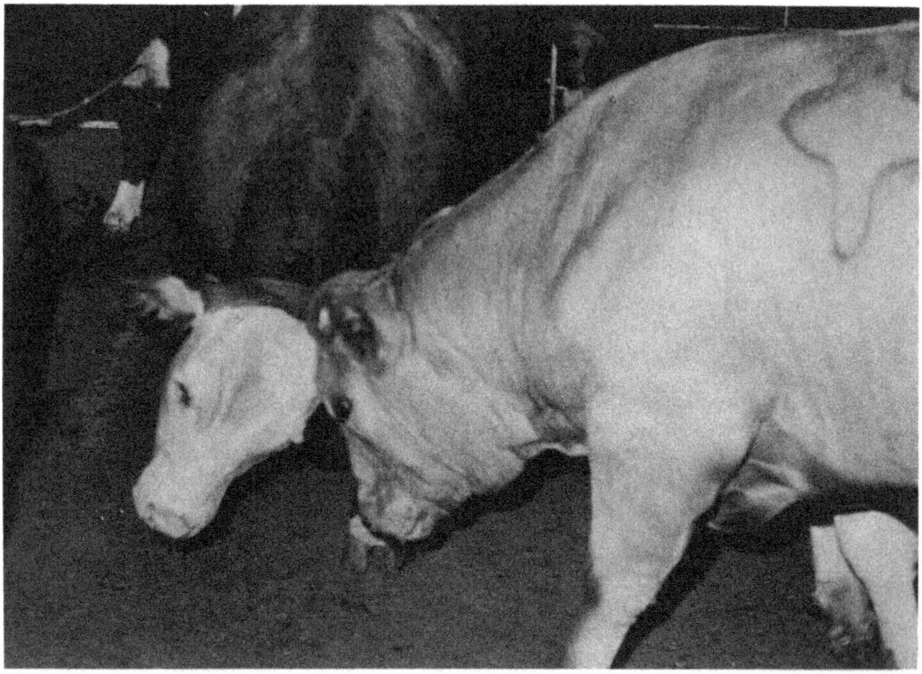

Figure 6. Fighting prior to slaughter is a major cause of dark cutting beef. When animals fight epine phrine (adrenalin) is secreted, which depletes the store of glycogen in the muscles. The big Charolais on the right disrupted the entire pen. Extremely aggressive cattle observed fighting in the packing plant holding pens should be removed to reduce stress.

REFERENCES

Addis, D.G. 1980. Preventive medication for feedlot replacement calves, Study II. In: Stress Calf Studies. Riverside and Imperial Counties, Cooperative Extension, University of California.

Addis, D.G., J. Dunbar, J. Clark, G.P. Lofgreen and G. Crenshaw. 1980. Effect of time and location of processing on feeder calves. In: Stress Calf Studies. Riverside and Imperial Counties, Cooperative Extension, University of California.

Ames, D.R. 1974. Wind chill factors in cattle and sheep. Special Publications SP-174, International Livestock Environment Symposium, Amer. Soc. Agr. Eng. St. Joseph, MI. pp 68-74.

Asplund, J.M. 1982. Shrink: The silent rustler. Beef. February, p 46.

Brownson, R. 1979. How to prevent cattle shrink. Montana Stockgrower. Reprinted in Beef Digest. February 1979, p 40.

Camp, T.H., D.G. Stevens, R.A. Stermer, and J.P. Anthony. 1981. Transit factors affecting shrink, shipping fever, and subsequent performance of feeder calves. J. Anim. Sci. 52:1219.

Grandin, T. 1982. Transportation of domestic animals, Symposium on Management of Feed Producing Animals. Purdue University.

Grandin, T. 1981[a]. Bruises on southwestern feedlot cattle. Paper presented at 73d Annual Meeting, Amer. Soc. Anim. Sci., July 26-29, 1981. (Abstr.).

Grandin, T. 1980. Bruises and carcass damage, Int. J. Stud. Anim. Prob. 1:121.

Grandin, T. 1980[b]. The effect of stress on livestock and meat quality prior to and during slaughter. Int. J. Study of Anim. Prob. 1:313.

Grandin, T. 1978. Transportation from the animal's point of view. Amer. Soc. Agr. Eng. Technical Paper No. 78-6013.

Hails, M.R. 1978. Transport stress in animals: A review. Animal Reg. Stud. 1:289.

Kilgour, R. 1978. The application of animal behavior and the humane care of farm animals, J. Anim. Sci. 46:1478.

Kilgour, R. and H. de Langden. 1970. Stress in sheep resulting from management practices. New Zealand Society of Animal Production proceedings 30:64.

Livestock Conservation Institute. 1981. Livestock Trucking Guide. By Temple Granding, South St. Paul, MN.

Marshall, B.L. 1977. Bruising in cattle presented for slaughter. New Zealand Vet. J. 25:83.

Mayes, H.F., M.E. Anderson, H.E. Huff, J.M. Asplund and H. B. Hedrick. 1980. Transport effects on carcass yield of slaughter cattle. Amer. Soc. Agr. Eng. Technical Paper No. 80-6509.

McFarlane, I. 1976. Rationale in the design of housing and handling facilities. In: M.E. Ensminger (Ed.), Beef Cattle Science Handbook. Agriservices Foundation, Clovis, CA 13:223.

Meischke, H.R.C. et al. 1974. The effect of horns on bruising cattle. Australian Vet. J. 50:432.

Muirhead, V.U. 1982. Interim report, an investigation of the internal and external aerodynamics of cattle trucks. Prepared for the Dryden Flight Research Center under Grant NAG4-8. University of Kansas Center for Research Inc., Lawrence, KS.

Pearson, A.M., R. Kilgour, H. deLangen and E. Payne. 1977. Hormonal responses of lambs to trucking, handling and electric stunning. Proc. New Zealand Soc. Anim. Prod. 37:243.

Ramsey, W.R. et al. 1976. The effect of tipping horns and the interruption of journey on bruising cattle. Aust. Vet. J. 52:285.

Reid, R.L. and S.C. Mills. 1962. Studies of the carbohydrate metabolism of sheep. Aust. J. Agr. Res. 13:282.

Rickenbacker, J.E. 1958. Causes of losses in trucking livestock. Marketing Research Report 261, Farmers Cooperative Service, USDA.

Rickenbacker, J.E. 1961. Loss and damage in handling and transporting hogs. Marketing Research Report 447, Farmer Cooperative Service, USDA.

Roberts, D.W. 1982. Yield in sheep meat processing. CSIRO Advances in Meat Science Conference, Brisbane, Australia.

Roose, J.C. 1974. Your stake in the $184,000,000 tangible farm to cooler loss. Proc. Livestock Conservation Institute, St. Paul, MN.

Self, H.L. and N. Gay. 1972. Shrink during shipment of feeder cattle. J. Anim. Sci. 35:489.

Seubert, T.J. 1982. Handling and transporting of bob calves and special fed veal calves. Official Proceedings, Livestock Conservation Institute, South St. Paul, MN.

Stermer, R.A., T.H. Camp, D.G. Stevens. 1981. Feeder cattle stress during handling and transportation. Amer. Soc. Agr. Eng. Technical Paper No. 81-6001.

Stevens, D.G. and T.H. Camp. 1979. Vibration in a livestock vehicle. Amer. Soc. Agr. Eng. Technical Paper No. 79-6511.

Stricklin, W.R. H.B. Graves and L.L. Wilson. 1979. Some theoretical and observed relationships of fixed and portable spacing behavior in animals. Appl. Anim. Ethol. 5:201.

Tennessen, T. and M.A. Price. 1980. Mixing unacquainted bulls: a primary cause of dark cutting beef. The 59th Annual Feeder's Day Report, Agriculture and Forestry Bulletin, University of Alberta, p 34.

Wythes, J.R., R.H. Gannon and J.C. Horder. 1979. Bruising and muscle pH with mixing groups of cattle pretransport. Vet. Rec. 194:71.

Part 6

PASTURE, FORAGE, AND RANGE

30

UNDERSTANDING RANGE CONDITION FOR PROFITABLE RANCHING

Martin H. Gonzalez

INTRODUCTION

In many parts of the world, improvement of the livestock industry is believed to be dependent on two primary factors related to animals: genetics and disease control. Another nonanimal consideration seems to be rain! It is common to listen to ranchers talking about a new bull they have or plan to import, about the last rain--or complaining about the lack of it (even when the normal rainy season is still months ahead--or about prices of livestock on the market and the latest government regulations. However, less often we find a group of ranchers here or in Mexico, discussing range conditions, the forage production on the ranch, or range improvements.

Perhaps because we are overgrazing and because our pasture and forage production is poor, we manage to put the blame on the rain, the market, or the government, we do not like to recognize that a great part of our attention should be devoted to the basic things on the ranch: soil, grass, and water.

In the U.S. and Canada, many ranchers might be familiar with the range condition concept and its implications, but in other countries the situation is different. Even when the common sense, the personal ability, and the experience of a producer is outstanding, he will require a good and basic knowledge of the condition of the range because it is the key to profitable ranching. And this is true, particularly, today, when the livestock industry needs to reduce the cost of production--and can by using desirable range plants--still one of the cheapest sources of forage.

SOIL DEVELOPMENT, PLANT SUCCESSION, AND RANGE MANAGEMENT

Range management as an art and as a science has its roots in ecological principles. A range user has to deal with the basic components of the ecosystem around him and his operation, and should understand how they function and how they interact. Profitable ranching requires an understanding of basic ecology.

Figure 1 will help us to understand how the basic ecological processes relate to range management, exemplified by a rangelands ecosystem.

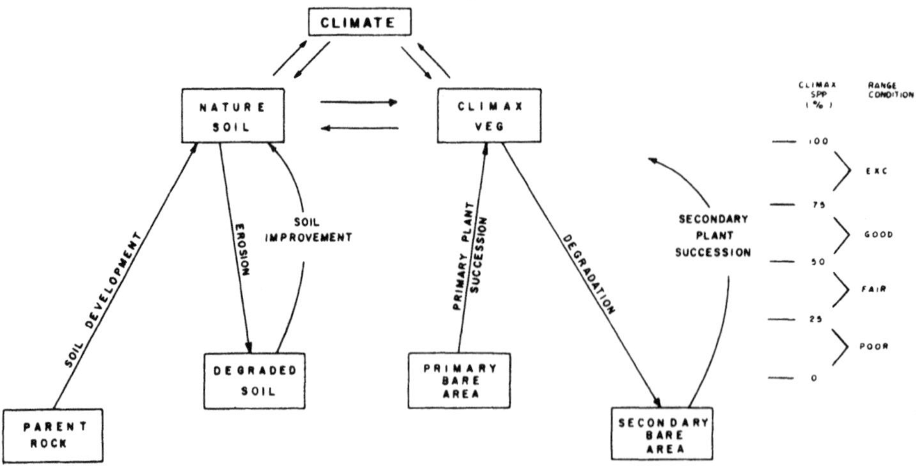

Figure 1. Relationship between soil, vegetation, climate, and range condition

Soil has developed for thousands of years from parent material. After a long process under various climatic conditions, it developed freely and reached maturity in its most productive phase. During that process, plant life had a parallel development, beginning from a primary bare area until reaching its climax stage (or maximum degree of development). This series of natural vegetative changes toward the climax stage are known as the primary plant succession. At this point, a fully developed soil and climax vegetation were in equilibrium with the climate for that particular area.

However, mainly due to man's actions, the soil has often been eroded, caused by degradation of the vegetative cover, and the pristine or climax condition has been destroyed. That which took centuries to build was destroyed by man in a much shorter period: by plowing of rangelands, by excessive fire, by timber exploitation, by over-grazing, by erosion, etc. It was then that the mature soil was converted into a degraded soil, and the climax vegetation was converted into a secondary bare area very far from its productive potential.

But man had not always been abusive of his resources. After he realized the serious damage caused to soil and vegetation and, of course, to the total productivity of an area, he started a series of soil improvement practices, together with some actions to accelerate the rehabilitation of the vegetative cover (secondary plant succession). These man-activated or "artificial" improvement practices (usually very expensive) were particularly needed in those areas where disturbances had been so severe that a spontaneous recovery was impossible.

And it is here, during different stages of this secondary plant succession, that most of our ranches are found today. The degree of disturbance they once experienced (or still are) and the degree of advancement within this succession, relate directly to range condition, according to the percentage of climax (desirable) forage species in the botanical composition found in the different pastures. It is impossible, under practical grazing conditions, to expect a range to remain in its primitive, climax state, and very few pastures can be found in excellent condition. There are many well-managed ranches in "good" condition, but most of the land, particularly in Latinamerican native ranges, would be characterized as in "fair" and "poor" condition. Many of these lands have the potential--under some grazing restrictions and good management--to improve and to recover productivity. However, on the lands, some agronomic improvement practices (soil conservation, brush control, range reseeding, etc.) must be used to reach their potential. We must pay this high price for the misunderstanding and mismanagement of our range resources.

QUANTITATIVE ECOLOGY AND RANGE MANAGEMENT

The Range Condition Concept

Range condition reflects the health of a range or pasture. It is based on the relation between the present vegetation and the vegetation that, potentially, a given site should have.

Dyksterhuis (1948) classified range plants in three categories according to their response to use by animals, and determined range condition based on the quantity of each of these species when sampling the range. These species are decreasers, increasers, and invaders. Figure 2 shows these relationships that every rancher should keep permanently in his mind.

Extensive field work has been done to classify plant species as decreasers, increasers, or invaders. Based on this work, range condition guides have been developed for every different site within any given vegetative type.

These include most of the plants that fall within each category and the percentage of each that is permitted in the botanical composition--based on what the climax vegetation should be. This permitted percentage, particularly for the increaser species, has been determined by experience and comparisons of the same species in different sites and under different grazing pressures.

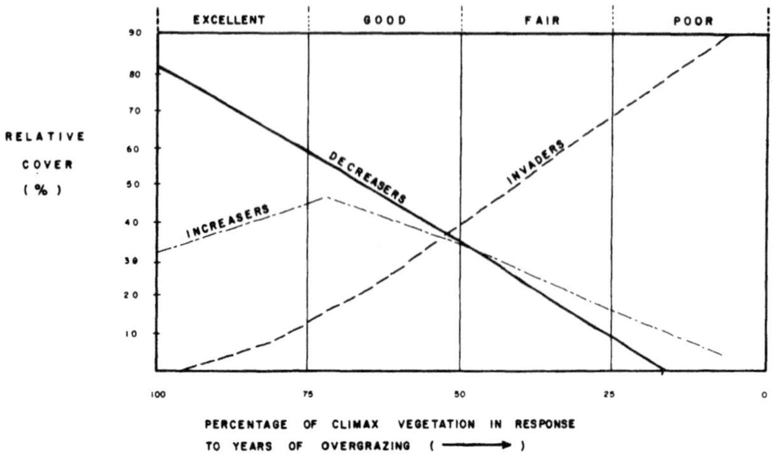

Figure 2. Quantitative basis for determining range condition (Dyksterhuis, 1948)

The decreasers are usually components of the climax community and are perennial, highly productive forage plants that are palatable and desirable. They are permitted in the percentages found in the botanical composition. The invaders are not permitted at all in the composition, so the percentage they accomplish for is substracted from the total. The increasers, most of which belong to the climax community, are permitted in varying percentages according to the range site and the associated species. The permitted percentage of each increaser species for the different sites has been calculated according to their response to grazing. This percentage has a maximum that will not interfere with the most desirable species--and a maximum that will not benefit the expansion of invaders. These changes in composition take place only when overgrazing occur over long periods, thus causing the deterioration of the range.

Dyksterhuis established a quantitative basis to determine range condition as follows:

Excellent Condition describes the botanical composition including more than 76% of climax or desirable forage plants. In other words, the vegetation is a mixture of decreasers and allowable increasers. There is almost no erosion and plant density is high. Almost no invaders are present. Litter is abundant.

Good Condition describes the biological composition that contains desirable species. The proportion of decreasers is less than that found in excellent-condition composition and few invaders may be evident. Litter is not as abundant but does occur.

Fair Condition describes the composition that has only 26% to 50% of the plants of climax or desirable species. In this condition, the signs of deterioration are more noticeable, particularly in marginal areas. Invaders appear in the same proportion as do decreases and increases; erosion has begun to be evident and density is generally low.

Poor Condition describes a badly degraded range with less than 25% of the plants in the composition of desirable or climax species. Forage production is very low. Erosion is a problem and invaders constitute a majority in the botanical composition. Almost no litter is evident. Spontaneous revegetation may not be economical under this condition and rehabilitation can be achieved only through range improvement practices.

We must realize that these conditions cannot be compared among different sites. Range site separation must be done before determining condition since their characteristics and potential are different.

Forage Production and Range Condition

The condition of the range and the amount of desirable forage produced are directly related for all types of range vegetation. However, from a production standpoint, a disclimax condition (just below the climax) may present a better diet for the grazing animals since some of the "increasers" in a given site could be more palatable or have a higher nutritive value than do some "decreasers" or climax species.

Figure 3 shows comparisons in the productivity of five selected rangeland types in northern Mexico, each for different condition: excellent, good, fair, and poor (Gonzalez, 1969).

It is observed that, even when there are substantial differences in kg/ha of usable forages between good an excellent conditions, there are greater differences between fair and good conditions and still greater differences between the poor and the fair conditions. This is true for all the vegetative types represented. Such large differences can serve as an index in estimating the effort, the expense and the time that may be needed to bring back the poorer conditions to a more productive stage.

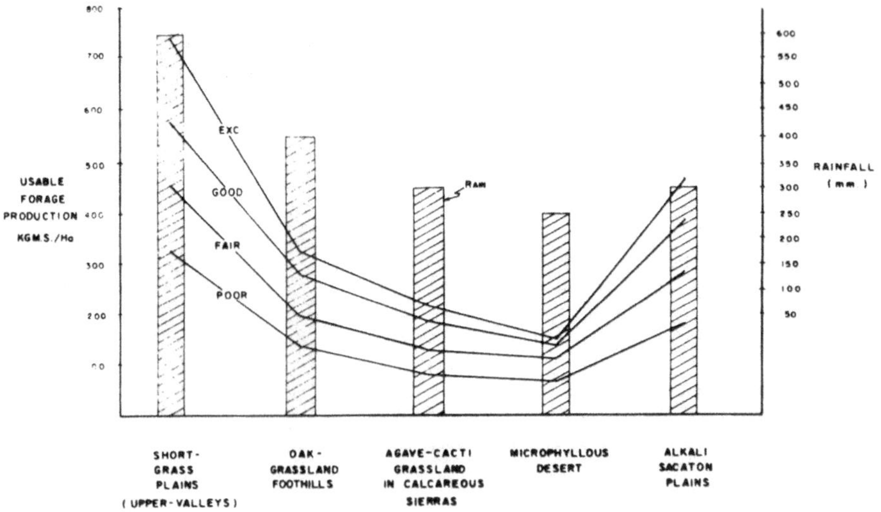

Figure 3. Forage production in five vegetative types under different range conditions in northern Mexico (Gonzalez, 1969)

RECOGNIZING RANGE CONDITIONS ON A RANCH

There are now sufficient range-condition guides for most of the rangeland areas of North America. For example, the Soil Conservation Service in the U.S. and COTECOCA-SARH in Mexico represent the agencies that developed such guides and that are using them extensively, along with many other government and educational institutions. It is not difficult to learn how to use these guides, but a basic knowledge of the terrain and the vegetation is needed.

Hoffman and Ragsdale (1974) from Texas A&M recommend for following steps in using the range-condition guides:

- Determine the range site.
- List the plants found on the site.
- Estimate the percentage of each plant that could be in the composition.

- Refer to the guide to determine the percentage of each plant that could be in the composition in the vegetational areas and site for which you are judging.

Carrying Capacity

Each guide includes the carrying capacity that is estimated for the different conditions in every site because range condition directly reflects the forage production in that particular site. A good condition means a stable, well-managed range; fair and poor conditions are indicators of heavy or severe use. On the other hand, if a site is in excellent condition, perhaps it is under-utilized. These guides provide an efficient, easy way to determine the carrying capacity (AU/ha) that the ranch should have--as compared with that under current management.

After the condition of the range has been determined, some adjustments may be necessary to keep the ranching operation flexible and to give the range the opportunity to recover. Adjustments may be necessary on:

- Livestock numbers (usually reducing them if fair and poor conditions are dominant).
- The present grazing system.
- The most economically convenient improvement practices to accelerate rehabilitation (brush control, range reseeding, etc.)--if condition is poor and hope for natural recovery is low.

REFERENCES

Dyksterhuis, E. J. 1948. Guide to condition and management of ranges based on quantitative ecology. Amer. Soc. Agron. App. Sec. Mimeo. p 25 (Abstr).

Gonzalez, M. H. 1969. Coeficientes de agostadero para el estado de Chihuahua. Memoria COTECOCA-SARH. Mexico.

Hoffman, G. O. and B. J. Ragsdale. 1974. How good is your range? Tex. Agr. Ext. Service Bull. Texas A&M Univ., College Station, Texas.

31

RANGE IMPROVEMENT PRACTICES AND COMPARATIVE ECONOMICS

Martin H. Gonzalez

Grazing lands of the world's rangelands have suffered abusive management that has, in turn, caused reduced productivity. This low productivity often is attributed to critical drought periods; however, the direct effect of a drought is to aggravate the poor management of the land.

The world demand for livestock products is increasing but the production from rangelands is decreasing because: faulty management has reduced the condition of the land and consequently the productivity; other demands for the land (agriculture, industrial, urban, highways) are deminishing the number of hectares available.

World evidence confirms the desertification of our ranges and the destruction of millions of highly productive hectares that have become unproductive and denuded.

A study conducted in Mexico (CFAN-CID, 1969), which included a survey in nine central and northern states and covered around 100 million hectares, indicated that overgrazing was evident in 85% of the land; light or advanced erosion was a problem in 87.5% of the overgrazed area, and 49.7% of the land was infested by undesirable plants, mostly brush. Overgrazing is a problem in almost all countries where degradation of rangelands has been so severe that economically it is impossible to expect a natural recover. Before the rangeland reaches this point, various range improvement practices, basically agronomic, are needed to facilitate and accelerate the secondary plant succession for returning those lands to a productive stage.

In this paper, emphasis will be given to the principles governing the most common improvement practices. More specific information for a particular problem may be obtained from neighboring experimental stations, the extension and university people, or the county agent.

MAIN RANGE IMPROVEMENT PRACTICES

Of the four elements applied in the rehibilitation of grazing lands (climate [rainfall], soil, vegetation, and grazing management), climate is the only one that man <u>cannot</u>

manipulate. However, manipulation of water from rainfall once it hits the ground is one of the most important aspects of range improvement. On the other hand, with correct planning and management of the other three elements, man may rehabilitate deteriorating lands. Some of the most common (and needed) improvement practices on rangeland are the following:

Water Conservation

Efficient use of rainfall enhances all the other improvement practices. When properly managed, even excess water can be beneficial. The principle in water conservation is to supply the forage species with moisture in such a way and at the right time so that the effects last the longest.

Water conservation can be managed in two ways: (1) by retaining each drop of rain where it falls and (2) by diverting excessive surface runoff to the highest-producing sites on the ranch. These two steps can be accomplished by either simple or complicated structures. However, the best way to conserve water for plant use is to have a good vegetative (grass) cover so that rain infiltrates into the soil and runoff and erosion are avoided.

Figure 1 shows how different grass covers on the rangelands of Chihuahua affect infiltration of water (Martinez, 1959). This figure averages data for tests on different short-grass ranges where bluegrama (Boutelouagracilis) was dominant. It was found that soils in an exclosure where they had been protected from grazing for 7 years had an infiltration rate 118% higher than bare ground, 63% higher than an overgrazed area, and 25% higher than a moderately grazed area. In turn, the soil that was moderately grazed absorbed 30% more water than the overgrazed site and 74% more than the bare area.

Table 1 shows the results of similar tests comparing different types and density of vegetative cover to conserve rainfall on different soils. On short-grass plains, sandy loam soil under excellent condition absorbed 450 mm more rain in a period of 105 minutes than a bare area; 270 mm more than range in poor condition, and 170 mm more than the soil with a cover in good to fair condition. The results were the same for the other two vegetative types--the oak-bunchgrass in the stony foothills and on the alkali flats with their heavy clay and deep soils (Sanchez, 1972).

Among the commonly used water conservation practices on ranches are contour furrows, range pitting (small, medium, or large pits), subsoiling, and low-retention fences. The conservation practice(s) to use depends on soil type, topography, slope, plant cover, costs, and storm intensity and frequency.

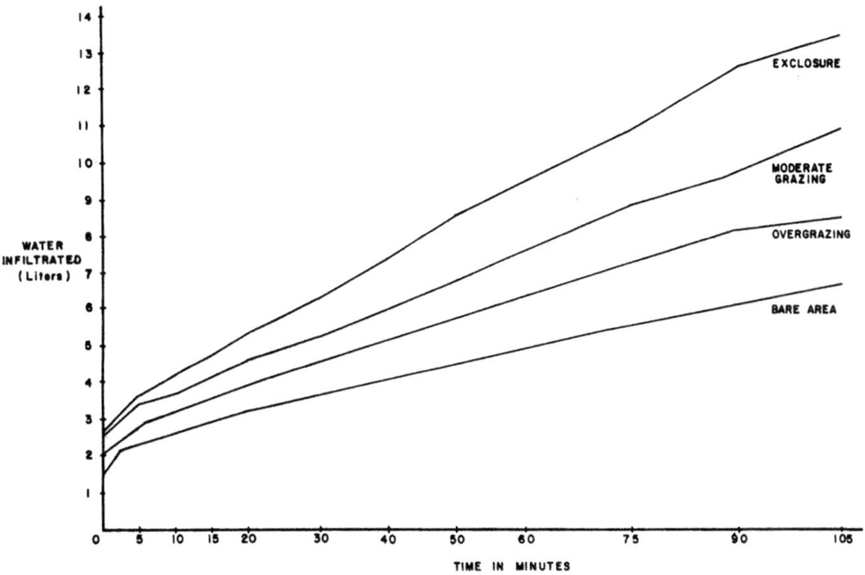

Figure 1. Water infiltration in short grass range under different covers. Chihuahua, 1964.

TABLE 1. INFLUENCE OF RANGE CONDITION ON WATER INFILTRATION[1] IN THREE VEGETATIVE TYPES AT LA CAMPANA, CHIHUAHUA

Range condition	Shortgrass plains	Oak bunchgrass foothills	Alkali flats bottom lands	$\overline{X}$
	mm	mm	mm	mm
Excellent	650	450	150	417
Good-fair	480	375	120	325
Poor	380	230	81	230
Bare area	200	190	73	154

[1] Mm of rain or equivalent absorbed in 105 minutes.

Soil Conservation

The most effective practice for soil conservation on rangeland is the same as for water conservation--a permanent cover of desirable plants. All soil conservation practices have a double purpose: to conserve water and to avoid erosion. Contour furrows, gully control by using different structures, terracing, and wind breakers are all effective soil conservation practices. As for water control, the practice to be selected will depend on the terrain. Heavy, sophisticated equipment for soil conservation is easily replaced by using some of the natural materials found on the ranch or in the pastures. For example, a gully can be controlled by filling an area with rocks, yucca plants, palmilla (sacahuiste) plants, old trees, or the undesirable vegetation nearby. Gullies can serve as dump sites for those materials. In a very few years the plant material, rocks, and soil eroding from upstream will compact, cause a dam to form, and erosion downstream will be reduced.

Control of Brush and Other Undesirable Plants

Unfortunately, most of the world's rangelands in use are infested, by varying degrees, with undesirable vegetation of which the woody species are the most common. This invasion is the direct result of the degradation caused by the different activities of man while using his grazing resources. Control of undesirable brush can be done by different methods: mechanical, chemical, and biological.

Mechanical control. Mechanical control is usually done with heavy or with light equipment. Heavy equipment includes dozing, chaining, disking, and heavy shredders or crushers. Light equipment includes the use of smaller tractors and some manual tools for shredding and disking. the equipment to use depends on the type of vegetation and its density, conditions of the terrain, availability of equipment, and costs. When using mechanical brush control, try to move or remove as little top soil as possible.

Research and rangeland experience all over the world demonstrate the competition between undesirable brush and grasses. The woody species of brush requires nearly four times more water to produce one kg of aerial growth and provides much lower forage quality than some of the best perennial grasses.

On Rancho Experimental La Campana, Chihuahua, Mexico, mechanical control of shrubs in a short-grass range invaded by *Acacia*, *Mimosa*, *Eysenhardtia*, and *Brickellia* resulted in a 102% range increase in forage production (Gomez and Gonzalez, 1978). Chemical control of "chaparrilo" (Eysenhardtia spinosa) produced 428 kg DM/ha more than the untreated areas, which had only 171.1 kg DM/ha. This represents 250% increment in forage production (Gomez and Gonzalez, 1976).

Chemical control. The chemical control of undesirable plants has developed intensively during the last two decades. The chemical products (herbicides) on the market for almost any brush-control program can be applied either by aerial spraying or ground spraying. Ground applications can be foliar, on the trunk or stumps, or directly on the ground in the form of pellets.

Herbicides work in different ways:

- contact herbicides are those that kill the plant or parts of the plant directly exposed to the chemical. This type is used on annual weeds. Examples: diesel oil, diquat, and paraquat.
- Translocated herbicides (also called hormonal or systemic)--are used in low concentrations. The toxic substance is carried through different parts of the plant by the plant's own liquids. Examples: 2,4-D, 2,4,5-T, silvex, dicamba, and picloram.
- Selective herbicides are the ones that kill a particular species or group of species without damaging others. If heavy dosages are used these may act as nonselective. This type of herbicide is the most widely used for range improvement because it does not affect grass or grass-like plants (monocots). Examples: 2,4,5-T picloram, TCA, and MCPA.
- Nonselective herbicides kill or damage all plants where applied. Example: AMS, amitrol, PCP, diesel oil, kerosene.
- Soil sterilants are generally in the form of pellets and are applied directly to the soil. When the pellets are diluted by rain, they are absorbed by the roots. Examples: Atrazine, bromocil, dicamba, diuron, and fenuron. The effect may be permanent or temporary, selective or nonselective. A detailed treatise on herbicidal plant control including products, methods, advantages, problems, etc., is found Range Development and Improvements (Vallentine, 1971).

Phenological stage of the plants, time of year, and meteorological conditions are three of the important factors to consider in brush-control programs.

Biological control. Another, but complicated, way to control undesirable plants and shrubs is through biological control. This can be achieved by insects or domestic or wild animals. There are insects that specifically live on certain plants, feeding from them, and eventually damaging them. Examples are the mesquite twig girdler (Oncideres rhodosticta) and two species of leaf-feeding beetles (Chrysolina gemellata and C. hyperícy) that attack the klmath weed (Hypericum perforatum).

Although biological control of shrubs and other undesirable plants by insects is complicated, the use of domestic animals is not. Studies in Central Chihuahua (Fierro, et al., 1979) used goats over a period of 3 years to defoliate five different woody species. Results indicated the goats' preferred shrubs to grasses. In moderate vs intensive browsing and in shredded and not shredded areas, intensive defoliation by goats caused a 36% mortality of "Chaparrilo" plants (Eysenhardtia spinosa); moderate defoliation killed 14% of the plants. A similar trend was observed for Acacia and Mimosa.

An additional benefit of using goats for biological control of noxious shrubs was their milk production--an average 30 ml per goat per day for 4 mo a year. The production of perennial grasses increased from 94 kg/ha to 360 kg/ha in the first year (1977) and to 950 kg/ha 3 years later with moderate browsing.

The main factor to consider if planning to control shrubs with sheep or goats is the type of vegetation. On many ranges invaded by shrubs and other plants that cattle will not eat, grazing systems combining small and large herbivores have been established. Even in some areas where toxic plants for cattle existed, changing to goats (or perhaps sheep) allowed those areas to be utilized.

Fire. The cheapest and most effective brush control tool is fire. However, there are certain rules that have to be followed: (1) the amount of fuel (grasses and associated vegetation) must be high enough to carry the fire at the desired intensity; (2) wind direction, velocity, and soil moisture content determine the season and time of day to use fire. Extreme care must be taken to avoid letting the fire run wild and damage adjacent areas. Vallentine (1971) in his book, Range Development and Improvements, describes in detail the procedures of burning for brush control and its consequences.

Numerous controlled burning experiments have been done to control undesirable vegetation. Glendening and Paulsen (1955) obtained 52% kill of young mesquites having basal diameters of 0.5" or less; however, only 8% to 18% of the taller trees were destroyed by fire. In other areas, only 9% were destroyed by burning (Reynolds and Bohning, 1956), but White (1969) reported a wild fire that killed 20% of the trees with moderate and severe burns. Cable (1965) reported that variable results were obtained from different fuel covers: 4500 kg/ha of fuel (ground cover) killed 25% of the trees when burned; areas with 2200 kg/ha of fuel killed only 8% of the trees. Other species damaged by fire were Ocotillo (Fouqeria splendens) with 40% to 67% killed; Sotol (Dasylirion texanum) had 97% dead plants, and creosote bush (Larrea tridentata) was reported with a "very high" percentage killed.

June fires killed 44% of the choya cactus (Opuntia fulgida) and 28% of the pricklypear (O. engelmanii) (Reynolds and Bohning, 1956).

Range Reseeding

Reseeding of denuded areas is an improvement practice that obtains fast results. However, many risks are involved in this type of operation, and extreme care must be exercised to establish and produce good forage.

Range reseeding has proven successful in many areas in the U.S., Canada, Mexico, and some Latin American countries. In arid and semiarid lands it is combined usually with some water catchment devices or structures in order to assist in the establishment of the new plants. Commonly associated with land clearing and brush control, reseeding is a must in many parts of the world.

The following are some points to help in understanding, planning, and managing a range reseeding operation.

What is reseeding? Reseeding refers to the artificial, not natural, revegetation of the range. It includes planting native or introduced species with one or several of the following objectives: 1) to improve plant cover/density, 2) to improve forage production, 3) to improve forage quality, 4) to improve efficiency of water use, and 5) to improve the diet of grazing animals.

Why reseeding? Reseeding is initiated for two reasons:

- To rehabilitate unproductive areas and put them into production again. Nongrazed areas, because of lack of forage, are a waste of money.
- To accelerate secondary plant succession in areas where natural revegetation is too slow or almost impossible.

When is reseeding necessary and justifiable? Reseeding is needed and justified:

- when percentage of desirable vegetation in the botanical composition is very low (below 18-20%) and plant density is rare.
- when the range site to reseed has the agronomic potential to respond to the treatment (topography, soil quality, etc.).
- when climatic and meteorological conditions are favorable and meet the minimum safe requirements: amount and distribution of rainfall, and frost-free period.
- when the proper seed is available.

How to reseed. There are some basic rules that one has to follow to minimize risks when trying to improve a range by revegetation:

- Choose the best adapted species, native or introduced, based on experimental evidence or on regional experience.
- Select a high quality seed with a high PLS (pure live seed) percentage; do not sacrifice quality for a low price.

- Seed at the beginning of the formal rainy season--when you are sure the rains will continue with regular frequency. Do not plant late in the summer because the plants will not have a chance to get established before the first frosts.
- Use the recommended seeding rates on a PLS basis.
- Try to use a grass seed driller that will assure the best distribution of seed on the ground. If no machinery is available and you have to broadcast, cover the seed lightly with branches from shrubs.
- Broadcast a mixture of seeds separately according to size, i.e., the small seeds (weeping live, clover, bluepanic) should be mixed with fine sand for a better distribution and planted separately from larger or fluffy seeds (side oats, crested wheat, buffel). Drillers usually have different types of boxes for different sizes of seeds and can be calibrated separately for the desired rate.
- Plant grass seeds no more than 1 to 1.5 cm deep in a firm seed bed. If possible, press the soil down lightly on the seeds to make contact between seed and soil. If this is not possible, and broadcast is used, cover lightly with a rake made of shrub branches.
- Do not fertilize at the time of planting since many weeds will take advantage of the fertilizer and will compete strongly with the planted species. It is better to wait until there is an established stand of the seeded species.
- Protect the seeded area from grazing at least during the first growing season. Depending on the density of the stand, the area could be grazed lightly in the second year. Remember: reseeding is expensive and risky, and proper management is necessary for a long-lasting stand.

Range fertilization. Recent increases in prices of fertilizers have limited their use on rangelands. However, areas receiving annual rainfall above 350 mm have responded significantly in terms of forage production, forage nutritive quality, and carrying capacity so that it might still pay to fertilize.

Nitrogen is the most commonly used fertilizer for semiarid and temperate rangelands; however, nitrogen plus phosphorous have demonstrated to be better in some rangelands. For instance, experiments in Central Chihuahua, Mexico, using different levels of N and P and their combinations, indicated that either N or P alone produces more forage than the nontreated areas, but the application of N and P together was much better. By applying 80 kg of N and 50 kg

P_2O_5/ha, forage production increased 132% the first year; crude protein per hectare changed from 31 to 107 kg and phosphorus per hectare increased from 0.42 to 1.46 kg. In this same experiment, 73% of the cost of fertilizers was recovered during the first year (Gonzalez, 1972).

In spite of the "energy crisis" and the high price for fertilizer, it is important to fertilize rangeland because of the growing demand for land and because of the need to intensify as much as possible by getting the maximum out of every hectare of range. That is why fertilizers have an important role in areas where ecological conditions are favorable for their use.

Type of fertilizer, apportionment/ha, time of application, and method of application are some of the aspects to consider once you have had a soil analysis and determined the need to fertilize.

PRIORITIES IN SELECTING IMPROVEMENT PRACTICES

Every ranch has its own problems and own needs for range improvement. All of the range improvement practices discussed are expensive, but each must be considered carefully when bringing rangeland back into production.

The following are some guides to help in the selection of which improvement practice(s) to use.

- First, evaluate the present range condition to determine if an improvement is necessary. Usually ranges in poor or fair condition need some help, but ranches in better condition may need only a change in management.
- Second, define the problem. Brush infestation? Low plant density? Poor botanical composition? Erosion? . . . Establish your priorities. Treat the best sites first because they respond better and faster to treatment. Define which improvement practice(s) to use and, if several are needed, establish your priorities.
- Third, in selecting the practice to improve your ranch, costs are basic. There must be an evaluation of costs in terms of the potential to improve forage production and pay back the expenses. One must be alert to prices and cost fluctuations.
- Finally, once the practice has been selected, plan how to do it: 1) what to use, 2) when to do it, and 3) who is going to do it. Before starting the program define the type of equipment and materials, the time of the year, the number of hectares to treat, and the people responsible for doing the job.

In analyzing all the parameters, a combination of two or more practices (if needed!) may be the most economically productive.

For example, figure 2 is a classic for understanding the priorities for improvement in northern Mexico in relation to the interaction of rain and vegetation. Of the 340 mm average annual rainfall, 40 mm, almost 12%, fall in the winter time when it is not used readily by plants. Of the remaining 300 mm about 100 mm are lost through evapotranspiration and surface runoff; only 200 mm are available for the vegetation. According to studies on over 100 million hectares of rangelands in Central and Northern Mexico (CFAN-CID, 1968), botanical composition in range pastures includes more than 50% of undesirable plants; therefore, at least 125 mm of rain are used by shrubs, forbs, and weeds, and the final 75 mm for desirable forage species--the equivalent of a mere 22% of the total rain recorded.

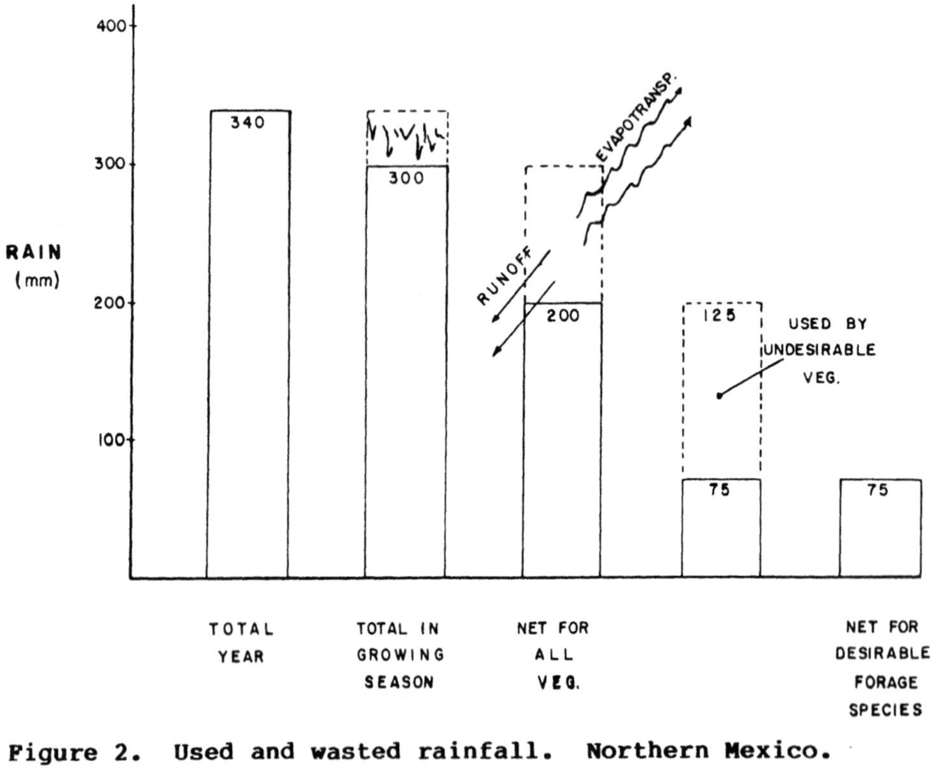

Figure 2. Used and wasted rainfall. Northern Mexico.

This simple rain-vegetation relation shows that range improvement practices should be directed towards: (1) obtaining a better conservation of water and (2) eliminating undesirable vegetation so the limited available rain is used by desirable plants.

ESTIMATED PRODUCTION INCREASE DUE TO IMPROVEMENT PRACTICES

Based on evidence accumulated by research at Rancho Experimental La Campana, in Chihuahua; CIPES, in Carbo, Sonora; and Vaquerias Experimental ranch, in Jalisco, as well as field data in different ranches in northern and central Mexico, the increase in range production due to several improvement practices is illustrated in table 2 (Gonzalez, 1977). This table includes averages of results of different vegetative types where improvement practices have been conducted. These indicate the benefits in a 4- to 6-year period (short-term) and the projected increases that can be obtained in 8 to 12 years (long-term).

TABLE 2. ACTUAL AND POTENTIAL INCREASE IN FORAGE PRODUCTION IN RANGELANDS OF CENTRAL AND NORTHERN MEXICO

		Potential	
	Present	Short-term (actual)	Long-term (projected)
Forage Prod. kg DM/ha	210	420	580
% increase due		100	176
Grazing systems and intensities		40	46
Soil and water conservation		15	40
Brush control		30	60
Range reseeding		15	30
Carrying capacity Ha/AU	20.6	10.3	7.5

Source: M. Gonzalez (1977).

In the short-term period, the forage production of the range increased 100% from an average of 210 kg DM/ha to 420 kg DM/ha. This increment was due to the adjustment of the carrying capacity of the ranches and the modification of their grazing systems (40%); to soil and water conservation (15%); to brush control (30%); and range reseeding (15%). These improvements increased the yearly carrying capacity of the range from 20.6 ha/AU to 10.3 ha/AU.

The projected increase in production for the long-term period was based on the continuation of adequate management and improved range conditions. For an 8- to 12-year period, it was estimated that forage production would reach 580 kg DM/ha or 176% above the present. Carrying capacity would be 7.5 ha--almost triple of the present one.

The range improvement will be reflected by increased livestock production, both in quantity and quality, because of better forage in the pastures.

REFERENCES

Cable, D. R. 1965. Damage to mesquite, lehmann lovegrass and blackgrama by a hot June fire. J. Range Mgt. 18:326.

Fierro, L. C., F. Gomez and M. H. Gonzalez. 1979. Control biologico de arbustivas indeseables utilizando ganado caprino. Bol. Pastizales, Vol. X No. 3 Rancho Exp. La Campana INIP-SARH, Mexico.

Glendening G. and H. A. Paulsen. 1955. Reproduction and establishment of velvet mesquite and relation to invasion of semidesert grasslands. USDA Tech. Bull. 1127.

Gomez, F. and M. H. Gonzalez. 1976. Evaluacion de cinco mezclas de herbicidas en elcombate de chaparrillo, Eycenhardtia spinosa. Bol. Pastizales Vol III, No. 2 Rancho Exp. La Campana INIP-SARH. Mexico.

Gomez, F. and M. H. Gonzalez. 1978. Efecto del chapeo mecanico en el incremento de un pastizal invadido por arbustivas Bol. Pastizales, Vol. IX, No. 5 Rancho Exp. La Campana INIP-SARH, Mexico.

Gonzalez, M. H. 1972. Aumentos en la produccion de carne con la fertilizacion de un pastizal. Bol. Pastizales Vol. III, No. 1 Rancho Exp. La Campana, INIP-SARH, Mexico.

Gonzalez, M. H. 1976. El potencial de las tierras no cultivables en la produccion de alimentos. Ingenieria Agronomica, Vol. II, No. 2, Col. Ing. Agr. de Mex. AC. Mex.

Martinez, J. 1959. Pruebas de infiltracion en un pastizal mediano-abierto. Tesis. Esc. de Ganaderia, Univ. de Chihuahua, Mexico.

Reynolds, H. G. and J. Bohning. 1956. Effects of burning on a desert grass shrub range in southern Arizona. Ecology 37.

Sanchez, A. 1972. Efecto de la condicion de pastizal en la infiltracion del agua de lluvia Bol. Pastizales, Vol. III, No. 3. Rancho Exp. La Campana INIP-SARH, Mexico.

Vallentine, J. F. 1977. Range Development and Improvements. Brigham Young Univ. Press. Provo, Utah.

White, L. D. 1969. Effects of a wild fire on several desert grass land shrub species. J. Range Mgt. 22.

32

WHAT TYPE OF GRAZING SYSTEM FOR MY RANCH . . . ?

Martin H. Gonzalez

INTRODUCTION

Ranchers have developed a growing interest in grazing systems during the last decade. The introduction of the new, "intensive" systems has received more attention from producers and researchers than the traditional, continuous, and extensive rotation systems we knew.

A never-ending controversy could evolve from a group of ranchers discussing the pros and cons of each one's favorite system. All of the systems have been used under different conditions and have their particular advocates but, at the same time, everybody knows that a specific system cannot be universally implemented. Each ranch has its peculiar objectives, ecological features, market situations, financial capabilities, and the personal preference of its owner. As Penfield (1982) wrote, "The most important element in a grazing system involves a commitment by you to make it work." And at the same time the rancher must be committed to working with nature not against it--a principle to be kept in mind when selecting a grazing system.

Both native rangelands and cultivated forage communities and their biological and economic ecosystems lead ultimately to the production of livestock products used by man. Most lands under livestock or wildlife production have unbalanced ecosystems in which one or more of the major constituents (plant, animal, or man) has a chronic or seasonal lack of nutrients, water, or other environmental requirement (Whythe, 1978). It is the responsibility of the soil and crop scientists to help overcome these deficiencies. But it is the rancher, the manager of the range, who has to decide whether the proposals of the scientists and the new "discoveries" are economically acceptable; the rancher, after all, is part of the ecosystem in which he lives with his domestic livestock.

TYPES OF GRAZING SYSTEMS

The number and variation of grazing systems is almost infinite. Every ranch has its peculiar characteristics--

even adjoining ranches with similar ecological conditions differ in objectives, management, and financial capabilities. For this presentation we will consider some of the most common grazing systems, knowing that within each of them the variations are numerous. Any rancher can identify with one of these models and design his own system.

Continuous Grazing

Continuous grazing is the constant use of forage on a given area, either throughout the year or during most of the growing period. This type of grazing does not always result in range decline (overgrazing). Light, continuous use results in range improvement, but returns per hectare are lower than those obtained from a four-pasture, deferred-rotation grazing (Merrill, 1980).

Under continuous grazing, stocking rates are set with only minor adjustments through the year. The number of animals does not vary greatly except during long drought periods (Kothmann, 1980). Under continuous grazing, stocking rate is the only variable the manager can adjust; this allows little flexibility in responding to drought seasons (Vaughan-Evans, 1978).

Rotation System

A rotation system moves animals from one pasture to another according to a fixed schedule. Within this system are the extensive (or non-intensive) and the intensive systems. In the extensive system, several pastures are grazed and one is resting for a certain number of months. The rest period is rotated so that the same pasture has the rest period during different months each year. In the intensive system, all the animals graze one pasture (for a short period) while all the other pastures rest.

Deferred rotation. Under deferred rotation, each pasture is deferred during each season (Ambolt, 1973). The design includes a fixed number of pastures for each herd of livestock. The goal is to improve the vegetation condition of the resting unit, but the length and intensity of grazing on the remaining units should not be so long as to cause range deterioration (Merrill, 1980). Carrying capacity is determined by the total land area in the system and should be set conservatively. Deferment periods usually vary from three to six months, but may be longer (Kothmann, 1980).

The purpose of such a system is two-fold: to improve the vigor and production of forage, which in turn increases the desirable forage species, the carrying capacity, and the animal response to the improved forage. The result is higher returns per unit area. Figure 1 shows the Merrill 4-pasture deferment-rotation grazing as a good example of this system.

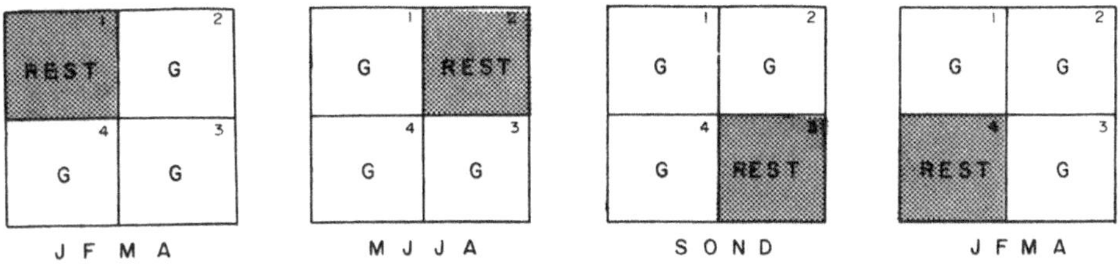

Figure 1. Merrill's four-pasture, deferred-rotation grazing system

Rest rotation. In the rest-rotation system, a pasture is not grazed for a full year and the rest period rotates among pastures. Deferments are provided for seed production and for seedling establishment. The number of pastures in the system may be from two to five (Hormay, 1970). The carrying capacity is based on that portion of the range that is grazed each year. Generally a much smaller percentage of the total area is available to grazing during the growing season (Kothmann, 1980).

These two basic nonintensive rotation systems sometimes are combined to take advantage of both the rest and the deferment on a varying number of pastures.

A disadvantage for some ranches is that pastures should be reasonably uniform in both size and carrying capacity. This often requires additional fencing and watering (figure 2).

Intensive systems. The two most common types of intensive grazing systems are the high intensity-low frequency, and the short-duration grazing. Both of these systems are based on grazing all the animals in one pasture while the other pastures rest. The difference between the two systems is the length of time given to the grazing and rest periods. The advantages of the long rest period are: it provides maximum vegetation improvement, eliminates seasonal rotations, and concentrates the livestock in one area which makes handling and working easier. The disadvantages of the intensive systems are: an increase in the need for water storage and daily output on individual pastures, generally lower livestock gains per head, and a possible parasite problem if sheep and goats are present (Merrill, 1980).

High intensity-low frequency (HILF)--When designing an HILF system, the forage species present are evaluated and a system that favors the plants we want to produce is designed. The time plants need for recovery from grazing determines the rest periods needed in the system (figure 3).

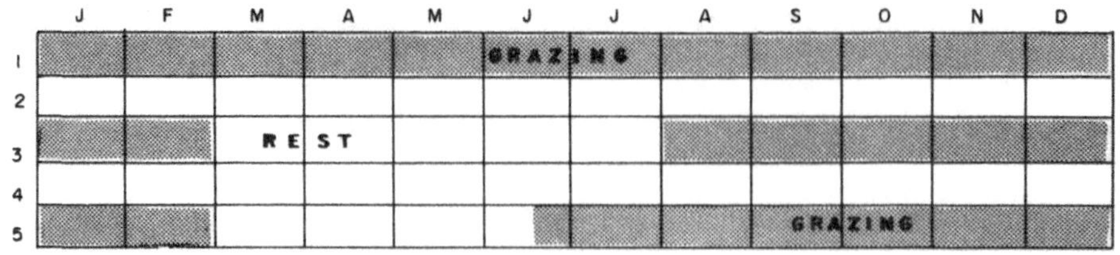

Figure 2. Rest rotation system with five pastures

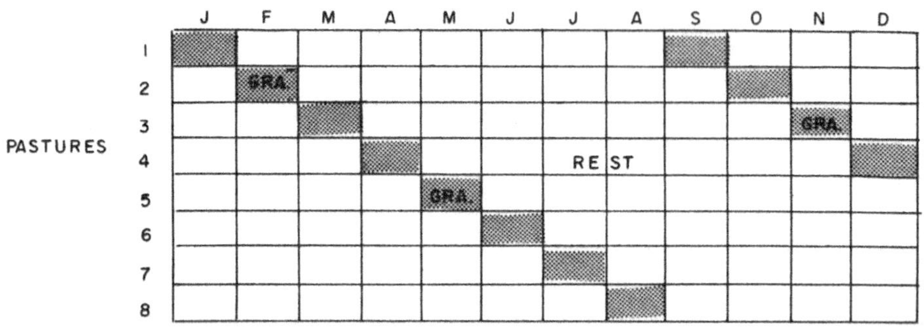

Figure 3. High-intensity--low-frequency grazing system

The HILF systems are based on the use of intensive grazing periods with relatively long rests. At least three pastures are required per herd. The grazing periods are generally more than two weeks, rest periods longer than 60 days, and grazing cycles over 90 days. Carrying capacity is based on the total land area in the system and moderation in length of grazing cycle to prevent an excessive decline in animal production.

Short duration--Like the HILF, this system also must have three or more pastures per herd, but the grazing and rest periods are shorter. Grazing periods should be less than 14 days (preferably 7 days or fewer and rest periods should vary from 30 to 60 days, but never exceeding 60 days (figure 4).

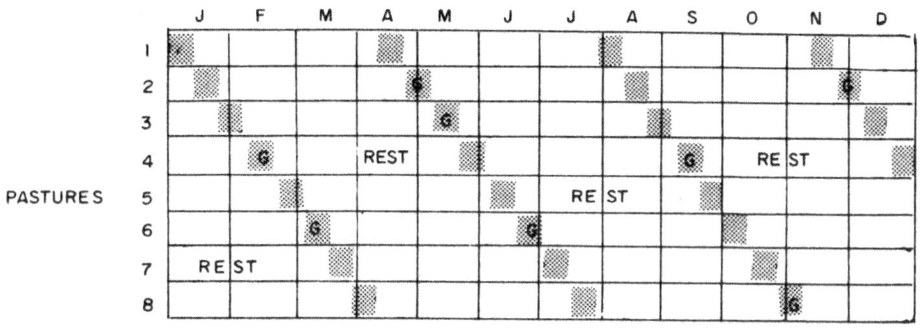

Figure 4. Short-duration grazing system

Grazing cycles are generally short enough to allow six or more full rotations per year (Savory, 1979). Carrying capacity is based on the total land area in the system. Because of the high degree of control over both frequency and intensity of defoliation (greater than under other systems), a higher degree of use in this system is possible without detriment to plant or animal production (Kothmann, 1980).

PARAMETERS FOR THE EVALUATION OF GRAZING SYSTEMS

Objectives

Apart from the specific objectives that a rancher might have, there are a number of general objectives for selecting his grazing system. Some of the most important are the following:

- To obtain a better livestock grazing distribution and, consequently, a more uniform utilization of the range or pasture.
- To maintain or improve forage density.
- To maintain or improve the botanical composition.
- To meet the nutritional needs of the grazing animals.

All of the objectives are affected by the major objective--to increase the overall efficiency of the operation.

Considerations

Thee are some basic premises that the rancher must consider when choosing his grazing system. The following deserve his attention:

- Viability of the system. Will all factors combined in its planning make it work?
- Cost of additional infrastructure (fencing, watering, pens, etc.) and labor involved.
- Minimizing the effect of adverse weather.
- Stocking rate.
- Pasture size.
- Location of water, salt, and supplements.
- The number of pastures in the system.
- The number of herds per system, if more than one system is to be used.
- The length of the grazing cycle and the rest period--the heart of the system.

Consequences

A well-planned and implemented grazing system will have some beneficial consequences. On the contrary, if the development of the system is not in agreement with the basic objectives and considerations, the results are likely to be poor. Some of the positive effects of a well-designed and managed system are:

Effects on the vegetation.

- The system will produce more forage.
- Seed production is increased.
- Botanical composition of pastures is improved.
- Chemical composition of forage plants (animal's diet) is improved.
- Over-all range resources are improved by the rapid increase in range condition.
- The amount of forage consumed by livestock due to nonselective grazing is increased.

Effects on the animal.

- Minimal or no stress on animals.
- Improved animal performance.
- Less handling of animals in general.
- Reduced supplemental feeding.
- Minimized labor costs.
- Higher individual weight gain.
- Better calving intervals.
- Increased stocking rate.
- Interrupted disease and parasite cycle.

A possible added benefit would be the increase in wildlife population and an improvement in their condition.

PLANNING THE SYSTEM FOR YOUR RANCH

Assuming that you have decided to develop your own grazing system or to adopt one of many models available, a logical sequence of action is recommended.

Inventory and study carefully your range resource.

- Vegetative types
- Range sites within vegetative types
- Your pastures:
  - a Number
  - b Size
  - c Forage production
  - d Season of use
  - e Present range condition

Reaffirm your objectives for:

- Type of livestock operation
  - a Cow-calf
  - b Feeder cattle
  - c Combination
- Type of grazing system
  - a Extensive rotation
    1. Deferred rotation
    2. Rest rotation
  - b Intensive rotation
    1. High intensity-low frequency
    2. Short-duration grazing

Design your grazing system according to:

- Your objectives
- Present number and size of pastures
- Projected number and size of new pastures
- Expenses involved in additional proposed infrastructure
  - a Fencing
  - b Water development
  - c Labor

Expand the management plan.

- For the range-pastures
  - a Grazing cycle
  - b Range improvement practices needed
- For the livestock
  - a Breeding and calving
  - b Nutrition program
    - 1 Range supplements for cows
    - 2 Possibilities of creep feeding
    - 3 For replacement heifers
  - c Sanitation program
  - d Marketing
- For monitoring
  - a Forage utilization
  - b Range condition and trend
  - c Animal production

1 Percentage calf crop
2 Weaning weights

d Market situations

We must emphasize that the best grazing system is the one best adapted to your particular interest and the conditions on your ranch. Even if economic resources and infrastructure are limited, a grazing system can be improved through better management. Work with what you have and follow the basics for the grazing system that best suits your ranching business.

REFERENCES

Hormay, A. 1970. Principles of rest-rotation grazing and multiple use land management. USDA, F. S. Training Test 4(2200).

Merrill, L. B. 1980. Considerations necessary in selecting and developing a grazing system. What are the alternatives? Proc. Grazing Mgt. Systems for S. W. Rangelands Symposium. The Range Impr. Task Force. New Mexico State Univ., Las Cruces, N. M.

Kothmann, M. M. 1980. Integrating livestock needs to the grazing system. Proc. Grazing Mgt. Systems for S. W. Rangelands Symposium. The Range Impr. Task Force. New Mexico State Univ., Las Cruces, N. M.

Penfield, S. 1982. Are you ready for a grazing system? Rangelands 4(1). Soc. Range Mgt.

Savory, A. 1979. Range management principles underlying short-duration grazing. Beef Cattle Sci. Handbook. Agr. Services Found., Clovis, Ca. 16:375.

Vaughan-Evans, R. H. 1978. Short-duration grazing improves veld conservation and farm income in the Oue group of I.C.A.'s Cenex reports. Mimeo. Rhodesia.

Wambolt, Carl L. 1973. Range grazing systems. Coop. Ext. Serv. Bull. 340. Montana State Univ.

Part 7

MEAT AND MEAT PROCESSING

33

CURRENT STATUS OF MEAT IN THE DIET AND ITS RELATIONSHIP TO HUMAN HEALTH

B. C. Breidenstein

Throughout the recorded history of man, and in all likelihood prior to that time, the primary diet/health concerns revolved around the need to obtain sufficient nutrients to maintain life. The priority status associated with consumption of meat and other animal products is well documented in the Old Testament of the Bible. Whether this status can be attributed to the perceived nutrient contribution to the diet or to taste appeal is not clear, but one could well speculate that it might have been a combination of those two reasons.

Nutrient deficiencies have been observed and recognized as such for several hundred years. The use of certain foods in the prevention or correction of these deficiencies has been commonly practiced. For example, scurvy, a symptom of vitamin C deficiency, was a common problem during long ocean voyages of European explorers and was found to be prevented by the consumption of citrus fruits such as limes, which are now known to be an excellent source of vitamin C. Nevertheless, the scientific study of man's nutrient needs and the food sources for satisfying those needs is relatively recent. Those "needs" of the human body are, however, not definable as a single quantifiable entity. Rather, age, sex, level of physical activity, ambient conditions, genetics, and general health factors are among the factors known to influence nutrient requirements. Further complicating the issue is the fact that certain nutrients exert a sparing effect on the need for other nutrients. The bioavailability of nutrients from different sources also varies. For example, the heme iron contained in meat is known to be one of the most bioavailable forms of the mineral. In addition, its presence in the diet makes the less available plant sources of iron more available to humans. On the other hand, consumption of an excess of some nutrients may result in metabolic waste of others.

The human not only has a threshold of minimum requirements, but in many cases, is also very tolerant of an excess beyond those requirements. It is generally recognized, for example, that approximately 200 mg of sodium are required daily for humans. Yet about 70% to 80% of the U.S. population is believed able to tolerate daily intakes of sodium of 10 to 20 times that amount without adverse effects. The magnitude of the range between the minimum requirement for any person and the intake level resulting in adverse effects is undoubtedly different for different nutrients--and is probably influenced by the same factors that affect the minimum requirement levels.

In spite of the continuous challenges to the desirability of its inclusion in the diet, per capita meat consumption in the U.S. has increased steadily during the current century. It is probably that, during at least the early part of that time period, its popularity in the diet was due largely to its taste appeal and secondarily to a perceived preferential source of nutrients. During the past several decades, however, meat has become increasingly recognized as a premiere dietary source of high-quality protein, B vitamins, and minerals. The reputation of meat as a source of these essential nutrients is rarely, if ever, seriously challenged.

Red Meat Consumption* Pounds Per Capita/Year

Decade	Beef	Veal	Lamb & mutton	Pork Exc. lard	Total red meat
Last half 1940s	63.5	10.2	5.7	69.5	148.9
Last half 1950s	75.3	8.1	4.3	66.0	153.7
Last half 1960s	98.7	4.8	4.3	63.3	171.1
Last half 1970s	117.2	2.9	2.5	64.1	186.7
1980	105.6	1.8	1.6	75.1	184.1

*Carcass weight disappearance.

Probably because the basal statistics are readily available, meat consumption figures are frequently expressed as total carcass weight of livestock slaughtered, divided by that year's total population. It is true that meat consumption per capita expressed on that basis increased during the current century, but obscured by such a figure is the change in carcass composition that has occurred in response to consumer demand. For example, prior to the wide availability of plant source oils swine were viewed as somewhat dual-purpose animals. They were prized not only for their ability to produce appetite-satisfying meat in a wide variety of products, but also for their ability to convert plant materials to fat, thereby representing a primary source of edible fats in the U.S. The rapid increase in the supply of relatively low-cost vegetable oils since World War II resulted in a depressed price for lard and the diminished status of swine as a source of edible fats. In recognition of this economic reality, the swine industry has, since the early 1950s, been very actively engaged in highly successful genetic, nutritional, and management programs designed to maximize muscle and minimize fat production per animal marketed. As a result of these efforts, lard production per head slaughtered in 1977 has been reduced to about half the level produced in 1950. This has been accomplished simultaneously with ever-closer trimming of fat from the cuts offered to the consumer.

While cattle and lamb fats have been, and remain, an important by-product of their slaughter, the reliance on them as converters of plants to fat has never been viewed as an economic asset as was the case historically for swine. However, consumers undoubtedly purchase all red meats primarily for their muscle content and consider the fat contained as an economic loss. Thus, the consumer pressure for increased leanness of meat products, which expressed itself forcefully in pork consumption in the early 1950s, became fully recognized in the beef and lamb consumption perhaps a decade later. As a result of this awareness on the part of the cattle and sheep industry, their response was similar to that of the swine industry. Even a casual observer of carcass composition must recognize the dramatic changes toward less fat and more lean that have taken place in the meat supply during the past three decades. As a result, if one assumes that little of the fat attached to retail cuts is actually consumed, then the true consumption of <u>meat</u> has increased to a greater extent than indicated by carcass weight disappearance data.

It is obvious from observing the per capita consumption, which has increased by more than 25% over the past three and one-half decades, that red meat is viewed by consumers as a desirable dietary component. Pork consumption has remained relatively constant over that time period while lamb and mutton and veal consumption has declined. Those declines, however, have been more than compensated for by increases of more than 80% of beef consumption.

The diet/health controversy, insofar as red meat is concerned, is not a new issue. During the first two decades of this century, the annual per capita consumption of red meat and lard declined by more than 27 pounds, with beef consumption accounting for nearly 23 pounds of that total. Thomas E. Wilson, president of the Institute of American Meat Packers speaking before the American National Livestock Association on January 12, 1922, said: "One of the outstanding factors operative for the last two decades has been the fostering and development by propaganda of an impression that meat is harmful to the health. In this connection, meat has been misrepresented in a damaging fashion and in a widespread way. The food value of meat has been mis-stated, its place in the diet minimized, and its healthfulness challenged. People are naturally sensitive to any propaganda relating to their health. They are quick to avoid foods said to be harmful. In this way the public, no doubt, has been materially influenced. Almost every other food interest has made invidious comparisons of its products with ours to the disparagement of meat. Many of these comparisons have not reflected the truth from a scientific standpoint." Characteristic of such information is an article from the April 1905 issue of The Ladies Home Journal entitled "Why I Do Not Believe in Much Meat," an exerpt from which is: "Nourishing diet, in the minds of most people, is a meat diet, which soon upsets the digestive organs and makes the person a martyr to rheumatism and gout. Women grow constipated and have cloudy complexions; men red-faced, or very thin, according to their various resisting powers."

Mr. Wilson suggested: "Scientific data wherewith to correct adverse propaganda should be collected, compiled, and disseminated showing the high value of meat in the diet. Such information should be circulated among dietitians, physicians, hospitals, teachers, home demonstrators, household editors, agricultural colleges and others."

In its January 21, 1922, issue, the National Provisioner reported on the creation of the National Live Stock and Meat Board stating that it "shall be created to conduct and direct an adequate educational campaign counteracting the widespread and insidious propaganda against the food value of meat and disseminating through all possible avenues correct information about meat in the diet."

Differences of opinion among scientists is to be expected. One of the valuable attributes of a scientist is the ability to think in independent and original terms. Such traits lead to the expansion of the frontiers of scientific knowledge and, as such, are a valuable asset of the scientific community and indeed to the community at large. It should not be surprising that such characteristics make it more difficult to reach even a consensus opinion, let alone a unanimous one, regarding such a complex issue as diet/health. Some recent examples of truly dedicated, knowledgeable, and well-meaning scientists reaching differing conclusions from observing a common body of knowledge reflect that difficulty. Dr. D. M. Hegsted of the Harvard School of Public Health in a press release on 2/14/77 regarding the second edition of dietary goals said: "The diet of the American people has become increasingly rich--rich in meat, other sources of saturated fat and cholesterol, and in sugar. It should be emphasized that this diet which affluent people generally consume is everywhere associated with a similar disease pattern--high rates of ischemic heart disease, certain forms of cancer, diabetes and obesity. These are the major causes of death and disability in the U.S."

Only some two years later in a 1979 report from the Surgeon General appears the statement that "the population of the United States has never been healthier" and "mortality rate for coronary heart disease has declined 20% during the last 20 years and is currently falling at 2% per year." The following year, 1980, the Food and Nutrition Board of the National Academy of Science said, "The American food supply on the whole is nutritious and provides adequate nutrients to protect essentially all healthy Americans from deficiency diseases" and "the excellent state of health of the American people could not have been achieved unless most people made wise food choices."

It should be recognized that a diet is a massively complex and variable combination of a multitude of nutrients and other components. Ingested by the human, which is an organism of an extremely complex and nonuniform nature, those dietary ingredients are broken down into simpler forms by various biochemical and physical means to exert their ultimate effect on the human organism. The oversimplification of the role of diet or even a single food entity on the health and well-being of the human leads to potentially dangerous conclusions. It is somewhat analogous to characterizing a single tree while not recognizing the existence of the forest of which the single tree is only a small part.

There are those in the scientific community who seem willing to make sweeping dietary recommendations on the premise that such changes involve an acceptable risk level for adverse effects and may possible result in benefits.

It is disturbing that such a segment of the scientific community seems to be gaining in influence, especially in the popular press. It would seem that challenges to the safety of our environment or our food supply whether or not properly founded in fact, are deemed newsworthy and hence receive widespread media distribution. On the other hand, a statement such as made by the Food and Nutrition Board in 1980 that "the excellent state of health of the American people could not have been achieved unless most people made wise food choices," is viewed as not very dramatic and hence receives little attention by the news media.

Although the diet/health controversy regarding meat has not subsided over the eight decades of this century, there has been, and continues to be, an increasing body of scientific knowledge reflecting the positive contributions of meat to the diet of man. Meat has long been recognized as an excellent source of high quality protein. A 3 oz serving of meat supplies more than half the daily human requirement for protein. The fact that it is a high quality protein simply means that it contains all of the essential amino acids in amounts appropriate for best use by the human. It therefore needs no supplementation from other proteins to be effectively utilized.

Meat is an excellent source of the B vitamins and ranks as the principal dietary source of most of them. Pork is an exemplary source of thiamin with a 3 oz serving supplying well over 50% of the U.S. RDA. A 3 oz serving of meat supplies about 15% of the U.S. RDA for riboflavin and more than 15% for niacin. Meat is also recognized as a good source of pyridoxine (B_6) and vitamin B_{12}.

Meat is a good dietary source of minerals, especially of iron. The heme iron in meat is highly available to the human and its presence also aids in the assimilation of the nonheme iron present in the diet from plant sources. Meat also supplies about 15% of the U.S. RDA for the trace mineral zinc.

Because meat contains such significant quantities of these important nutrients in relation to its caloric contribution of about 200 calories, it is described as a food of high nutrient density.

There are many human conditions that the scientific community and/or the popular press have purported to be related to the diet. It is the intent of this discussion to limit such considerations to those having implications for the livestock and meat industry and likewise perceived to be important by the public-at-large. Dietary implications regarding obesity, cardiovascular diseases, hypertension, and cancer are discussed.

Probably obesity is a condition that excites the least controversy. The Food and Nutrition Board contends that obesity, or excess fatness, is the most common form of malnutrition in the Western nations of the world. It has a multiple etiology and is influenced by neurohumoral, endocrine, metabolic, and social factors. It is generally recognized that, in many persons, obesity is associated with significant increases in morbidity and mortality from such diseases as hypertension, diabetes, coronary heart disease, and gall bladder disease, and that mortality from these diseases is reduced with weight reduction. Obesity results from the failure, for whatever reason, to balance energy intake with energy expenditure. Such a balance is difficult to achieve when energy expenditure is low, as is generally true for the adult population of the United States.

Because meat is of high nutrient density, it fits very well into either a weight-reduction diet or a diet low in energy and designed to maintain body weight at the desired level. Dr. Maria Simonson of John Hopkins has compared response to vegetarian vs meat diets in weigth-reduction programs. She reports that those on a meat diet 1) lost weight somewhat more slowly, though not significantly so, 2) were more consistent in weight loss, 3) had fewer drop-outs and less cheating, 4) had no feelings of hunger, 5) had far fewer physical and psychological problems, 6) had no anemias, 7) improved their work productivity and work efficiency (by 13%), and 8) showed excellent weight maintenance for up to one year for those who achieved goal weight. To obtain the necessary nutrient intake in diets of reduced energy level, it is obvious that nutrient-dense foods, such as meat, must play a prominant role.

Arteriosclerosis and its complications, i.e., coronary artery disease, stroke, and peripheral vascular disease are the leading causes of death in the U.S. Cardiovascular disease is the leading health problem accounting for about 50% of the deaths in this country. Mortality rates from cardiovascular disease, as is true for other degenerative diseases, increase sharply with increasing age. Several risk factors for cardiovascular disease have been identified through epidemiological studies. Among these is hypercholesterolemia or high levels of cholesterol in the blood serum. Risk factors are those factors found to be statistically associated with an increased incidence of the disease and cannot, without independent evidence, be considered causative agents of the disease. In any event, the identification of high serum cholesterol levels as a risk factor has led to concern about cholesterol and fat, especially saturated fat, in the diet. Cholesterol is found primarily in foods of animal origin with a 3 oz serving of lean meat containing about 75 mg. The human synthesizes between 800 and 1500 mg of this essential metabolite per day. Dietary cholesterol is poorly absorbed by man, with only 10% to 50% of dietary intake actually absorbed.

Diet modification with respect to level and kind of fat and the amount of cholesterol has been shown to alter serum lipid and lipoprotein concentrations of subjects in metabolic wards under rigid dietary control. A high intake of saturated fat as a percentage of calories is a major factor in elevating serum cholesterol and LDL (low density lipoproteins) levels and the reverse is true for a high intake of polyunsaturated fat. Studies among free-living subjects, however, indicate that such dietary modification is only about 60% as effective in altering serum cholesterol and lipoprotein levels as was the case for the controlled-metabolic-ward studies. Obviously, other factors still to be identified, in addition to diet, influence serum lipid values of free-living persons in an as yet unpredictable manner.

Intervention trials utilizing diet modification to alter the incidence of coronary artery disease and mortality in middle-aged men have generally been negative. It has not been proven that lowering serum cholesterol and LDL levels by dietary intervention will consistently affect the rate of new coronary events. The Food and Nutrition Board "recommends the dietary fat content be adjusted to a level appropriate for the caloric requirements of the individual" and "that sedentary persons attempting to achieve weight control may be well-advised to reduce the caloric density of their diets by reduction of dietary fat." Regarding dietary cholesterol, the Food and Nutrition Board says, "no significant correlation between cholesterol intake and serum cholesterol concentration has been shown in free-living persons in this country" and "the board makes no specific recommendations about dietary cholesterol for the healthy person."

A genetic predisposition for hypertension is believed to exist in the U.S. for about 15% to 20% of the population. While sodium has been implicated in this condition, an association between blood pressure and salt intake has not been demonstrated within selected U.S. populations.

Nevertheless, dietary sodium reduction is a frequently made recommendation for those persons with elevated blood pressure.

Sodium appears to exist in red meats in a reasonably constant relationship to protein. Red meat contains about 3.5 mg of sodium per gram of protein. Thus a 3 oz serving of fresh meat having 16% protein would contain about 48 mg of sodium. It can, therefore, be used very effectively in diets restricted in sodium.

Salt (NaCl) has been widely used for centuries as a preservative for meats in addition to its flavor-enhancement properties. Salt is the primary contributor to the sodium content of processed meats. At this time, the meat industry has no economically viable preservative alternative to the use of sodium chloride. Research is currently in progress to develop alternative compounds, combinations, etc., to permit reduction in the amount of sodium used. It is probable that some reduction in amount will be possible as a result of such research, with potential reductions estimated at about 25%.

Processed meats, such as ham, bacon, and a myriad of different sausages, have provided a multitude of dietary variations and have added greatly to our food enjoyment. They have greatly enhanced the overall palatability of the American diet. It seems that the first requirement of a suitable diet is frequently overlooked; namely that it be palatable and otherwise appealing to the consumer. Processed meats have much to contribute in that regard as well as making a substantial contribution to nutrient requirements.

The issue of diet/cancer was recently addressed by a special committee of the National Research Council reported on June 16, 1982, in a document entitled Diet, Nutrition and Cancer. As acknowledged by that committee, the extent of the relationship between diet and cancer is not precisely understood at this time. That committee further stated: "It is not now possible, and may never be possible, to specify a diet that would protect everyone against all forms of cancer. Nevertheless, the committee believes that it is possible, on the basis of current evidence, to formulate interim guidelines that are both consistent with good nutritional practices and likely to reduce the risk of cancer."

Two of those interim guidelines have potential for direct impact on the red meat industry: 1) the consumption of both saturated and unsaturated fats be reduced in the average U.S. diet from the current level of about 40% of calories to about 30% and 2) the consumption of food preserved by salt curing (including salt pickling) or smoking be minimized.

Regarding dietary fat, it is of interest to note that the per capita consumption of animal fats has declined over the past 70 years by about 7%, but the per capita consumption of vegetable fats has increased by 180%.

As consumers consider reduction in caloric intake, they should be fully aware of the nutrient contributions made by the fat-containing foods they consume. Animal-source foods are often nutrient-dense, providing significant quantities of nutrients in relation to calories contributed. Eliminating or reducing such foods because they contain fat may result in a deficiency of other essential nutrients provided by those foods.

To further illustrate this point, the U.S. Department of Agriculture's Nationwide Food Consumption Survey 1977-78 has shown that women from 20 to 50 years of age did not receive the Recommended Dietary Allowances for the following nutrients: vitamin B_6 (78% of RDA), folacin (71%), iron (77%), and zinc (69%). This short-fall occurred with a dietary pattern in which 42% of the calories originated from fat. It seems prudent to suggest that any change in food consumption aimed at reducing dietary fat should be made only after consideration of its effect on other essential nutrients, especially those already perilously low.

The committee's recommendation that the consumption of salt cured or smoked food be minimized appears to be based primarily on epidemiological data. It is presumed by the committee that such foods contribute to the incidence of esophageal and gastric cancer, especially gastric cancer. This recommendation is made despite the acknowledgement by the committee that gastric cancer in the U.S. is low and is decreasing. It has been described by others as "almost a vanishing disease." That declining incidence has occurred during a time period in which per capita processed-meat consumption has increased by about 50%.

The transfer of associations revealed by epidemiological studies form extremely diverse life styles would seem to be tenuous at best. In this case, populations in China, Japan, and Iceland that consumed diets containing salt-cured or smoked foods were referenced as having high incidences of esophageal and stomach cancer. The body of the committee's report identifies salt-pickled vegetables as being widely consumed in China and states that they are frequently contaminated by fungi. The committee acknowledges that the N-Nitroso compounds (carcinogens) contained in such foods were believed to originate with the fungus contaminants. To suggest that this food consumption pattern and its association with a high incidence of stomach cancer is useful in establishing recommendations for the U.S. population regarding processed-met consumption would seem, at best, questionable.

Red meat continues to be a vital contributor to the satisfaction of the human nutrient requirements in the U.S. The recommendation to consume a balanced and varied diet, including foods from all food groups, remains the most effective approach to good nutrition. The red meat industry takes justifiable pride in its contribution to the current health condition described by the Surgeon General of the United States, "The population of the United States has never been healthier."

Finally, the Food and Nutrition Board of the National Academy of Sciences, since 1941 the nation's recognized authoritative voice regarding nutrition/health, expressed its concern in 1980 over excessive hope and fears in many current attitudes toward food and nutrition. The Board said, "Sound nutrition is not a panacea... Good food that provides appropriate proportions of nutrients should not be regarded as a poison, a medicine, or a talisman, IT SHOULD BE EATEN AND ENJOYED."

34

PRINCIPLES OF MEAT SCIENCE AFFECTING LIVESTOCK PRODUCERS: MICROSTRUCTURE

Donald M. Kinsman

INTRODUCTION

The adage "We are what we eat" is a truism for the human race as well as it is for the livestock we produce. If we wish a high-protein result, then we consume a high-protein food; if we wish high energy, we consume that source, assuming each is available to us at a cost we can afford. Fortunately, we generally ingest a balanced diet from many sources, and we usually eat those things we like most, such as meat. Meat supplies a complete protein, chiefly from muscle; ample energy, from the accompanying fat; vitamins, especially the B complex; and minerals, most certainly iron. In addition meat can provide palatability which is described as a combination of appealing tenderness, juiciness, and flavor, plus the satiety value or "lasting value" that provides "staying power." The fact that meat is highly digestible (93 - 97%) makes it especially valuable and a highly efficient source of assimilative nutrients. This is all documented, but we do need to remind ourselves, the nation, and the world that it is a nutrient-paced product. As our consumers become more and more remote from the production scene, it becomes increasingly important that we call these facts to their minds.

Furthermore, if we are to continue to produce this commodity, we must have an appreciation of how it is manufactured. Our urban citizens may believe that it grows in cellophane wrappers. We, of course, have watched it grow from "conception to consumption" and know the care and costs involved. But do we really know how it grows? Certainly we know that feed goes in one end, waste products out the other, and the difference is used for maintenance and growth. But what are these growth units? How are they formed? What do they contribute to the end-product?

MUSCLE TISSUE

The chief component of meat is muscle, which the consumer generally refers to as lean. There are three types of muscle:

- Striated (sometimes called voluntary or skeletal) which represents the greatest amount of body muscle.
- Nonstriated (also referred to as smooth, visceral, or involuntary muscle) is found only in the body linings, such as the digestive system.
- Cardiac or striated-involuntary muscle is peculiar to the heart only.

We will confine our remarks to the striated muscle. Without becoming too involved with the ultrastructure of muscle, let us take a broad overview of this tissue. There are more than 600 muscles in the animal body, which vary decidedly in size, shape, activity, and function. They are attached to bone by tendon and, of course, have a blood supply and nerve supply.

Starting with the smallest physical structure, the myofilament is composed of the proteins actin and myosin, which are vitally involved in muscle activity - contraction and relaxation - in the living animal. These, in combination, form myofibrils, which course through the muscle fiber or muscle cell in conjunction with the cell fluid or sarcoplasm, nuclei, and other cellular inclusions (figure 1).

Numerous individual muscle fibers (cells) in turn combine as muscle bundles, and these bundles or fasciculi compound to form the approximately 600 individual muscles attached to the skeleton. Yes, "mighty beasts from little myofibrils grow."

CONNECTIVE TISSUE

Each of these units--muscle fibers, muscle bundles, and muscles proper--have connective tissue associated with them. As the term implies, this tissue serves to connect, envelope, or strengthen and give elasticity to these structures. There are three types of connective tissue:

- Collagen is the most abundant. It is white and transparent and gelatinizes when exposed to moisture and heat. Thus, it will be tenderized when meat is cooked by moist heat.
- Elastin is also prevalent but, unfortunately, does not tenderize easily. It is yellow in color and dense and elastic in consistency. It is best seen in the backstrap or neckleader (ligamentum nuchae) of carcass meats, especially in the neck and chuck.

- Reticulin is composed of small fibers that form networks around cells. They may be precursors to collagen and/or elastin.

Although connective tisue is not readily apparent, it is associated with muscle structure and is a major factor influencing meat tenderness. Fortunately, conective tissue, collagen in particular, is an excellent source of protein.

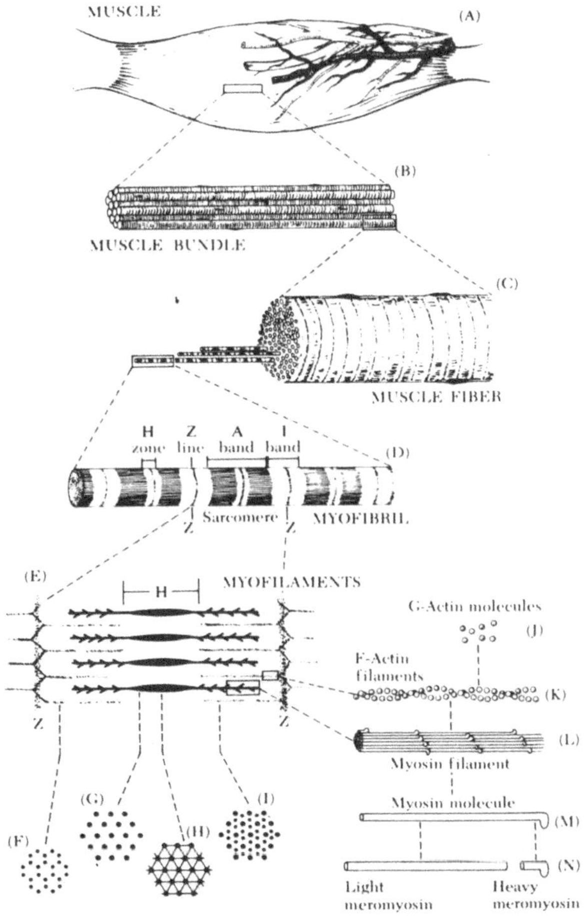

Figure 1.

Source: <u>Principles of Meat Science</u> by J. C. Forrest, E. D. Aberle, H. B. Hedrick, M. D. Judge, and R. A. Merkel. W. H. Freeman and Co. Copyright © 1975.

ADIPOSE TISSUE

In polite society, we refer to fat as adipose tissue. It is made up of fatty acids of two general types, saturated and unsaturated. Meat animals produce both types of fat, with ruminants incorporating more highly saturated fatty acids such as myristic, palmitic and stearic, whereas the monogastrate produces more unsaturates like oleic, linoleic and linolenic fatty acids. Thus, cattle and sheep have harder, more saturated fats, and hogs produce softer, less saturated fats (table 1). In total, lamb has a ratio of about 58% to 42% saturated to unsaturated fatty acids, while beef runs approximately 55:45, and pork 45:55. These facts all have ramifications relating to quality, consumer acceptance, processing, and shelf-life; however, time does not permit discussion of these topics.

However, we are most concerned with the amount of deposition of fat on and in the carcass and its influence on cutability and quality. Fat is laid down first internally, around the kidneys, heart, and pelvic region. Next, it is deposited subcutaneously to form covering or finish. Simultaneously, but slightly later, it is being infused between the muscles as seam fat (intermuscular fat) and within the muscles as marbling (intramuscular fat).

This sequence causes us much concern if it is necessary to accept all the internal and external fat that appears to be necessary to realize the degree of marbling preferred. Fortunately, once identified, we can select genetically for marbling or grade without producing wasty carasses to obtain these characteristics. Thus, from a quality or palatability standpoint, it is reassuring to know tht we can produce desirably finished animals that will "grade" without being excessively wastey.

HOW DOTH YOUR MUSCLE GROW?

Visualizing now the make-up of muscle, let us consider its growth and development on a unit basis and as manifested in the living animal and its resulting carcass (table 2). Growth is basically an increase in size. It may be accomplished in muscle by either, or both, of the following:

- Hyperplasia is the production of new cells and is largely determined prenatally. The number of muscle fibers (cells) does not appear to increase significantly after birth.
- Hypertrophy is the enlargement of existing cells. The greatest increase in muscle size occurs postnatally as the muscle fibers increase both in diameter and length. The diameter increase is caused by a proliferation of the myofibrils within the muscle fiber (cell), possibly by as much as 10 to 15 times from birth to

TABLE 1. FATTY ACID AND TRIGLYCERIDE COMPOSITION OF SOME ANIMAL FAT DEPOSITS (PERCENTAGE BY WEIGHT)

		Pig			Cattle		Sheep	
Component	Chicken	Subcutaneous Outer	Subcutaneous Inner	Perirenal	Subcutaneous	Perirenal	Subcutaneous	Perirenal
Fatty acids								
Lauric	-	trace	trace	trace	0.1	0.2	0.1	0.1
Myristic	0.1	1.3	0.1	4.0	4.5	2.7	3.2	2.6
Palmitic	25.6	28.3	30.1	28.0	27.4	27.8	28.0	28.0
Stearic	7.0	11.9	16.2	17.0	21.1	23.8	24.8	26.8
Arachidic	-	trace	trace	trace	0.6	0.6	1.6	2.6
Total saturated	32.7	41.5	47.3	49.0	53.7	55.1	57.7	59.5
Palmitoleic	7.0	2.7	2.7	2.0	2.0	2.2	1.3	1.9
Oleic	20.4	47.5	40.9	36.0	41.6	40.1	36.4	34.2
Linoleic	21.3	6.0	7.1	11.8	1.8	1.8	3.5	4.0
Linolenic	-	0.2	0.3	0.2	0.5	0.6	0.5	0.6
Archidonic plus chipandonic	0.6	2.1	1.7	1.0	0.4	0.2	0.6	0.8
Total unsaturated	67.3	58.5	52.7	51.0	46.3	44.9	62.3	41.5

TABLE 2: APPROXIMATE COMPOSITION OF MAMMALIAN SKELETAL MUSCLE (PERCENTAGE FRESH WEIGHT BASIS)

	Percent		Percent
Water (range 65 to 80)	75.0	Non-Protein Nitrogenous Substances	1.5
Protein (range 16 to 22)	18.5		
Myofibrillar	9.5		
myosin	5.0	Creatine and creatine phosphate	0.5
actin	2.0		
tropomyosm	0.8	Nucleotides	
troponin	0.8		
M protein	0.4	(adenosine triphosphate [ATP], adenosine diphosphate [ADP] etc.)	0.3
C protein	0.2		
α-actinin	0.2		
β-actinin	0.1	Free amino acids	0.3
Sarcoplasmic	6.0	Peptides (anserine, carnosine, etc.)	0.3
soluble sacroplasmic and mitochondrial enzymes	5.5	Other nonprotein substances (creatinine, urea, inosine monophosphate [IMP] nicotinamide adenine dinucleotide phosphate [NADP]	0.1
myoglobin	0.3		
hemoglobin	0.1		
cytochromes and flavo-proteins	0.1	Carbohydrates and non-nitrogenous substances (range 0.5 to 1.5)	1.0
Stroma	3.0		
collagen and	1.5	Glycogen (variable range 0.5 to 1.3)	0.8
recticulin elastin	0.1		
other insoluble proteins	1.4	Glucose	0.1
Lipids (variable range 1.5 to 13.0)	3.0	Intermediates and products of cell metabolism (hexose and triose, phosphates, lactic acid, citric acid, fumaric acid, succinc acid acetoacetic acid, etc.)	0.1
Neutral lipids (range: 0.5 to 1.5)	1.0		
Phospholipids	1.0	Inorganic Constituents	1.0
Cerebrosides	0.5	Potassium	0.3
Cholesterol	0.5	Total phosphorus (phosphates and inorganic phosphorous	0.2
		Suflur (including sulfate)	0.2
		Chlorine	0.1
		Sodium	0.1
		Others (including magnesium, calcium, iron, cobalt, copper, zinc, nickel, manganese etc.	0.1

Source: Principles of Meat Science by J. C. Forrest, E. D. Aberle, H. B. Hedrick, M. D. Judge, and R. A. Merkel. W. H. Freeman and Co. Copyright © 1975.

maturity. In hogs, the maximum proliferation, and therefore muscle fiber size, is at about 5 months of age. Individual muscles vary in their rates of growth. The larger muscles, hind limb and back generally, have the greatest rate of growth. However, the greatest influence is the number of muscle fibers initially present to be hypertrophied. In turn, the diameter expansion, is influenced by a number of factors such as species, breed sex, age, nutritional level, and exercise. The muscle fibers of sheep are smaller than those of cattle and swine; intact males have muscle fibers larger in diameter than those of castrates, which have larger fibers than do females. Generally, muscle fiber diameter increases with maturity, nutritional plane, and physical activity.

We should be able to control some of these factors to develop maximum muscle while minimizing unnecessary fat; we have an opportunity to produce the most of the best for the least. Like much theory applied to practice, such production may be more easily talked about than done. Nonetheless, using this knowledge, proper records, wise selection, and culling, adaptation of genetics, nutrition, and environmental manipulation, we should be able to put much of this information to work and thereby improve our production of the "right kind" of livestock for the marketplace of the future. The goal then should be, in most basic terms, to increase muscle fiber size and numbers and to decrease adipose tissue cell size and numbers!

The time is upon us for action to feed our nation and much of the world on the most nutritious single foodstuff we have to offer and known to man - red meat!

We should be aiming for Choice quality grade and yield grade 1 to meet the demands of the future. It can be done, it is being done, it must be done. We can refute the saying of "No waste, no taste." Let's maintain the quality we desire, but eliminate the inefficient waste fat that we have tolerated in the past. This challenge I leave you!

35

PRINCIPLES OF MEAT SCIENCE AFFECTING LIVESTOCK PRODUCERS: GROWTH AND DEVELOPMENT

Donald M. Kinsman

INTRODUCTION

In mathematics, "The whole is equal to the sum of its parts." In nature and with all biological materials, everything is interrelated and, under normal conditions, the harmonious blending of all cells, organs, and systems produces the final organism. Understanding these micro and macro structures is important in the production of all animals, especially those we are raising for meat purposes. Certainly we need to be cognizant of the end-product of this chain of events to be certain that we are producing the kind of product that will sell, that the consumer desires and demands, and that can be produced and marketed as efficiently and economically as possible for a reasonable profit.

WHAT'S UNDER THE HIDE?

Let us first examine this end-product by its components, as viewed from a production perspective. In essence, meat is the edible tissue of animals. That which we ingest is principally muscle, but it is accompanied by some form of fat, either intramuscular, subcutaneous, or intermuscular. Connective tissue in the muscle and fat is also consumed. Although we do not normally consume bone, it is the framework on which the muscle and fat develop and is frequently included as part of the roast or steak. How these basic components grow and develop is a function of many interrelated factors--with genetics, nutrition, and environment serving as major influences. Table 1 depicts the heritability estimates of the growth and carcass characteristics of cattle, sheep, and swine. Nutrition certainly plays a major role in the growth and development of our meat animals, and figure 1 portrays the influence of plane of nutrition on the rate of skeleton, muscle, and fat growth in swine. Environmental effects are more difficult to measure, including weather, temperature, climate, and stress, but we have the greatest control over management.

TABLE 1. HERITABILITY ESTIMATES OF GROWTH AND CARCASS CHARACTERISTICS OF CATTLE, SHEEP, AND PIGS

Characteristic	Species	Heritability estimate[a] approximate average
Weaning weight	Cattle	25
	Sheep	33
	Pigs	17
Postweaning rate of weight gain	Cattle	57
	Sheep	71
	Pigs	29
Feed efficiency in feed lot	Cattle	36
	Sheep	15
	Pigs	31
Carcass grade	Cattle	48
Tenderness	Cattle	61
	Sheep	33
Fat thickness	Cattle	38
	Sheep	20
	Pigs	49
Longissimus muscle cross sectional area	Cattle	70
	Sheep	48
	Pigs	48

Source: Principles of Meat Science by J. C. Forrest, E. D. Aberle, H. B. Hedrick, M. D. Judge, and R. A. Merkel. W. H. Freeman and Co. Copyright © 1975.

[a]Estimates were compiled from various sources.

These all combine to yield a growth curve as shown in figure 2. Note that this curve starts slowly, rises gradually, and then escalates very rapidly at an early stage before leveling off at its eventual mature state. It is during this rapid escalation period that we need to capitalize on the animal's greatest feed and growth efficiency in producing maximum muscle with minimum fat to meet the market demands.

NOTHING IS STATIC

The term "homeostasis" expresses the maintenance of a physiologically-balanced internal environment. This is virtually a system of checks and balances that permits the

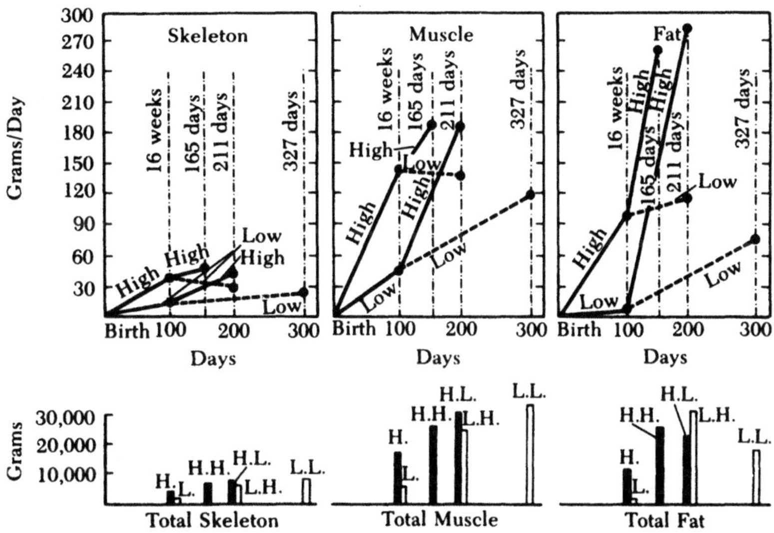

Source: **Principles of Meat Science by J. C. Forrest, E. D. Aberle, H. B. Hedrick, M. D. Judge, and R. A. Merkel. W. H. Freeman and Co. Copyright © 1975.**

Figure 1. Effect of plane of nutrition on rate of skeleton, muscle, and fat growth in the carcasses of pigs with live weights of up to 91 kilograms. Key: H = high plane of nutrition; L = low plane of nutrition; HL = high plane, then low plane (to 91 kg); HH = high plane (to 91 kg); LH = low plane, followed by high plane (to 91 kg); LL = low plane of nutrition (to 91 kg). [Data from McMeekan, C. P., "Growth and Development in the Pig, with Special Reference to Carcass Quality Charcters; III. Effect of the Plane of Nutrition on the Form and Composition of the Bacon Pig," J. Agr. Sci. 30, 511 (1940).]

animal to cope with the stresses that tend to alter that environment. This mechanism is controlled by the nervous system and the endocrine glands. Fortunately, they usually maintain the harmony and equilibrium that permits adaptability of the animal to its oft-changing environment. However, change is always occurring in any organism. Certainly, amounts and proportions of the basic meat components change in meat animals during both prenatal and postnatal growth and development, and the stage at which these factors change is of importance to all of us as producers of the nation's meat supply. Let us study figure 3 and recognize the changes in the percentage of bone, muscle, and fat in beef carcasses during growth. Bone has the least percentage change, decreasing from about 25% of the carcass in the neonatal animal to about 13% at maturity. Muscle percentage of

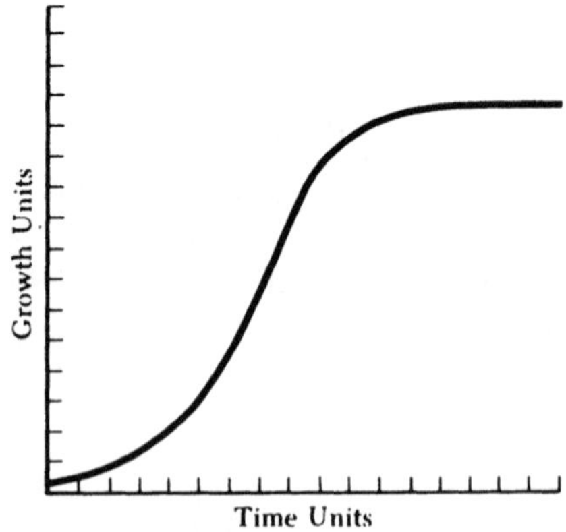

Source: <u>Principles of Meat Science</u> by J. C. Forrest, E. D. Aberle, H. B. Hedrick, M. D. Judge, and R. A. Merkel. W. H. Freeman and Co. Copyright © 1975.

Figure 2. A Typical Growth Curve

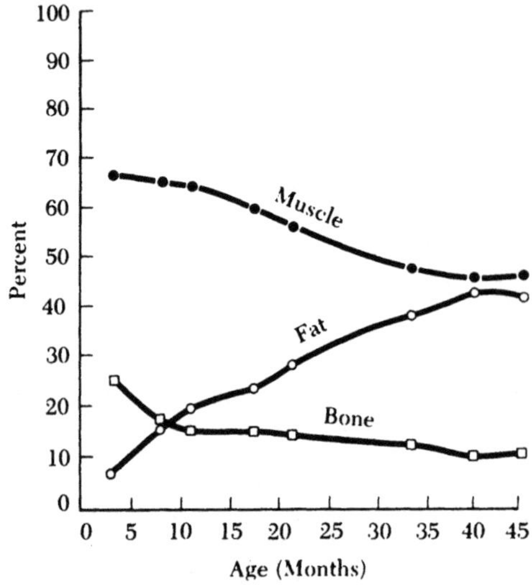

Source: <u>Principles of Meat Science</u> by J. C. Forrest, E. D. Aberle, H. B. Hedrick, M. D. Judge, and R. A. Merkel. W. H. Freeman and Co. Copyright © 1975.

Figure 3. Changes in the percentage of bone, muscle, and fat in beef carcasses during growth. [Data from Moulton, C. R., P. F. Trowbridge and L. D. Haigh, "Studes in Animal Nutrition: III. Changes in Chemical Composition on Different Planes of Nutrition." Mo. Agr. Exp. Sta. Res. Bul. 55 (1922).]

the carcass decreases as fat increases. Thus, muscle represents approximately 68% of the newborn animal and may decrease to 50% or less as fat approximates that proportion. On an age basis, fat is about 30% (the maximum fat permitted in ground beef) at 20 to 24 months of age in beef cattle. Thereafter, it climbs steadily, approaching 40% to 50% at maturity. This is, of course, dependent upon nutritional planes. Furthermore, we need to concentrate on those stages that affect the final product and then interrupt it at the optimum point of greatest value to producer, packer, processor, and consumer.

Where do you believe that point or segment lies? To carry that thought further, we may wish to relate the percentages of muscle and bone to various percentages of fat in the carcasses of beef and sheep to determine the influence of carcass fat on muscle, in particular, and to bone, secondarily (table 2).

TABLE 2. APPROXIMATE PERCENTAGES OF MUSCLE, AND BONE PLUS TENDON, IN BEEF CATTLE AND SHEEP CARCASSES HAVING VARYING PERCENTAGES OF FAT

Fat	Muscle	Bone plus tendon
8	66	26
12	62	26
16	62	22
21	61	18
26	59	15
32	54	14
37	49	14
42	46	12

Source: Principles of Meat Science by J. C. Forrest, E. D. Aberle, H. B. Hedrick, M. D. Judge, and R. A. Merkel. W. H. Freeman and Co. Copyright © 1975.

WHAT IS THE OPTIMUM "FINISHING' POINT?

At what point is it best to "finish" our livestock? There is no single, uniform answer as we consider efficiency, economics, profits and the multitude and variability of our markets. However, if we are producing for USDA Choice, quality grade and yield grade 1 or 2 then we must harvest these meat animals while young and tender, with sufficient quality to satisfy the goals of palatability and cutability. This generally implies slaughter of beef cattle at 15 to 20 months of age, and of hogs and lambs at 4 to 8 months of age--animals with adequate genetic, nutritional, and environmental background to provide the tenderness, juiciness, and flavor with minimum waste to meet these precise goals.

It can be done, it is being done, and it must be done if we are to continue to provide this highly-nutritious, heavily-demanded, reasonably-priced basic foodstuff for America's dinner table with decreasing energy and costs. Go forth and do likewise.

36

NEW TECHNOLOGY IN MEAT PROCESSING AND MARKETING

B. C. Breidenstein

For purposes of this discussion, the term "new" is used to describe innovations that have either evolved over the past 20 years or appear now to be in an emerging evolutionary state. Technological innovations are usually evolutionary in nature rather than revolutionary, and observations over an extended time frame tend to place events in a more appropriate perspective. Some of these innovations are discussed next.

METHODS OF ANIMAL DISPATCH

Methods of livestock dispatch have undergone marked changes during the past two decades. These changes were motivated, not so much for their impact on the resulting meat entity, as for compliance with a desire to render the dispatch process more humane.

In the case of cattle, they had historically been rendered unconscious by a sledgehammer blow to the pate prior to bleeding. That was an effective means of rendering them unconscious but was subject to operator error in aim or to sudden movement of the animal after the swing was begun. Thus, occasionally, the animal was injured but not rendered unconscious by the blow. It was generally recognized that the humanitarian aspects of the process could be improved upon. The method of general preference that has emerged is the use of a gunpowder charge to propel a hammer-like device against the pate. It is activated by a "trigger" that is depressed upon contact with the forehead of the animal. Thus, the "aim" of the operator is greatly improved and the possibility of injury as opposed to rendering unconscious is greatly diminished.

In the case of both hogs and sheep, the method of general choice for rendering the animal unconscious is electrical stunning. In spite of several years of experience and considerable experimentation with regard to voltage, amperage, and duration of electrical energy application, certain adverse effects upon muscle tissue are purported to result from improper stunning procedures. These adverse effects include capillary hemorrhages resulting in blood spots. Such spots may appear on the muscle surface and although possible to remove them during processing, the result is a rather obvious reduction in product value. The spots may, however, occur deep in the muscle tissue so that they are not visible during processing. Such deep hemorrhages must surely have an adverse impact on the consumers' perception of value.

BEEF DRESSING METHODS

Innovations in beef dressing methods have had a significant impact on the industry over the past two decades. Cattle dressing through the late 1950s and into the 1960s was largely accomplished on stationary skinning beds where the job was essentially done on an individual animal basis, rather than on an assembly-line basis that has characterized the hog kill and cut for decades. More recently, on-rail dressing procedures have had a dramatic effect on the area of kill floor required per head, through-put per unit of time, and man-hours required per head killed. Because of these and other factors, individual plants are able to handle a much higher capacity per unit of time. The on-rail dressing approach has led to a totally different perspective regarding the economics of scale and has resulted in appreciably lower kill costs than would have been possible with the old stationary-bed systems.

A phenomenon called "cold-shortening" has been a recognized factor in postmortem conditioning of meat in which tenderness of meat is adversely affected. The term "cold-shortening" is relatively new as an apt description, but the condition itself is not new. It is now recognized as a postmortem shortening of muscle fibers during the onset of rigor mortis and is believed to be energized by the residual glycogen of the muscle. That energy is believed to be released by adenosine triphosphate (ATP). Upon complete disipation of the energy source for the muscle contraction, the muscles relax and bring an end to the shortening phenomenon. In the case of typical postslaughter chilling/handling, that relaxation requires four to seven days for its completion, at which time muscle tenderness is similar to immediate postslaughter muscle.

The use of electrical stimulation shortly after slaughter has been found to induce rapid and rather violent muscle contractions that quickly dissipate the glycogen, thereby eliminating the energy source for further muscle contraction. Consequently, the cold-shortening phenomenon does not occur and muscle tenderness is maintained at the immediate postslaughter state. Typically, electrical stimulation improves tenderness. Biological variances affect both the rate and extent of cold-shortening resulting in increased variability in tenderness among carcasses handled in a similar fashion postslaughter. Electrical stimulation has been found to improve uniformity of tenderness among beef carcasses of similar visible quality and postmortem handling. In addition, electrical stimulation appears, for unknown reasons, to hasten the "set-up" of carcasses. That phenomenon is a recognized requirement for maximized grade achievement. Upon complete set-up, the muscle is firm and marbling is most prominently visible, thereby facilitating grading at an earlier time postmortem than would otherwise have been the case. Lean color uniformity of the rib eye also influences carcass grade. A phenomenon called "heat ring" results in a darker color for the outer (skin side) region of the rib eye. Such a condition in the absence of electrical stimulation is more common in thinly finished carcasses than in those having a more abundant fat cover.

It has been theorized that the darker color results from rapid chilling because of the greater prevalence of this condition in cattle carrying minimal fat as an insulation on the outer surface of the muscle. It has been postulated that the more rapid chill slows down the biochemical reactions in that area and that this reduced reaction rate is expressed in a visibly darker color of lean. In any event, electrical stimulation has been observed tc reduce the incidence and(or) the severity of this "heat ring" phenomenon.

The form in which beef was distributed from the kill/chill plant to retail stores began to undergo change beginning in the mid 1960s. Prior to that time, beef was delivered to the individual retail stores primarily in the form of quarters, with limited supplementation of requirements in the form of primal cuts. Thus, each site required the skills and expertise necessary to convert that carcass form to the individual retail cuts in the kinds and amounts required to meet that market's need.

Under those conditions, the back room of the retail market also required workers with sufficient physical strength to lift and transport the ungainly beef quarters weighing 125 to 200 pounds from the storage cooler to the processing area. The difficulty of maintaining the most capable personnel in each individual retail market resulted in some sacrifice of control and adherence to a prescribed merchandising plan. Additionally, the reliance on "cattle" as the beef raw material in each individual store tended to limit merchandising programs that emphasized the anatomy of cattle, rather than the consumers' wishes in that particular market area. Balance in the movement of the product was thus achieved in many cases by reducing the broader market prices on slow-moving cuts to a price level that reflected the lack of demand for them in that narrowly defined market. Conversely, to maintain a profitable operation, the more popular cuts were priced at unduly high prices compared to the price level that reflected their demand in the broader market. Thus, it seems probable that relatively low price levels on slow-moving cuts created a lower-value perception by the retailer for beef carcasses than would have been the case for a broader market outlet, i.e., a nationwide market. It also is probable that the compensating higher prices ascribed to those cuts in high demand in a narrowly defined market tended to reduce demand for them.

The advent of boxed beef in the industry emerged as a significant factor during the late 1960s. Essentially, it involves the breakdown of carcasses into subprimal cuts that are cut, boned, and trimmed to conform to either a single or very limited number of end uses. For example, rib eyes or loin strips are boxed for broiling steaks and chuck clods and chuck rolls are boxed for boneless pot roasts, etc. With boxed beef as the purchase form, it became possible for the retailer to identify cost much more definitively for each retail cut and to make his purchase pattern conform much more closely to the consumer demand in his defined market area.

In addition to improved merchandising flexibility, boxed beef offers more opportunities for improved value/cost during the distribution process. Since boxed beef is generally vacuum packaged in film having both a high-moisture barrier and a high-oxygen barrier, losses through dehydration and trimming of oxidized, dehydrated, and discolored muscle surfaces are minimized. The removal of a significant portion of fat and bone at a central location instead of at an individual retail store represents significant improved value for beef carcasses. The centralized location for the removal of those portions of fat and bone also reduces freight costs per pound of retail cuts sold. Boxed beef with its uniform symmetrical cartons can be mechanically handled to lower costs and reduce the physically demanding part of the job for both the distributing warehouse and the retail store. Because of the film/vacuum packaging, the product is protected against bacterial contamination during the distribution process. Storage space required for boxed beef is about 40% less than that required for a comparable amount of beef stored as quarters. Because each carton is identified as to item and weight, inventory control is simplified.

Because regulatory control of water content of many processed meats is based on a protein multiplier, as well as the processors' desire for finished products of uniform composition, the piece-to-piece variance in the introduction of water-carried ingredients is an important consideration. The more uniformly such solutions are introduced into the meat, the more closely one can set targets to the regulatory limits. Thus, reduced variance leads to improved yield. Significant progress is being made in the pumping machines used to introduce such solutions into the meat.

To conform to consumer desires, it is often necessary to combine two or more pieces of meat in such a way that they give the appearance of a single muscle unit. With the consumer demand for reduction of both subcutaneous and intermuscular fat, it is frequently necessary to "muscle bone" as a means of removal of the intermuscular fat. As a result, the muscle units may be considerably smaller than desired. Contained within the muscle cell is a natural binder or adhesive that is activated by heat. So long as it remains within the cell, however, it cannot function as a binding agent. These intracellular proteins, primarily myosin, are salt-soluble and can be extracted by the application of mechanical energy and (or) by cell membrane rupture. The application of mechanical energy is commonly achieved by use of mixers, tumblers, or stirring devices in vats. The term commonly applied to the process is "massaging" or "tumbling" or sometimes both. Frequently, a combination of cell rupture and massaging are employed. The widespread adoption of such techniques in the meat processing industry occurred during the decade of the 1970s and still continues. Systems are available that combine the pumping, cell rupture, and mechanical agitation within a single totally closed system. Such a system may also combine all these operations under vacuum. The application of vacuum to a meat mass increases its volume, creates a presumably looser muscle structure, and distributes the curing solution in the meat mass more rapidly and uniformly.

The finished product, appearing as a single, solid muscle mass, is formed by placing the meat mass in some kind of a shaping, cooking container and subjecting it to heat. The bonding together of the individual pieces is a heat-induced phenomenon commencing at a temperature of about 145°F. Temperatures above 175° to 180°F do not further enhance the "bind."

The subject of restructured products has received considerable attention in the trade press in recent months. The concept of restructured products is not particularly new--ground meat formed into patties is one form of restructuring. What is new are new applications for existing equipment and new equipment to produce products having a different texture and mouth feel than ground products. The principles of restructuring in this newer sense involve the extraction of intracellular proteins to the meat-mass surface where, upon heating, the particles adhere to each other forming a solid, more nearly muscle-cut-like-structure than is true for ground products.

Restructuring basically involves particle-size reduction, mechanical agitation, usually the addition of salt and phosphates to extract the intracellular proteins, forming or shaping, freezing and portioning. Great variations in texture and/or mouth feel are possible, depending upon raw material used, extent and/or nature of particle-size reduction, amount and time duration of mechanical energy applied, kind and amount of additives used, etc. Thus, a broad spectrum of products can be produced through the restructuring process. Much good technology exists, but the amount of meat surface area exposed during the process makes the product more susceptible to oxidative changes affecting color and flavor than is true for whole muscle cuts. It is, therefore, essential that technology be properly applied if consumer satisfaction is to be maximized.

The use of liquid smoke as an alternative to direct application of wood smoke became much more widespread in the 1970s. The primary stimulus was probably the need to minimize atmosphere discharge from cooking/smoking operations as a means of compliance with the Clean Air Act. Additionally, the advances made in suitability of liquid smoke and the desire for a higher level of process control also were contributing factors. In any event, liquid smoke is now widely and successfully used as a part of meat-processing systems. The use of additives to enhance functional properties has been practiced as long as meat processing has been done. As an example, various forms of phosphates have been used for the purpose of water retention and altering the mouth feel of finished products. Not only do phosphates aid in yield control, but products containing phosphates tend to have less free moisture accumulation in consumer packages. The absence of free moisture in the package greatly improves consumer acceptability. Recent regulatory changes of the USDA now permit the use of phosphates in a broader array of products including some sausage products. Ascorbate is another additive that has been used for some time but has had increased emphasis placed on it in recent years. Its presence in bacon has been shown to inhibit the formation of nitrosamines.

Processing of prerigor meat, as opposed to traditional kill, chill, and heat process is receiving increased interest and attention. As energy costs increase, a frequently asked question is: why start with a carcass at a temperature of about 100°F, expend energy for refrigeration to reduce its temperature to about 30°F or 40°F, then expend energy to heat it back through that temperature range to its ultimate processing temperature? That is obviously an oversimplification, but it has resulted in increased efforts to examine the issue of prerigor processing of those carcass portions that are expected to be processed further prior to sale or distribution.

PACKAGING MATERIALS

Packaging materials for primary containers have undergone dramatic improvement over the past two decades. The barrier properties of films to both moisture and oxygen have been sharply improved as well as the film resistance to the rigors of distribution. Only quite recently have films been used in which the product could be cooked and also serve as the primary packaging container. In conditions of water cook, such as boiled hams, product exposure to the cooking water results in leaching of meat components and curing ingredients. Protein loss thereby represents a loss of yield potential not only of the protein but of the water as well, because of the protein multiplier. Controlling yields thus becomes a matter of estimating cooking losses so that the pump can be adjusted to compensate for it. The new cook-in-

-the-container packaging does not allow exposure of the meat product to the water--thus cooking loss (one of the previously estimated variables) is eliminated.

The film package has recently appeared in retail markets as a replacement of the metal can as a primary container; however, it is too soon to make any judgments regarding its success. Its primary advantages would appear to include lower container cost, product visibility through the transparent film, and less hazard in the home upon opening. The primary question would appear to be container integrity and ability to withstand the rigors of distribution expected of products in metal containers.

RETAIL STORE DELIVERY

A major shift in retail-store delivery of meat and other products has occurred over the past two or so decades. The use of distribution centers by supermarket chains, cooperative groups, etc., has permitted the combination of products for delivery to greatly reduce the number of deliveries to the individual stores. In earlier years, each different meat supplier sent delivery trucks to each store as frequently as required. The distribution center now takes quantity deliveries from each supplier, warehouses them, and combines all suppliers in accord with each store's orders to make a single delivery. Such a system has resulted in significant distribution efficiencies.

During the late 1960s and early 1970s, beef fabrication facilities were often a part of a distribution center. Generally, these were justified on the basis of improved quality control and meeting perceived requirements as compared to the alternative of purchasing boxed beef from a slaughterer /fabricator. Many such operations have been discontinued. One can speculate as to the reasons. Most assuredly, product specification standardization among boxed-beef suppliers has vastly improved since the childhood of boxed beef in the late 1960s. Secondly, the ability of the boxed-beef industry to adapt to a true market requirement has largely accommodated most "unique" retailer requirements. Finally, many of the features of economy that characterize boxed beef packaged at the source of supply were not captured if naked carcass beef was transported to a distant facility to be fabricated.

The use of piggy-back refrigerated trailers once appeared to be a very viable meat-shipping mode. It seemed to offer the economy of rail transport, the flexibility of truck delivery, and the advantages of mechanical refrigeration. Experience in the industry, however, was that it was entirely too unreliable to meet the needs for transport of a valuable, perishable product such as meat. The over-the-road truck/trailer has evolved as the method of choice for the transport and delivery of the vast majority of meat products. Air transport has been championed for meat from time to time but has never accounted for any significant proportion of the volume of meat product delivery. It would seem possible that air transport might be useful in the delivery of ready-to-cook fresh meat items for export. Such a situation could conceivably arise from demand for high-quality American meat products overseas, i.e., to satisfy tourist requirements.

COMPUTER APPLICATIONS

As in almost all elements of life, the computer revolution has had a dramatic impact on the meat industry. The necessity for timely information has always characterized the meat industry. Its perishability and narrow margins per unit of volume make it essential that management information regarding costs, yields, inventories, shipments, etc., be available at frequent intervals, i.e., daily or more frequently. The computer has made such things possible, though probably few companies have fully exploited the computer for such purposes.

Least-cost sausage formulation has been a widely used computer capability for at least two decades. However, the ever increasing computer capacities and capabilities have led to ever more sophisticated models that provide an expanded examination and interpretation of data.

In the early 1970s, the National Livestock and Meat Board assembled industry representatives and assumed the leadership role in the development of Uniform Retail Meat Identity Standards. This activity resulted in the publication of a book by the Board in which those standards are set forth. That effort has been widely accepted as a basis for improving the consumer information contained on the label as well as reducing the confusion resulting from geographic variation in the names by which each retail meat cut is known.

Beyond their original intent, however, the presence of those widely accepted standards has provided the opportunity for the adoption of computers by the meat department of retail stores. Universal product coding of catch-weight packages is now a feasible use of the computer. Optical scanning makes it possible to track meat sales and other operational parameters of the retail meat department much more comprehensively than was previously possible. Numbers have been assigned to the standard cuts described in the Uniform Retail Meat Identity Standards thus greatly facilitating the use of the computer.

To that end, the Food Marketing Institute and the National Livestock and Meat Board have cooperated in sponsoring educational seminars attended not only by the representatives of the meat department but also by his coworker, the computer systems analyst, and/or programmer. Such a joint approach for each to better understand his coworkers' problems and needs is expected to lead to a smoother and more effective transition to the computer age.

The presence in each retail store of minimanufacturing operations has long been recognized as something less than maximally efficient. Decentralized preparation and packaging of retail cuts have been on the horizon for decades. As new packaging and preservation technologies evolve, they have been seized upon as possibly the missing element required for that dream to become a reality.

Freezing, in combination with advances in packaging, has been tried several times with, in most cases, less than satisfactory results in the U.S. Some centralized retail cutting operations appear to be successful in Europe. In the U.S., however, some vacuum-packaged subprimal cuts, prepared and packaged using the boxed-beef concept, have been successfully marketed at retail. Such products are available as beef, pork, and lamb cuts in many areas of the country.

Finally, the productivity of American agriculture is the envy of the world. It is a real challenge for our meat-processing distribution and marketing industries to achieve comparable proficiency. The extent to which we are successful in meeting that challenge, thereby reducing marketing costs and/or improving upon the perceived value of our products, plays a very important role in the economic health of this great industry.

Part 8

COMPUTER TECHNOLOGY

37

AGNET: A NATIONAL COMPUTER SYSTEM FOR CATTLEMEN

Harlan G. Hughes

In the early 1970s, two professors at the University of Nebraska conceived the idea of an agricultural computer system designed specifically for farmers and ranchers. They developed the computer system now known across the country as AGNET--The Agricultural Computer Network. In 1977, the governors of five states (Nebraska, South Dakota, North Dakota, Montana and Wyoming) jointly funded a pilot project to test if farmers and ranchers in their respective states would use a computer system to make better management decisions. AGNET has now developed so that over 400,000 calls a year are being made to the AGNET computer. AGNET is, indeed, a management tool for agriculture.

Wyoming now has computer terminals in all 23 county extension offices, and county extension agents are now receiving training on how to use these terminals with their farmer and rancher clientele.

AGNET is one of three computers in the Wyoming computer center. This operator controls the AGNET computer from the central station. If we have done the job right, the operator should not have to do much. Due to the speed of the computer, we prefer that the machine do as much of its own operation as possible. This operator, however, can and does take over control of the machine whenever necessary.

AGNET is a mass storage system. Behind the dark windows in AGNET storage units are stacks of phonograph-like records used for storage of data and programs. All of AGNET's programs are stored on disks so that when you type in the name of a program, the computer can immediately go to the appropriate disk and find the requested program. We do not have to wait for an operator to mount a tape or to do any manual intervention. AGNET is one of the largest mass storage systems in the world.

Farmer advisors on the AGNET payroll are very special persons to AGNET. George, one of the advisors, is a real character whom I wish everyone could met. George's role is to help make sure that what we have on AGNET will work for farmers and ranchers. I have heard George say, "Harlan, that is the dumbest #%&"* thing I have ever heard!" Or I have heard George say, "That may be well and fine in your ivory tower, but out on the farm we do not have that kind of data." We have two half-time farmers on the AGNET payroll and they play a very important and unique role in the design and operation of the total AGNET system.

AGNET is equipped so that we can have over 200 telephone calls coming in to the computer at one time. We are now averaging a phone call into AGNET every four minutes, seven days a week, 24 hours a day. That is over 400,000 phone calls a year.

The AGNET computer in Nebraska is located in the basement of the State Capital. By design AGNET is not on a university campus computer (and probably never will be on a campus computer) because of our computer needs and demands. A university computer is set up for research and administrative data processing. We need a service-oriented computer center that can consult us before the system is changed or shut down. Our users are not computer science PhD's and become frustrated with computer down time or off time. AGNET is often our user's first contact with a computer and since they are paying for the computer time, we place some stringent demands on the computer center.

THE FULL PARTNER STATES AND THE STAFF

In 1977, five states previously mentioned became full partners in the AGNET system. The best way to describe a full partner state is to say that each state has a member on the AGNET Board of Directors.

In July of 1980, the state of Washington joined as a full partner state and in October 1980 the state of Wisconsin joined AGNET. In July of 1981, Wisconsin withdrew, which leaves six full-partner states in the AGNET system, but other states currently are considering partner status.

The concept behind AGNET is to share the development and operating costs among the full-partner states. There are approximately 17 people on the AGNET payroll. Of the 17 people, Wyoming is paying for two. Each state pools its resources with the other states so that each state can take advantage of the total efforts of the total 17 people.

I have a goal in life and it is to dissolve state boundaries when it comes to information dissemination and use. We are proving that states have information and computer programs to share across state lines, and as we share our extension resources, the winners are our clientele.

AGNET PROGRAM LIBRARY

The six partner states in the AGNET system have developed the world's largest agricultural and home economics computer-program library in the world. Today there are over 200 programs available to AGNET users. With a library of this size, no one is expected to use or even know how to use all the programs.

Our goal is not to have users able to use all the programs in the library, but rather to have a large enough library so that every user can find at least one program of interest. This large smorgasbord of programs means that users should be able to find several programs of special interest. Appendix A provides a partial list of the programs available on AGNET. I have grouped the programs by subject matter to facilitate user interests.

The AGNET library has been put together with approximately 35 man-years of programming effort. In addition, each program development is supervised by a subject matter extension specialist who is responsible for the content of the program. Each subject-matter program is owned by the subject matter specialist and not by AGNET.

AGNET is exceptionally well equipped for the livestock producer. There are livestock ration-formulation programs available for range cattle, feedlot cattle, hogs, sheep, and poultry. There are programs available that will let you simulate on paper what your cattle will do in the feedlot given a description of your cattle and the ration that you are going to feed them. There are livestock budgets and planning prices stored in selected programs.

AGNET also has programs for the crop farmer. Machinery-cost calculators and crop budgets are available. In addition, there are whole farm or ranch budgeting programs designed to help you make long-run business investment decisions. There are many, many more programs designed to help you make better management decisions.

HARDWARE USED TO ACCESS THE AGNET LIBRARY

Touch-tone telephone. The first computer terminal that I installed in a county extension office in 1972 was a touch-tone telephone. It cost us $14 per month. We used the number pad to send information to the computer and the computer sent back the information over the special loud speaker attached to the phone. We would send in the input numbers by typing them into the telephone. The computer would talk back and say, "Answer number 1 is 420." We printed the answer onto a preprinted form that explained the interpretation of the number. This touch-tone terminal served us very well as a low-cost computer terminal. Industry still uses this type of small, low-cost terminal.

Execuport terminal. It soon became evident that we would like to have terminals in our county extension offices that would print out the computer information. We now have five of the Execuports in the Wyoming AGNET inventory. These cost $1,400 for reconditioned terminals.

Texas Instrument 745 Terminal. We have installed small portable TI-745s in most of our extension offices. The TI-745 weighs 13 pounds, has a clamp-on lid and a handle. It is the size of a small briefcase and weighs about half as much. Wyoming agents transport their terminals all over their counties. The TI-745 costs approximately $1,400 new.

North Dakota's CRT Terminal. Terminals come in all sizes and shapes. The Animal Science Department at North Dakota has a CRT terminal with a TV screen where one can read the data. It also has a printer that can be used to generate a printed copy of the output when desired. These dual-purpose units cost more money, but the flexibility is convenient and does reduce paper costs.

A Decwriter terminal. A Decwriter terminal is used by the Department of Agriculture in Alberta, Canada. I used their terminal to check my electronic mail. Alberta Agriculture subscribes to the AGNET system. This terminal costs around $2,000.

Teletype 43 terminal. My secretary and I use Teletype 43 terminals. Obviously this is the terminal that I like best. The TT-43 gets used more hours than any of our terminals and is virtually a maintenance-free terminal. The only problem is that it is not portable. It weighs 45 pounds and has the terminal plus the telephone coupler and the paper to move. This terminal also costs approximately $1,400.

Terminal with TV screens. We have one special terminal that drives 23-inch TV screens for demonstration and teaching purposes. These are the same TV screens that you see in airports with flight schedules. We use these screens so that clientele and students can see exactly what we type on the terminal and exactly what the computer sends back to the screens. These screens have helped to promote AGNET in Wyoming. The screens work so well that I will not give a demonstration without these screens. The special terminal and the two screens cost approximately $4,000; therefore, we have only one in Wyoming.

MICROCOMPUTERS FOR FARM AND RANCH

Let's now boil this all down. What does it mean for you on the farm or ranch?

Agriculture is going to have some serious challenges in the 80s. During the 60s and 70s your challenge was production, but the challenge in the 80s is going to be financial management.

Yes, the computer has the potential to improve your financial management. Let me make a prediction. Those of you that will be farming in 1990 will be using computers. Those of you that do not want to use a computer will not be farming in 1990. I often hear, "No damn computer is going to tell me how to run my ranch!" I predict that that person won't be farming in 1990. Many will have retired and others will have gone out of business. Computers are going to become commonplace on U.S. farms and ranches during the 1980s. Producer owned microcomputers can be useful in relation to AGNET. As I travel around the country talking to farmers and ranchers, I hear them expressing interest in three applications of microcomputers. The three applications are:

- Business accounting.
- Herd performance reporting.
- Financial management.

In an accompanying paper, I have discussed microcomputers, their use, and purchase.

Information Networking on AGNET

If you have a telephone coupler for your micro, you can access the following from AGNET:

- Current commodity market prices.
- Current USDA, Foreign Ag and Wyoming news releases.
- Agricultural outlook and situation reports.
- Western Livestock Market Information Project livestock analyses.
- Hay for sale.
- Sheep for sale.
- Certified pesticide applicators in Wyoming.
- People interested in judging county and state fairs.
- Horticultural tips during the summer.
- Home-canning tips during the canning season.
- Emergency information such as drought tips, Mount St. Helen's emergencies, etc.

You can even use your micro to access the UPI and AP news services for news stories dealing with, for example, the Farm Bill or "beef." The AP and UPI news services are available from two commercial time-sharing companies. You can do all this today with your micro if it has a telephone coupler on it.

Marketing Information On AGNET

We are putting about 17 different daily market-price files on AGNET, including the futures opening and closing prices. We have Chicago, Kansas City, and Minneapolis futures going onto AGNET. In addition, we have both national and selected local cash markets. We are reporting local feeder-cattle sales in Wyoming, Northeastern Colorado, and Western Nebraska. Local grain and cattle markets are being put on weekly for Nebraska. Feedlot reports for the major cattle feeding states are going periodically. Export data is also going on weekly. AGNET is becoming a major source of market information for agricultural producers.

This appears to be the major reason for most of our producer subscriptions to AGNET. They want current market information.

SUMMARY

In Wyoming we are using the AGNET system to provide Wyoming farmers and ranchers with their first contact with computers for:

- Record keeping such as beef herd performance.
- Problem solving for computer-aided decision making.
- Information networking such as daily market information.
- Electronic mail to speed up the delivery of research and extension information to clientele.

Computerized Management Aids (CMAs) are not new to agriculture. They are just new to the west. Leading midwest farmers have been using CMAs for over 10 years.

APPENDIX A

PARTIAL LIST OF AGNET PROGRAMS AVAILABLE

Livestock Production Models on AGNET:

BEEF	Simulation and economic analysis of feeder's performance.
BHAP/BHPP	Beef herd performance program and beef herd analysis program.
COWCULL	Package to help determine which dairy cow to cull and when.
COWGAME	Beef genetic selection simulation game.
CROSSBREED	Evaluates beef crossbreeding systems & breed combinations.
FEEDMIX	Least cost feed rations for beef, dairy, sheep, swine, & poultry.
FEEDSHEETS	Prints batch weights of rations including scale readings.
RANGERATION	Ration balancer for beef cows, wintering calves, horses & sheep.
SWINE	Simulation and economic analysis of feeder's performance.
TURKEY	Simulation and economic analysis of turkey's performance.
VITAMINCHECK	Checks the level of vitamins & trace minerals in swine diet.
WEAN	Performance testing of weaning-weight calves.
YEARLING	Performance testing of yearling-weight calves.

AG Engineering Models on AGNET:

BINDRY	Predicts results of natural air & low temp. corn drying.
CONFINEMENT	Ventilation requirements & heater size for swine confinement.
DRY	Simulation of grain drying systems.
DUCTLOCATION	Determines ducts to aerate grain in flat storage bldg.
FAN	Determination of fan size and power needed for grain drying.
FUELALCOHOL	Estimates production costs of ethanol in small-scale plants.
GRAINDRILL	Calculates the lowest cost width for a grain drill for your farm.

PIPESIZE Computes most cost-effective size irrigation pipe to install.
PUMP Determination of irrigation costs.
SPRINKLER Examines feasibility of installing sprinkler irrigation.
STOREGRAIN
Cost analysis of on-farm and commercial grain storage.
TRACTORSELECT
Assists in determining suitability of tractors to enterprise.

Crop Production Models on AGNET:

BASIS Develops "historical basis" patterns for certain crops.
BESTCROP Provides equal return yield & price analysis between crops.
CROPINSURNACE
Analyzes whether to participate in crop insurance program.
FLEXCROP Forecasts yields based on amount of water available for crop growth.
IRRIGATE Irrigation scheduling.
RANGECOND Calculates the range condition and carrying capacity.
SEEDLIST Lists seed stocks for sale.
SOIL LOSS Estimates the computed soil-loss (tons/acre/ year).
SOILSALT Diagnoses salinity & sodicity hazard for crop production.
SOYBEANPROD
Demonstration soybean production management model.

Home Economic Models on AGNET:

BEEFBUY Comparison of alternative methods of purchasing beef.
BUSPAK Package of financial analysis programs.
CARCOST Calculates costs of owning & operating a car or light truck.
DIETCHECK Food intake analysis.
DIETSUMMARY
Summary of analysis saved from DIETCHECK.
FIREWOOD Economic analysis of alternatives available with wood heat.
FOODPRESERVE
Calculates costs of preserving foods at home.
MONEYCHECK
Financial budgeting comparison for families.
PATTERN Helps select a commercial pattern size & type for figure.
STAINS Tells how to remove certain stains from fabrics.

4-H and Youth Models on AGNET:

CARCASS	Scoring & tabulation of beef or lamb carcass judging contest.
FAIR	Scoring and tabulation of judging contests.
JUDGELIST	List of judges available for fairs and contests.
PREMIUM	Compiles and summarizes fair premiums.

Farm and Ranch Planning Models on AGNET:

BUSPAK	Package of financial analysis programs.
CALFWINTER	Analyzes costs and returns associated with wintering calves.
COWCOST	Examines the costs and returns for beef cow-calf enterprise.
CROPBUD	Prints out select Wyoming crop budgets.
CROPBUDGET	Analyzes the costs of producing a crop.
DAIRYCOST	Analyzes the monthly costs and returns with milk production.
EWECOST	Analyzes the costs & returns of sheep production enterprise.
FARMPROGRAM	Analyzes USDA Acreage Reduction Program.
GRASSFAT	Analyze costs and returns associated with pasturing calves.
LANDPAK	Package of programs to assist in land management decisions.
LSBUDGETS	Designed to print out stored livestock budgets.
MACHINEPAK	Machinery analysis package.
PLANPAK	Package of programs designed to help analyze and plan aspects of the business.
PLANTAX	Income tax planning/management program.

Information Networking on AGNET:

CONFERENCE	A continuing dialogue among users on a specific topic.
EWESALE	Lists sheep for sale.
FAS	Prints trade leads & commodity reports provided by USDA-FAS.
GUIDES	Prints available reports of reference material information.
HAYLIST	Lists hay for sale.
MAILBOX	Used to send and receive mail.
NEWS	Latest notifications about programs and user-related information.
NEWSRELEASE	A program for rapid dissemination of news stories.

WHO IS — Retrieves name and company affiliation of individual users.
WYOPROGS — List of specialized Wyoming programs available only to Wyoming users.

Market Price Retrievals, Plotting, and Forecasting Models on AGNET:

CASHPLOT — Prints a plot of selected cash prices.
CORNPROJECT — Projects avg U.S. corn price for various marketing years.
MARKETCHART — Prints various charts on selected future and cash prices.
MARKETS — Various market reports and specialists' comments.
PRICEDATA — Prints selected historic cash and/or futures prices.
PRICEPLOT — Designed to plot market prices in graphic form.

Miscellaneous Programs on AGNET:

EDPAK — Demo programs illustrating computer-assisted instruction.
FILLIN — A "fill in the blank" quiz routine.
GAMES — Package of game programs.
INPUTFORMS — Prints available input forms.
JOBSEARCH — Matches abilities and interests to occupations.
MC — A multiple choice quiz routine.
MICROPROGRAM — Lists programs for microcomputers.
TESTPLOT — Standard analysis of variance.
TREE — Summarization of community forestry inventory.

38

RANCHER-OWNED MICROCOMPUTER SYSTEMS: WHAT'S AVAILABLE

Harlan G. Hughes

In the fall of 1977, Radio Shack started advertising the TSR-80 microcomputer for Christmas. This was the beginning of general-public awareness of the personal microcomputer. Another highly advertised microcomputer is the Atari which can be hooked up to a regular TV set, but the Atari is a game computer and, to my knowledge, has no agricultural programs available yet.

CURRENT MICROCOMPUTERS FOR RANCH AND FARM USE

There are two levels of microcomputers being considered by farmers. For the lack of any other terminology, I will use Level I and Level II as the classifications. Level I micros are the lowest cost and most popular systems. The three most common Level I micros in agriculture are the Radio Shack, Apple, and Pet Commodore.

Level I Hardware

Radio Shack Models I, II, & III. The Animal Science Division at the University of Wyoming has a Model I Radio Shack microcomputer. As is typical of most microcomputers, it has a keyboard, a TV screen, a disk unit, and a printer. Dr. Schoonover from Wyoming has developed a herd performance program for the Radio Shack Model I and III microcomputers. This program keeps track of the cow/calf information that ranches have been keeping on 3- x 5-inch cards. Once the data is inside the computer, management reports can be quickly printed out to help the rancher determine the cows to keep and the cows to cull. The same herd performance program that Dr. Schoonover has on the Radio Shack microcomputer is also on the AGNET system. We have several Wyoming ranchers currently using these herd performance programs.

The Radio Shack Model II has the disk drive built into the unit. Radio Shack refers to this as their small business machine.

The Radio Shack Model III has two disk drives built into the unit and presents pictures and graphs of your data.

The Model III can present a bar graph to show how ranch profits have changed the last 5 years. It has been suggested by some ranchers that a graph is purely academic if it represents ranch profits since profits have disappeared rather than changed.

Apple Computers. AGNET has an Apple computer with which we have one of our Teletype 43 AGNET terminals as a slave terminal. With proper connections, you can use your existing terminal as a slave printer on your microcomputer. Also, if you have a black and white TV you can back it up as the CRT on the Apple (and other brands as well). The resolution is not quite as clear as a regular monitor, but it is a cheaper way to get set up with a microcomputer.

Dr. Menkhaus, at the University of Wyoming, uses his Apple microcomputer in his Price Analysis class to teach undergraduate students how to use microcomputers.

The newest Apple is the Apple III. It has been out for about a year, but has had some technical troubles that has set its acceptance back. The Animal Science Division at Wyoming cancelled its order for the Apple III and ordered the Apple II Plus. This fast-growing company moved into a new product and forgot something called "quality control."

Pet Commodore Microcomputer. The Pet Commodore is being used by Alberta Agriculture in Canada and the Ag Economics Department at Wyoming. The Canadians have written a fair amount of agricultural software for the Pet and have been willing to share it with Wyoming so that we do have several decision aids for our Pet Commodore.

Word Processor on Screen. Micros also can be used for word processing. You can buy word processing programs for almost all micros that will let you use your micro to generate printed materials like letters and reports.

Word processing allows you to electronically add words, delete words, add paragraphs, move paragraphs, etc. When you have your paper like you want it, you can print out the letter or paper on the computer's printer. I now write all my papers on the word processor.

While word processing will not be a big thing for many farmers or ranchers, it might be of value to those of you that are officers of farm organizations. Dave Flintner, President of Wyoming Farm Bureau, could surely use word processing in his Farm Bureau business.

Level II Computers

The more common level II microcomputers that farmers are considering are: Northstar, Vector Graphics, Superbrain, Hewlett Packard, and Cromenco. There are also other brands but they tend to be less popular.

Northstar Microcomputer. One Level II microcomputer that is fairly popular is the Northstar. Country Side Data out of Utah is selling agricultural software for the Northstar computer.

Vector Graphics Microcomputer. Another Level II microcomputer is called the Vector Graphics. Homestead Computers out of Canada has several software packages for the Vector. In addition, Loren Bennett in California has a dairy-ration package for the Vector.

This microcomputer and others can be equipped with a "professional" printer that is used for word processing. If we had a letter typed with this type of printer, I could convince you that the letter was typed by my secretary on her IBM electric typewriter. Professional printers sell for around $3,000; however, if you are going to do word processing, a professional printer is preferred.

One purebred cattleman has a professional printer on his micro. He uses the word processor to write individual letters to his purebred cattle customers. He keeps a list of potential customers inside his computer. When he has a bull for sale, he then uses the word processor to generate and address personal letters to each customer. Each customer thinks the cattleman personally typed the letter to them. In reality, his microcomputer merged the names into the standard letter stored in the micro. This cattleman argues that this is a very cost-effective way to advertise his purebred cattle. The key is the professional printer and the word processing software.

Superbrain microcomputer. A Superbrain is used by South Dakota AGNET with disk drives that are built into the cabinet. This is extremely nice when you move the microcomputer around.

Hewlett Packard. Hewlett Packard recently announced the HP-125 as their small business machine. HP long has a reputation of producing high quality products, and we believe this is also true for their microcomputers. To date, I am not aware of any agricultural software available for the HP machines.

Cromenco Computer. Cromenco microcomputer is configured to be a fairly powerful microcomputer, yet there are several empty slots for future additions to meet your expanding needs. The Level II machines are considerably more flexible than the level I machines.

Comparing Level I and II Microcomputers

There are several differences in the Level II micros as compared to the Level I micros. The key differences are: 1) basic language compilers that are faster than Level I interpretors, 2) 80 character screens that make VISICALC and communications easier to use, 3) more standard operating systems such as CP/M (this means it is easier to exchange programs from one machine to another), 4) more error diagnostics for software and hardware, 5) and the S-100 buss (for more hardware exchangeability).

Hardware Accessories

Data cassette. In the past, we used cassettes for data and program storage. In fact, you can use your kids' cassettes and their tape recorder on your micro to record data and programs. While this is a very cheap storage device, by today's standards it is too slow and inflexible.

Floppy disk. The technology that has made microcomputers of value to agriculture is the floppy disk--a phonograph record with a paper covering around it. Instead of recording music on the disk, the micro records data and computer instructions on the disk. The floppy disk now provides the microcomputer with mass storage capability. Dr. Schoonover can store data for 500 cows in the beef program on one of these disks. If you have 1,000 cows, you simply use two disks. In fact, you can have as many of these disks as you want on the shelf. You just pull off the shelf the disk that you want and put it into your microcomputer.

Hard disk. The newest storage technology is the hard disk. Inside this little box is the ability to store 5 million characters of data. You could store all the management information that you would ever need or generate on your farm or ranch on one hard disk. Most farmers or ranchers do not have this kind of data storage need. The purebred cattleman I know with a Vector Graphics machine keeps all his pedigree information for his cow herd on the hard disk. He can go back to 1932 with his pedigree searches. He feels that the microcomputer has helped his purebred business out considerably.

Instructional Aids

A Radio Shack Teaching Center on our campus has 15 microcomputers hooked up to a sixteenth computer. The sixteenth computer can monitor the other 15 computers. Wyoming's Agricultural Extension Service needs one of these to bring 15 ranchers or farmers in for computer training. You learn more about microcomputers by hands-on experience than from lecture or books. Many high schools and vocational technical schools have such instructional centers but the university extension services are behind.

MICROCOMPUTERS FOR FARM AND RANCH

Let's now boil all this down--what do microcomputers mean for you on the farm or ranch? Agriculture is going to have some serious challenges in the 80s. During the 60s and 70s your challenge was production, but the challenge in the 80s is going to be financial management. And the computer has the potential to improve your financial management. Let me make a prediction. Those of you that will be farming in 1990 will be using computers, and those of you that do not want to use a computer will not be farming in 1990. I often hear, "No damn computer is going to tell me how to run my farm!" I predict that that person won't be farming in 1990. Many will have retired and others will have gone out of business. Computers are going to become commonplace on U.S. farms and ranches during the 80s.

As I travel around the country talking to farmers and ranchers, I hear them expressing interest in three applications of the microcomputer. The three applications are:

- Business accounting.
- Herd performance reporting.
- Financial management.

Top producers are recognizing that they need to keep better books. They are looking to the microcomputer as a means to make bookkeeping easier and more flexible. They want current cashflow situations several times during the year. Today's profit margins do not allow the management errors that you could get by with in the 70s.

Top ranchers know the benefits of good cow-calf records and they have been keeping them on the 3- x 5-inch cards; however, it takes a lot of time to sort them into useful management reports. A herd performance system fits well onto a microcomputer and once the data is in the computer, management reports can easily be printed out. We even know of one rancher that takes his micro right out to the scales and enters the calf weights as they are weighed. When the last calf is weighed, he pushes the button and identifies the cows to be immediately culled. By not having to wait for culling data, this rancher argues that the dollar amount saved from not rounding up cattle the second time will pay for his microcomputer.

Bankers are requiring more and more financial information before they will make loans to producers. Top producers are starting to see the potential of being able to use the microcomputer to help generate these needed reports: financial statements, profit and loss statements, cash flow projections, five-year plans, etc.

<u>VISICALC - a financial management tool</u>. One of the most powerful financial management tools available is VISICALC. It is designed so that you don't have to be a programmer to program your own financial management programs. There is nothing equivalent on AGNET! Since I don't know how to describe in words what VISICALC can do, I sug-

gest that you stop into a computer store and ask for a VISICALC demonstration.

Disk oriented system. In order to have sufficient capacity to handle your agricultural applications, producers should buy a disk-oriented system. It should contain:

- Dual-disk drives.
- A good 80-column printer.
- 32K to 48K memory (the horsepower of the computer).
- 80-column screen (preferred over a 40-column screen).
- Telephone coupler.

The system will cost between $4,000 to $5,000 for the hardware and about $2,000 to $3,000 for programs (software) for your farm or ranch.

Telephone coupler. One of the extremely useful attachments that you can purchase for your microcomputer is a telephone coupler. This will allow you to use your micro as a terminal to large mainframe computers such as AGNET, TELEPLAN, and CMN. You can call the mainframe on the telephone and type in your information on your micro's keyboard and have the output printed out on your micro's printer. The cost of a phone coupler is around $300 and you can access:

- Current commmodity market prices.
- Current USDA, Foreign Ag, and Wyoming news releases.
- Agricultural outlook and situation reports.
- Western Livestock Market Information Project livestock analyses.
- Hay for sale.
- Sheep for sale.
- Certified pesticide applicators in Wyoming.
- People interested in judging county and state fairs.
- Horticultural tips during the summer.
- Home-canning tips during the canning season.
- Emergency information such as drought tips, Mount St. Helen's emergencies, etc.

You can even use your micro to access the UPI and AP news services such as news dealing with the Farm Bill and "beef." The AP and UPI news services are available from two commercial time-share companies. You can do all this today with your micro if it has a telephone coupler on it.

HOW TO BUY A SMALL COMPUTER

What should a farmer and rancher do if he is thinking about buying a small computer?

There are two newsletters that I recommend that you subscribe to on computers in agriculture. Successful Farming publishes one newsletter for $40.00 per year. They make useful evaluations of hardware and agricultural software.

The second newsletter is published by Doane-Western Agricultural Service out of St. Louis, Missouri. Their subscription rate is $48.00 per year. If you are seriously considering a microcomputer, I strongly recommend that you subscribe to one or both of these newsletters.

The second thing I recommend that you do if you are considering purchasing a computer is attend one of the computer seminars that are being held around the country. Almost every state extension service is holding these seminars specifically for farmers and ranchers interested in learning more about microcomputers and the potential agricultural applications. Contact your local county extension agent or extension advisor for information on these seminars.

Books and magazines on microcomputers and how to use and program them are also helpful when selecting a microcomputer. I strongly encourage farmers and ranchers who are thinking seriously about purchasing a computer to get one or two magazines or books on microcomputers. Farmers and ranchers read several agricultural-related magazines, so why not read at least one computer-related magazine.

I personally subscribe to BYTE. It is a good magazine to read to find out what kind of hardware is available and to learn the jargon of computers.

I also subscribe to the Personal Computing magazine. It has stories written by people who are familiar with microcomputers for people like you and me who are not familiar with microcomputers.

SUMMARY

Microcomputers are the new farm- and ranch-management tools and innovative producers are buying them. More and more farmers and ranchers are going to own one or more microcomputers.

If you buy a microcomputer, be sure and buy the telephone coupler so that you can access the agricultural information networks being set up across the country. You will need to spend around $4,000 to $5,000 for a microcomputer with enough horsepower and flexibility to do your farm or ranch applications. I assure you that we are going to see considerably more farm and ranch purchases in the next five years.

39

SIX STEPS FOR A CATTLEMAN TO TAKE IN BUYING A COMPUTER

Harlan G. Hughes

INTRODUCTION

Today's low profit margins and high interest rates place a premium on a cattleman's management-information system. Automation of that system lends itself to the microcomputer. Microcomputers represent a relatively new farm and ranch-management tool that farmers and ranchers are investigating. Purchasing one may prove to be one of the few profitable equipment purchases of the 1980s. One study indicates that as high as 64% of the producers interviewed were planning to buy a microcomputer as a management tool in the next five years. Twenty-seven percent indicated they would purchase a microcomputer in one to two years. These producers ranked business record keeping as the number one management function they wanted to perform on the microcomputer. The preparation of financial balance sheets and income and cash-flow statements ranked second. Breakeven analysis of individual enterprises and crop-production records ranked as the third and fourth management functions, respectively.

An Alberta, Canada, study of producers owning microcomputers indicated they were using the microcomputers for 1) farm planning, 2) financial record keeping, 3) physical record keeping, and 4) analysis of records (cash flow, breakeven analysis, and costs of production).

What kind of microcomputers do producers own? Sixty percent of the Canadian producers owned Radio Shack and the rest owned Apple, Pet Commodore, Vector Graphics, and others.

A recent Successful Farming magazine survey indicated that 46% of the respondents owned Apples, 34% owned Radio Shacks, 4% owned IBMs, 4% owned Commodore or Pet and the remaining percentage covered all other brands.

SIX STEPS FOR A COST-EFFECTIVE INVESTMENT

A producer-owned microcomputer should pass the same cost/benefit analysis as any other machinery investment.

Costs can be easily identified and documented; however, the benefit of improved management is considerably more difficult to document. What is clear, however, is that benefits received depend heavily on the preparation that the cattleman makes before purchasing the microcomputer.

Step 1

Before purchasing a microcomputer, study your management-information needs. Collection and analysis of management information requires time and money. You cannot afford to collect management information that you do not use or need. Some questions that you should ask are: What are the most important and significant decisions that I need to make? What information is needed to make these decisions? Can the generation of the needed information be scheduled? Can a microcomputer make this information collection easier? Studying your information requires some time and effort. It may well be worth your time to hire a consultant or visit with your university extension service and get a second opinion.

Step 2

Identify computer programs (software) that are available that might meet your management-information needs. As a cattleman you have four potential ways that you can obtain needed software. You can (1) buy it from a commercial vendor, (2) obtain it from the extension service, (3) hire it custom programmed, or (4) program it yourself.

If the software needed is available from a commercial vendor, this may well prove to be the most satisfactory method of acquiring software. Sometimes, however, you'll need software that is not available from a commercial vendor. The local extension service may have what is needed. Occasionally the only viable alternative is to hire a program custom-programmed or to program it yourself. Unless you have special training or a lot of spare time, I cannot recommend that you program the software on your own. Obtaining software tailored to your specific needs will be the most difficult and time-consuming task.

Step 3

Determine the hardware specifications required to execute the needed software. The size of the business affects the volume of management information needed and this, in turn, determines the size of hardware needed. Microcomputers come in different sizes (memory units), have different storage capabilities on the diskette (floppy disk), and have different add-on capabilities (80 column screens, upper and lower case characters, computer languages, CP/M operating systems, telephone modems, word processing software, etc.) Again, it is recommended that you contact a consultant or the extension service. Computer dealers are not necessarily the best information sources for determining specific hardware needs. Generally, they promote what they have to sell.

Step 4

Contact local hardware dealers and determine the viable hardware alternatives. Cattlemen should use the same criteria that they would use for any other equipment purchase: dealer knowledge of his own hardware, quality of the service department, apparent financial stability of the dealer's business and, in general, compatability with the dealer. Since cattlemen have purchased equipment before, they should feel reasonably comfortable with this step.

Step 5

Estimate the cost/benefit of the proposed computerized management-information system. A dealer can tell the purchaser exactly what the hardware will cost; and the cattleman already should have an estimate of what the software will cost. Remember that the cattleman-buyer can take investment credit and depreciation on computer hardware just like any other piece of machinery.

The clerical cost of collecting and processing the management information should also be included. This frequently is your time or that of your spouse. Collecting and typing data into the computer is time-consuming and boring. You might even consider hiring a person to be specifically responsible for the data processing of the management information.

While determining the cost/benefit, cost of the total management information system should be projected. A Michigan State University study indicates that it may cost $500 to $600 a year to process a producer's business records through his own microcomputer. Again, an outside consultant can be useful.

Estimating the dollar benefit of having a computerized management-information system is difficult for most cattlemen to do. Today's high costs of production and high interest rates do not leave much margin for management errors. Just preventing one management error a year may well pay for the microcomputer system. As could the ability to experiment with a decision on paper before implementation.

Step 6

The final step is to make the decision whether to set up a computerized management-information system. You should consider talking to other cattlemen that already own microcomputers. Many states are offering educational seminars for ranchers and farmers to learn more about how microcomputers can enhance a producer's decision-making process. The final decision rests with you the individual. There is no blank recommendation that will fit all situations. Microcomputers can, however, be an effective management tool.

Microcomputers are becoming a more common management tool for cattlemen. Innovative producers are purchasing microcomputers to enhance their personal management-information systems. This article summarizes six recommended steps that you should go through in making the decision to purchase a microcomputer. If these six steps are followed, you will have a higher probability for a successful experience with your first microcomputer.

REFERENCES

Engler, Verlyn, E. A. Unger and Bryan Schurle. 1981. The potential for microcomputer use in agriculture. Contribution 81-412-A. Department of Agricultural Economics, Kansas State Univ.

Nott, Sherrill. 1979. Feasibility of farm accounting on microcomputers. Agricultural Economics Report No. 336. Michigan State Univ.

Successful Farming. 1982. Successful Farming farm computer news. A special survey summary. Successful Farming.

1981. A survey of on-farm computer use in Alberta. Alberta Farm Management Branch, Olds, Alberta.

40

DIRECT DELIVERY OF MARKET INFORMATION THROUGH RANCHER-OWNED MICROCOMPUTERS: A RESEARCH REPORT

Harlan G. Hughes, Robert Price,
Doug Jose

Ranchers needs for marketing information have changed dramatically since the early 1970s. Increasing price variability, rapid inflation, higher interest rates, and closer ties to world supply-and-demand conditions for agricultural commodities have resulted in increased needs for short, intermediate, and long-run marketing information. Also, ranchers continually have fewer market outlets available so that they must do a better job of marketing their product. The net result is that many ranchers are unable to adequately evaluate marketing alternatives and, thus, are often unable to make good marketing decisions.

In late June 1981, Cooperative Agreement Number 12-05-300-522 was signed between the USDA Extension Service and the Colorado State University Cooperative Extension Service on behalf of the Western Livestock Marketing Information Project to give ranchers decision assistance. The agreement was to conduct a pilot study concerning the feasibility of direct electronic delivery of marketing and management information to farm and ranch families.

Ranchers base marketing decisions on information from both internal and external sources. Accounting records, herd performance records, and budgets are examples of internal information used. Market news, outlook reports, price forecasts, weather forecasts, and research reports are examples of external information. Internal and external information are required for almost all short, intermediate, and long-run marketing decisions.

Needs for short-run market information commonly relate to selling decisions. There are sometimes substantial risks associated with selling agricultural commodities today rather than waiting a few days, or vice versa. Short-run decisions are relatively simple to analyze in a budgeting sense as the costs are readily predictable. The difficult element is the probability of price increases and decreases.

Typically, university and government outlook specialists have not provided short-run market information. It has generally been left to the commodity brokerage firms and other private organizations to provide short-run market information. These sources tend to discount the risk and uncertainty aspects.

Intermediate-run needs for market information relate to such decisions as purchasing of stocker and feeder cattle, crop selections, fertilizer application, feed choice, and other decisions that do not result in immediate revenue. These decisions are generally more complex as the information needed to evaluate possible outcomes is more complicated and has more chance of error. University and government outlook specialists generally have been most active in providing intermediate-run information.

Examples of long-run market-information needs include land purchases, irrigation development, machinery selection, cattle herd expansion, and the construction of livestock production units. These decisions, although not made as frequently as the previous types, require significant information to allow for success of a farm business. Although farm management economists have devoted much time and effort to investment analysis, outlook specialists in the university and government realm generally have concentrated very little on this long-run arena.

Because of variability in agricultural prices and production, as well as high financing requirements, producers may risk bankruptcy before profits from an investment can be realized. Long-run market information can also be useful in assessing the amount of risk that a specific producer can afford when making investment decisions.

A comprehensive marketing-information system, used properly, could play a major role in stabilizing or increasing net ranch income during the 1980s. In the coming years, ranchers are going to need more marketing information, delivered faster, and available in an easy-to-use form. Computers can and should play a major role in such a marketing-information system and the associated educational needs. The rapid development of electronic technology also presents an exceptional opportunity for the Cooperative Extension Service to assume an even greater role in the delivery of timely market information.

DELIVERY OF MARKET INFORMATION IN THE WEST

The problem of delivering timely market information in the western U.S. is compounded by the vast geographical dispersion of producers. The extension specialists and county agents must travel extensively to accommodate the needs of farmers and ranchers. Most newspapers carry very little, if any, current market data. Farm magazines are major sources of intermediate-run market information, but timeliness of that information does not meet the standards necessary for decision making in today's economic environment.

The Western Livestock Marketing Information Project (WLMIP) was created over 25 years ago in recognition of the void that existed in the delivery of timely market information to livestock producers in the West. The proven record of WLMIP as a major source of useful intermediate-run market information for the region has been well documented (WLMIP, 1977; Bolen, 1949). However, the changing complexities of the livestock market, combined with the rapid growth in computer technology, present the need and opportunity for WLMIP to expand services. These opportunities include direct delivery of market information at the producer level, as well as increased service to professional economists and others in the West. It also presents the opportunity for WLMIP to go from almost exclusive emphasis on intermediate-run market information to expanding short-run information.

AGNET--AGRICULTURAL COMPUTER SYSTEM

AGNET is a time-sharing computer network headquartered in Lincoln, Nebraska. There are over 2500 subscribers to the network with a total yearly connecttime of over 75,000 hours. This averages out to 8.5 users per hour concurrently on a 24-hour 7-day-a-week basis. AGNET is being utilized for problem-solving and information networking. The system is very "user-friendly" and is designed for use by people with no computer background. Ranchers are allowed to subscribe to the AGNET System by paying variable costs associated with operating the system.

PREVIOUS STUDIES

In 1979 a survey was sent to state extension service administrators inquiring about the priority of marketing extension programs. (Watkins and Hoobler, 1980). Thirty-seven of the 44 state administrators returning the survey placed extension marketing programs in the range of "important" to "of highest importance." The following summary statement was taken from the report:

- "It is recommended that each state Cooperative Extension Service administration, in cooperation with their marketing specialists and representative clientele groups, examine the results of this national study, analyze their state's specific needs, determine where a cooperative effort is needed with other states, and develop plans for renewing and/or initiating programs to effectively manage the problems."

Brown and Collins (1978), University of Missouri, conducted a national study in 1977 on the information needs of large commercial farms. Their study revealed that:

- Commercial family farmers and ranchers perceive marketing information as their number one need.
- Extension and universities were rated the most important source of production technology, but only of minor importance as a source of marketing information.

- Farmers, agribusiness, extension, and the agricultural media all expressed the belief that marketing information is critical now and will continue to be critical in the future. They also agreed that present sources of market information are inadequate.

A joint USDA/NASULGC study (1968) committee recommended in 1968 "that extension increase its emphasis on marketing and farm-business management while reducing the percentage of effort in husbandry and production." The study goes on to say, "Extension should gradually shift towards giving more in-depth training to producers and to wholesaling information through supply firms."

Most extension marketing-program-appraisal studies generally include recommendations for experimentation with the latest electronic and computer innovations. For example, New York dairymen in a 1977 telephone survey felt that extension could improve its effectiveness by placing more emphasis on the use of the computer as an educational tool. (Ainsle et al., 1977).

In spite of the emphasis placed more than 10 years ago on changing extension priorities, little progress has been made to implement these program shifts. This is mainly due to extension administrators' reluctance to changing priorities of their extension programs. We are hearing the same priority requests coming from producers today as we did a decade ago. This paper reports on one pilot project that attempted to respond to some of these priority requests.

PILOT PROJECT

The state of Wyoming piloted a basic electronic market information system on the AGNET computer network during 1978-79 (Skelton, 1980). Four objectives of the pilot system were:

1. To collect price information of interest to Wyoming producers.
2. To provide county extension offices with the ability to retrieve market information that allowed them to put together today's, yesterday's, last week's, last month's, or last year's markets of interest for use by their producers.
3. To provide simple, down-to-earth interpretations of what market prices and associated outlook mean to Wyoming producers.
4. To provide price forecasts for extension personnel to use with producers in planning.

The Western Livestock Marketing Information Project piloted some initial work in computerized market-information delivery in 1980. Major livestock reports (Cattle on Feed, Hogs and Pigs, etc.) were placed on AGNET, complete with analysis and interpretation. WLMIP was instrumental in substantially increasing listings of producers with hay for sale and making these listings available to areas hardest hit by the drought of 1980. In addition, WLMIP served as a clearing house for drought conditions in many areas of the western plains region. This information was collected and transmitted throughout the region and forwarded to the office of the Secretary of Agriculture in Washington, D.C.

OBJECTIVES OF THIS STUDY

In an effort to provide an evaluation of electronic delivery of market information, a cooperative agreement was signed between the Colorado State University Extension Service on behalf of WLMIP and the USDA Extension Service. Subsequent cooperation was obtained from the University of Wyoming and the University of Nebraska. Four of the six objectives of the pilot study reported in this report are:

1. Research and develop mechanisms for direct producer access to AGNET via farmer-owned microcomputers and computer terminals.
2. Add current livestock market news information on AGNET for retrieval by producers and others.
3. Improve the documentation of marketing information and other pertinent information on AGNET and make it available to producers.
4. Evaluate the effectiveness and efficiency of this new delivery system in providing useful information to farmers.

RESULTS

Objective 1: Research and develop mechanisms for direct producer access to AGNET by farmer-owned microcomputers and computer terminals.

The technology of communication between computers of different brands and types is an involved science of its own. Different hardware requires different communication protocols and procedures. Much additional work is needed in this area that is receiving a lot of interest at the current time.

The importance of networking between microcomputers and mainframe computers housing networks such as AGNET is becoming increasingly obvious. As an example, of the 12 producers that participated in the pilot study, 9 accessed AGNET through the use of microcomputers. The other 3 used "dumb terminals" for communication. It is the authors' opinion that microcomputers will become more the norm in producer hardware than dumb terminals. The reason is that the microcomputer can also be used to solve on-the-farm types of production and marketing problems, handle production and accounting records, and handle other applications.

Current technology is readily available to enable a microcomputer to operate as a dumb terminal in communicating with AGNET. Generally, all that is required is a modem (telephone coupler) and software for the microcomputer, which is generally included with the hardware coupler. Such packages for microcomputers generally run in the price range of $500 or less. However, the technology involved in making a microcomputer into an "intelligent terminal" with AGNET is more complex. A high degree of interest in this type of software appears evident throughout the western region.

The important of operating a microcomputer in an intelligent mode with AGNET arises from the tremendous potential savings in telephone costs. A large part of the user's time during any terminal session is now spent typing in the needed information to respond to the AGNET questions. If such files could be developed on the microcomputer before the telephone call is actually made to AGNET, much of the telephone cost could be eliminated.

The authors have experienced substantial savings in telephone costs (50% or more) when using microcomputers as intelligent terminals. The capability to access AGNET as a central warehouse for information, download the information to the microcomputer, hang up the telephone, and work with the information that has been accessed results in even more cost savings.

In summary, it is a relatively easy procedure to turn a microcomputer into a dumb terminal for communicating with AGNET. It becomes a little more difficult to operate in the intelligent-terminal mode, but software has been developed in this study that makes this possible for most brands of microcomputers. The idea of interfacing farmer-owned microcomputers and a regional computer, such as AGNET, could be one of the most significant thrusts in extension service activities for computer applications to agriculture in the coming years.

Objective 2: Add current market-news information on AGNET for retrieval by producers and others.

This objective has been pursued heavily since the beginning of the pilot project. The system has been expanded so that 17 different market-price files are going onto AGNET daily. In addition, weekly and monthly analyses are going onto AGNET. During the six-month study period, 19,873 market-price files were retrieved by all AGNET users.

In addition to providing information for its current market value, most of the information included in the price files is captured by the computer and put into historical data files. By building such a data bank, files are in place for retrieval by the user in various programs for management and marketing decisions. Most of the captured information is already available for use in various retrieval and charting programs. However, retrieval programs for AGNET market information are still under development.

In an effort outside the scope of this study, the Foreign Agriculture Service (FAS) of USDA has begun a test using AGNET markets for distribution for much of their information. The response from users of the FAS information has been very enthusiastic, and several new users have subscribed to AGNET just to receive the FAS information.

The addition of market news and other information on AGNET will be a continuing process. Feedback from participating county agents and producers during this pilot study resulted in several files being added. Additional feedback on the final end-users' evaluations points to the need for even more types of files.

Market information that is currently being placed on AGNET is almost exclusively done by volunteer labor. Therefore, relatively little money is allocated to staffing explicitly for placing market information on AGNET. Several staff hours weekly are being devoted from numerous offices to place the information on AGNET.

By relying so heavily on manual labor to provide the information to the computer network, costs are magnified and the chances for errors arise. USDA Extension Service is working with Agricultural Marketing Service for direct electronic transfers of market information from AMS to the AGNET computer. If such a system could be put in place, cost savings for staff time would be tremendous and the timeliness of the availability of the information could be much improved.

Objective 3: Improve the documentation of marketing information and other pertinent information on AGNET and make it available to producers.

One of the developments coming out of this pilot study has been an AGNET Market Information Users Guide. The guide is intended to be just that, a guide to help the new user know what type of market information is available and how to access that information. In addition to documenting the market information, management decision tools are also referenced in the guide with a brief explanation of how to access and use those tools.

AGNET is a very "user-friendly" computer system. The major part of any documentation needed by the user is accessible directly from the computer with the use of "HELP" commands built into the system. Such HELP commands are unique to AGNET. Consequently, most of the needed documentation and aids are available at any time during a terminal session and preclude the necessity of having a manual available for reference while the user is online.

Objective 4: Evaluate the effectiveness and efficiency of this new delivery system in providing useful information to farmers.

The study tested two methods of market-information delivery. The first method was actual direct delivery to farmer-owned microcomputers or computer terminals located on the farm or ranch. The second method tested was "wholesaling" market information through county agents or trained agricultural professionals. These agricultural professionals used marketing bulletin boards in the county agent's office or in the financial institution center and then used frequent mailings of market information from these offices to selected producers in their area. This course of delivery was used mainly to acquaint producers with the type of information that could be obtained from the computer network and to test their responses to see if the information delivered was useful. The authors were also interested in seeing whether, after receiving the information in this manner, producers would be more interested in obtaining their own computer hardware for direct delivery of the information.

EVALUATION OF USERS DIRECTLY ACCESSING INFORMATION

An evaluation form was sent to the producers who were directly accessing the information from their own hardware. Evaluation forms were returned from 12 direct-access users. It should be noted that this group is a representative subset of farmers and not all AGNET users.

Users were asked to evaluate six general types of market information they could access from the computer. The results of that evaluation are listed in the following table:

Evaluation By Direct Users

Information	Very useful	Slightly useful	Not useful	No response	Total
Futures prices	3	4	1	4	12
Cash prices	5	3	1	3	12
Commentary & interpretation	6	3	0	3	12
News releases	2	5	1	4	12
Retrieval programs	3	2	1	6	12
Conferences	1	4	0	7	12

The files that contain commentary and interpretation of factors influencing the market were very well received by the users who were accessing them directly. Various comments received on this question included: "Good insights." "Comments really helped to get a feel for the market." "More of this type of information needed."

The files on various cash prices were also very well received by the users who were directly accessing the information. AGNET is very unusual in that several files contain localized information for various areas within a state that is not available anywhere else in a condensed, summarized form. Comments included such things as: "We need more local prices on the system." "Used these the most." "Often AGNET is the only source of this information." "Excellent."

The files on futures prices were not perceived by the end users to be as useful as the two previously discussed categories. One of the main reasons for this is that AGNET only offers each day's open and close of the futures. Producers who are active in the futures markets find that they need more current quotes, which they obtain from their farm radios or from their brokers. Also, not many producers use the futures market. Some comments on the futures included: "Would be better if we had a detailed report on weekly futures price movement." "Information didn't fit our area completely as there were no sugar futures." "Out of date by the time the producers really need this information." "Useful if picked off daily."

The retrieval programs were not used as heavily as the authors hoped they might be. One of the main reasons was that perhaps the users did not feel they were sufficiently versed in the correct technical aspects to use the program. Typical comments for this information included: "Did not use." "What are these?" "We need a lot more information on how to run these programs." "I liked these very much and accessed them regularly."

NEWSRELEASE items were also not rated very highly by the end users. This was not too surprising as the NEWSRELEASE program on AGNET is generally considered to be more consumer oriented, although there is much good useful information for livestock and grain producers. Typical comments included: "Very few used." "Checked only on occasional basis." "Some good, some bad." "Especially liked the ones on economic issues." "Some were excellent."

The lowest-rated information source by the end users was the electronic CONFERENCES. Again, this was not too surprising. CONFERENCES are of more use to people other than farmers. It is the responsibility of an individual AGNET user to link with the electronic CONFERENCES. Although the authors had sent out the procedure for doing this via U.S. mail, it is doubtful that many of the users took the time to go through the procedure to link up to the CONFERENCES. Typical responses included: "Did not use." "So what?" "Helped sometimes." "Need more information on how to use." "Not enough conferences sales or prices."

The users were asked to evaluate the timeliness of information delivered by AGNET. The response broke down as follows: very timely--5; average timeliness--5; too late to be useful--1; no response--1.

The users, who were all paying their own computer and telephone costs, had a very high expectation of when the information should be available on the computer. Many times the information for a given day would not be available until the following morning because of the manual transfer of the information onto the system. Once again, this points out the high desirability of automatic linkages with the AMS teletype system so that the information can be available much more quickly. Typical responses to the timeliness question included: "Most information was available from other sources at lower costs like newspapers and radio; however, this service shines in the fact that information is available on demand." "Many times it is hard to check information everyday. Why not put on a program that records daily futures-prices information and then on Friday evening we could pull them off for our records."

Users were asked to report costs. Very few had kept records of their costs, but those who did report indicated that \$50 to \$75 a month was a normal combined telephone and computer cost for accessing the AGNET information. A good share of the users responded that they did take advantage of nonprime-time telephone and computer costs by calling early in the morning or late in the evening.

Only three users indicated any problems from trying to access the information on AGNET. They also indicated that a workshop on operating technique would have been helpful. Most indicated that they felt it was quite easy to use the system. However, six respondents indicated that a workshop on how to apply the information being received from AGNET would be extremely useful.

Users were asked to give suggestions for improving AGNET delivery of market information and whether they felt it was worthwhile to continue providing information across the system. Most of the respondents who indicated that additional information would be desirable were looking for more localized cash prices and more commentary with specific projections for what the markets might do in the future. The overwhelming response was that the direct delivery of market information was extremely worthwhile and that the project should be continued. Only two users indicated that they did not intend to continue accessing AGNET information regularly.

Although very few respondents put a dollar value on the information received, the majority indicated that the cost-benefit ratios for accessing the information were highly favorable.

SUMMARY

The need for better information to be used by agricultural producers in making agricultural marketing decisions has been well documented. The thrust of this study has been to evaluate the feasibility of direct electronic delivery of this needed market information. The development of mechanisms for direct producer access to AGNET via farmer-owned microcomputers and computer terminals was one of the main objectives.

The best evaluation of this project lies in the large increase in retrievals of market information. During the time period of the study there was over a three-fold increase in the number of times that AGNET was accessed for market information.

Of the cooperators in this study, nearly three-fourths accessed AGNET through the use of microcomputers, while the remainder used dumb terminals for communication. It was found that operating a microcomputer in an intelligent mode with AGNET becomes increasingly desirable due to the tremendous potential in telephone savings. Savings of 50% or more resulted from using microcomputers in this manner.

It is relatively easy to turn a microcomputer into a dumb terminal for communicating with AGNET. It becomes much more difficult, however, to operate in the intelligent-terminal mode. This study uncovered software that makes this possible for most brands of microcomputers. The idea of interfacing farmer-owned microcomputers and a regional computer such as AGNET could be one of the most significant thrusts in extension service activities for agricultural computer applications in the future.

The pilot study was successful in providing current market news information to AGNET for retrieval by producers and others. A wide variety of new files has been made available in the MARKETS section on AGNET. Many of these files were the direct result of this pilot study. These files and others will continue to be available on AGNET.

Feedback from the final end user indicated the need for several more types of files, particularly of a regional type. The timeliness of the market information provided on AGNET was a concern to the producers involved in the study. Although the majority of the participants felt that the material was very helpful to them, they also expressed a desire for more timely information. This end-user evaluation points toward a critical need for direct electronic transfers of market information from AMS to the AGNET computer.

In summary, this pilot study provided much needed background information on the electronic delivery of market information. This study found that there is, indeed, a demand for the direct electronic delivery of marketing and management information to farm producers. There exits a distinct opportunity for the extension service to assume an even greater role in the delivery of this timely market information. In addition, the study provided the documentation of the need for increased development of computer applications to agriculture in information networking and evaluation of marketing alternatives. This information provides a base from which the extension service can evaluate and plan their activities in the computer arena for the future.

REFERENCES

Ainsle, et al. 1977. An evaluation of cooperative extension dairy programs. Specialist Report. Cornell University.

Bolen, Kenneth R. 1979. Economic information needs of farmers. Report of ESCS and SEA/Extension Study.

Brown, Thomas R. and Arthur Collins. 1978. Large commercial family farms information needs and sources. A Report of the National Extension Study Committee.

Skelton, Irvin. 1980. Wyoming agricultural extension service accomplishment report for FY-1980.

Watkins, Ed and Sharon Hoobler. 1980. Report of ECOP Subcommittee on agriculture forestry, and related industries extension marketing program and priorities survey. SEA/Extension.

USDA. 1968. A people and a spirit. Report of the joint USDA/NASULGC study committee on cooperative extension. Colorado State University.

WLMIP. 1977. Evaluation of the western livestock marketing information project. Report of WLMIP Technical Advisory Committee Survey of Users.

Part 9

FARM AND RANCH MANAGEMENT AND PRODUCTION SYSTEMS

41

OPPORTUNITIES FOR INCREASING PRODUCTION EFFICIENCY IN INTENSIVE CROP-SHEEP PRODUCTION SYSTEMS

Hudson A. Glimp

INTRODUCTION

The production systems discussed in this series vary both between and within ecosystems. Our system is largely unique to our farm, yet most of it could be applied to all farms within our ecozone and many production practices apply to other ecosystems as well.

Central Kentucky receives an average of 45 in. of rainfall annually. The average date of the last killing frost is around March 25, with the average first killing frost around October 15. Growth of winter-hardy crops may begin as early as March and will not normally cease until approximately December 1. The land is gently rolling to rolling, with generally shallow soils, but the soils are well-drained and highly productive. Approximately 65% of the rainfall occurs in the spring and early summer, with another 20% during the fall months.

Crop production on our farm can be characterized by the following 5-year average yields; corn at 135 bu per acre; wheat at 40 bu per acre; soybeans (full season) at 40 bu per acre; soybeans (double crop behind wheat) at 30 bu per acre; alfalfa hay at 5 tons per acre; and burley tobacco at 2700 pounds per acre.

We are in the land-use business with all that this implies--the most efficient use of land resources through crops that are productive and, hopefully, profitable to produce. We must also use production systems that conserve and improve the soils entrusted to us during our brief tenure on earth as farmers. Economics may dictate short-term variations in our farming program, but the long-term objective must always be to protect and improve those resources entrusted to us. Pasture in intensive cropping areas of the U.S. is often viewed as "growing grass on land not good enough to plow." We still have economists in the area that value pasture at zero. In our view, pastures and forage crops are a vital and competitive part of our integrated farming program. Sheep are currently the most efficient and profitable method of harvesting and merchandising the forages produced on our farm.

INTENSIVE CROP/PASTURE ROTATION SYSTEMS

As stated previously, soil conservation is a major concern in our farming program. The topography, rainfall, and soil types do not lend themselves to continuous cropping by conventional farming methods. No-till and minimum-tillage farming practices were developed in Kentucky through necessity dictated by these factors.

Our basic crop rotation program includes no-till corn, followed immediately in the fall by wheat. Wheat harvest in early summer of the second year is followed immediately by no-till soybeans; thus we harvest three cash crops in two years and have relatively continuous ground cover for protection against erosion. After three or four cycles of this rotation (6 to 8 years) pasture will be established through broadcast seeding of forage crops in winter wheat during February and March. Following wheat harvest, the pasture is then established and will remain for approximately three years. Established pastures include orchardgrass, timothy, or tall fescue as grasses--and red or ladino clover as legumes. Alfalfa is also a valuable crop in the area but is used almost exclusively for hay production.

In addition to soil conservation, the above crop rotation system results in increased soil fertility. The pasture phase contributes to soil organic matter through residue and animal manure and results in reduced fertilizer needs when rotated back into no-till corn.

WHY SHEEP?

Our rotation scheme would not be possible if we did not have a profitable method of harvesting and merchandising forages during that phase of the cycle. Planting all of the forage-crop phase of the cycle to hay crops for harvest and sale might achieve the soil protection part of our objectives but would not meet our soil-building objectives. Sheep are needed in our farming program for several reasons:

- Sheep are profitable to produce. Sheep production generally has been profitable for the last 20 years, and there are no reasons to believe this trend will not continue. In terms of potential for improvement, current production levels could be improved by 50% or more by fully utilizing existing technology. We have increased productivity on our farm during the last 6 years from 100 pounds of lamb per ewe to 125 pounds per ewe per year. We have also increased stocking rates from 3 ewes per acre to 4.5 ewes per acre, thus increasing pounds of lamb produced per acre from 300 pounds to over 560 pounds. We are certain that increases of this magnitude are possible with what we know today.

- Sheep fit our farm operating plan. We have surplus winter labor, which is used in our winter lambing program. When cropping begins in the spring, sheep go to pasture and require minimum supervision. Burley tobacco is a major crop for us, with seven large tobacco barns located on the farm. These barns are empty during the winter months and meet almost all of our housing needs for sheep during the lambing season.
- We like sheep. I cannot overemphasize this component of the profitable sheep-production equation. Sheep require hard work, and there are few shortcuts to doing the job right. Many people are willing to accept substantially less profit and raise cattle simply because they do not like sheep or are not willing to commit themselves to the extra work.

IMPROVED PASTURE MANAGEMENT SYSTEMS

It should be emphasized that we are doing nothing unique or particularly innovative. We are merely applying those practices that have proved to be practical and profitable in other regions of the world. The integration of these components into our production system has created new interest in the potential of sheep in our area and has convinced us that we have much room for improvement.

Improved grass-legume pasture mixes. Improved varieties of orchardgrass and timothy, when combined with legumes such as red or ladino clover, have resulted in dramatic increases in forage quality and productivity. Pasture-management systems that harvest these crops at optimum stages of development and nutrient content have also resulted in increased pounds of forage--and in turn, of--lambs produced per acre. We also are quite interested in other improved forages, particularly perennial ryegrasses and subterranean clovers that may be adaptable to our area.

High-intensity/lower-frequency grazing systems. Probably the greatest increases in both pasture and animal productivity have been due to our conversion to this pasture-management concept. The growth cycle of the grasses and legumes used in our pastures is 30 to 35 days from closely clipped to first-flower stage. Since maximum nutrient content and growth rate of forages are during the preflowering growth stage, our pasture rotation scheme must fit this roughly 30-day cycle. The grazing scheme we use provides enough sheep to closely graze an entire area in a short period of time, then removes the animals from the area to permit regrowth. This regrowth is not permitted to reach the flowering stage of plant growth.

We have developed a pasture-rotation scheme, using 4 to 6 pastures. This translates into roughly 4- to 7-day grazing on each pasture, with 21 to 24 days between grazing periods. If the animals cannot clean the pasture during this grazing period, then it must be clipped. This rotation cycle has greatly reduced our internal parasite problems with sheep. The exception to this scheme is during the extremely rapid growth period from mid-April to mid-June. It is not possible to stock at levels adequate to utilize pasture growth during this period. We then go back to using 2 or 3 pastures in our rotation scheme and cut hay from the remaining pastures. Once those pastures from which hay has been cut have recovered, the sheep are then moved to these pastures and hay harvested from the remaining fields in the unit. During an exceptional year, such as 1982, two cuttings of hay may be obtained from many of the pastures. In most years we will also select certain pastures in the system for harvesting grass or clover seed. Our pastures will normally provide all hay needs for the sheep plus an average half ton of hay per acre for sale.

Conversion to this system would not have been possible without the use of electric fence. Conventional fencing is too expensive to consider in converting to this grazing scheme. The temporary nature of electric fencing is also compatible with our total cropping system. When a field is in pasture, for example, we may have 4 fields of 20 acres but the electric fence will be removed to convert to one 80-acre field for more efficient crop production.

Strategic grazing crops. Another highly profitable practice we have implemented is the use of turnips as a late-fall/early-winter grazing crop. During the past 4 years we have averaged 1150 sheep-grazing days per acre, with an average gain of 20 lb per ewe, from grazing turnips from mid-November to January. We could no doubt get more production from grazing early, but this is a period that our pastures are no longer productive and supplemental feeding would be necessary for late gestation ewes.

We have been able to establish turnips in pasture sod by using a power-till seeder and one pint of paraquat per acre to control pasture growth until the turnips are established. Fifty pounds of actual nitrogen per acre is applied at seeding time. The seeding rate for turnips is 3 to 4 lb per acre when planting no-till in sod and 1 to 2 lb per acre in a prepared seed bed. Turnips should be planted approximately 60 days before first killing frost.

Total cost for seed, fertilizer, chemicals, machinery, and labor amounts to approximately $30 per acre. This converts to less than $.03 per ewe per day feed costs during a critical nutritional period. Prior to using turnips during this period, we were feeding an average of 2 lb hay and 1.0 lb corn per day which would cost over $.10 per ewe per day.

The best seed available at this time is the conventional garden variety of purple-top turnips. We have tried New

Zealand varieties of turnips and swedes (rutabaga) and have also tried Japanese radishes, but have found no significant advantages for these crops or varieties over the purple-top turnip.

EARLY WEANING OF LAMBS

No lambs are weaned until pasture growth begins in early April, so lambs born in late December and January are not actually "early weaned." However, all lambs 60 to 70 days of age in early April are weaned. This practice is continued on a biweekly basis until all lambs are weaned.

This permits grazing the best and cleanest pastures with lambs, which may be followed by ewes in a "clean-up" grazing program. This system has been easier for us to manage than the "forward creep" grazing system pioneered in Great Britain.

Lambs are given free-choice access to creep feeders from approximately 1 week of age until weaning. These creep feeders become self-feeders on pasture postweaning. With the high crude-protein content (17% to 20%) of spring grass-legume pastures in our area, little more than whole or shelled corn is needed in the post-weaning concentrate feed. The lambs tend to consume approximately 80% forages and 20% concentrates when weaned at 50 to 60 lb and shift to approximately 30% forages and 70% concentrate by the time they reach slaughter weight of 100 to 110 lb.

USE OF CROP RESIDUES AND CROP GRAZING

Crop residues can be an important part of intensive crop-sheep production systems. Approximately 50% of the nutrients are left in the field when a corn or soybean grain crop is harvested. Even soybean residues have provided up to 200 ewe grazing days per acre for midgestation ewes in November on our farm.

Many areas of the U.S. plant winter wheat early enough in the fall to provide sufficient growth for significant fall and winter wheat grazing. Areas further north, such as our area in central Kentucky, do not get adequate fall growth but can get 2 to 4 weeks of spring grazing without affecting wheat yields.

42

AN ECONOMIC ANALYSIS OF LAND USE IN AN INTENSIVE CROP-SHEEP PRODUCTION SYSTEM IN KENTUCKY

Hudson A. Glimp

INTRODUCTION

This paper briefly describes a comparative economic analysis of the farm operated by my partner William Balden and myself near Danville, Kentucky. Table 1 presents the average acreages, and the percentage of gross income, net income, and of labor for the various crops on our farm. The figures represent averages over the past 5 years, and vary widely from year to year.

TABLE 1. ACREAGES, INCOME, AND LABOR COSTS, AND RETURNS BY ENTERPRISE*

Farm enterprise	Acres	% of gross income	% of net income	% of labor	% of net/% of gross	% of net/% of labor
Tobacco	50	33	50	50	1.52	1.00
Pasture (sheep, hay, seed)	400	24	33	29	1.38	1.14
Corn	450	16	5	7	0.31	0.71
Soybeans	400	12	5	6	0.42	0.83
Wheat, barley	480	15	7	8	0.46	0.88

* The above data represent 5-year averages for Blue Meadows Farm and may vary widely from year to year.

The basic sheep operation includes approximately 1100 crossbred commercial ewes that are bred to Suffolk rams, 400 Rambouillet ewes bred to Finn X Dorset crossbred rams to produce replacement females, and 100 Suffolk purebreds that produce our own rams plus limited breeding stock sales. In addition, the farm will normally import 400 to 1000 yearling ewes from Texas each year for resale as breeding stock.

A major portion of the corn crop and all of the soybeans, wheat, and barley are produced for seed. In addition to sheep, income from pastures is produced by hay and clover and grass seed. A commercial seed cleaner located on the premises processes small grain, soybeans, clover and grass seeds, while the hybrid seed corn is grown under contract to a neighbor in the hybrid seed-corn business.

INTEGRATED DIVERSIFICATION

It is important to emphasize the diversification of our farming program, which we consider essential for survival in today's agricultural economy. It is our strong belief that integrated diversification, rather than specialization as advocated for the past 20 years, is essential.

Our various farm enterprises are integrated in terms of their distribution of labor requirements, facilities, and equipment needs, and must fit into the long-range farm-utilization plan. Integration vertically through sales of seed stocks from the grain crops and sheep, and to an increasing extent direct consumer sales of lamb and mutton, continue to be important factors in our long-range plans. The enterprises should be diversified in terms of cash-flow requirements and income periods and should be noncompetitive for markets.

As shown in table 1, tobacco is the major crop produced on our farm and contributes 50% of our net income from only 50 acres. The sheep, hay, and seed income ranks second, with 33% of net income. During the 1970s the food and feed-grain crops were more important but are not profitable to produce at current prices. It is our fervent hope that prices for these will recover and their contribution to our farm economy will become more positive.

One of the tenets preached during the recent past, in addition to specialization, has been to get bigger--borrow more money and buy more acres and get larger equipment so the farmer can "more efficiently" utilize his labor. However, the grain crops to which this applies are currently losing money, while those crops (tobacco, sheep, hay) requiring the least equipment and the most manual labor are the most profitable to produce. These enterprises have resisted the "substitution of capital for labors" progression that has occurred in most other agricultural enterprises. A farmer has limited capital that he can supplement with his knowledge and labor to convince his banker to lend additional money for financing his farming operation. He must be careful to apply his labor resources, as well as his financial resources, to those enterprises that will yield the greatest return on his investment.

With tobacco and sheep producing over 80% of farm net income from less than 60% of gross income, one might legitimately ask why we do not increase production of these enterprises. First, the pounds of tobacco we can produce is strictly controlled by allocation. The 1600 ewes are about the number we can manage in the tobacco barns and other facilities available for sheep. We are not willing at this time to invest in specialized buildings for sheep production. In addition, although we like to work with sheep, the 1600 ewes in our production system provide about all the work we want.

CAN PASTURE COMPETE ECONOMICALLY?

"But this land is too good for pasture!" are the words I continue to hear throughout the grain belt. I would agree with this statement if all we are going to do is turn a few old ewes or cows out in the back pasture and check on them once a month. Intensive, pasture-management systems, however, can be competitive with grain crops for profitable land use.

The other question most frequently asked is, "Can I buy a farm and pay for it through crops produced from that land?" For the purposes of the discussion, the model used in tables 2 and 3 is a farm we purchased 6 years ago for $1200 per acre that has current annual payments of approximately $150 per acre. Our actual land costs are approximately $80 per acre since only about one-fourth of our land has payments at this level; the rest are being leased for $35 to $65 per acre.

Table 2 shows a net return to capital after land payment of $13.34 per acre when pasture is sold through sheep and hay in our system. The net return to capital and labor, which is the figure of interest in a family farm structure where the family provides the labor, is $89.84 per acre after land payments. In other words, 100 acres of pasture carrying 450 ewes at these levels of productivity would be a modest family unit that should meet land payments plus provide approximately $9,000 net income to the family.

Table 3 shows a net loss to capital per acre from corn production of $17.75 per acre, or less $4.00 return to capital and labor for a family farm unit. These figures were based on average corn values for the last three years of $2.75 per bushel. Corn currently is selling for $2.25 per bushel and is projected to decline further. These figures clearly illustrate the dilemma of the young corn farmer trying to make land payments and provide a living for his family. This clearly spells disaster for our farm economy in the grain-belt states.

Sheep require more labor, with 18 hours per acre required for sheep versus 5 hours per acre with corn. Sheep may be dirty and smell bad and may be hard work requiring long hours on cold winter nights during lambing, but we like the smell of money! These data may or may not be typical, but they clearly illustrate what can be done with sheep in an intensive crop-livestock production system.

TABLE 2. EXPECTED COSTS AND RETURNS FROM SHEEP WITH 4.5 EWES PER ACRE

EXPECTED RETURNS:	
Lamb @ 125 lb per ewe x $0.65 per lb x 4.5	$365.63
Wool @ 11 lb per ewe and lambs x $1.10 per lb x 4.5	54.45
Cull ewe @ 0.16 ewe x $40 per ewe x 4.5	28.80
Surplus hay @ 0.5 tons per acre x $70 per ton	35.00
Total Returns Per Acre	$483.88
EXPECTED COSTS	
Pasture expenses (fertilizer, mowing, fence, etc.)	$14.00
Corn @ 4.5 bu/ewe x $2.75 per bu x 4.5	55.69
Mineral and feed supplements @ $2.40 per ewe x 4.5	10.80
Hay (nonland costs) @ $18 per ton x 0.2 ton per ewe x 4.5	16.20
Power and fuel @ $2.00 per ewe x 4.5	9.00
Ram costs @ $2.00 per ewe x 4.5	9.00
Veterinary and drugs @ $2.50 per ewe x 4.5	11.25
Marketing and promotion @ $3.50 per ewe x 4.5	15.75
Shearing @ $4.80 per ewe x 4.5	21.60
Labor @ 4 hours per ewe x $4.25 per hour x 4.5	76.50
Buildings and equipment @ $15 per ewe x 10% x 4.5	6.75
Ewe depreciation @ $6.00 per ewe x 4.5	27.00
Taxes	4.00
Operating capital interest	43.00
Land charge	150.00
Total Costs Per Acre	$470.54
Net Returns to Capital Per Acre	$13.34
Net Returns to Capital and Labor Per Acre	$89.84

TABLE 3. EXPECTED COSTS AND RETURNS FROM 130 BUSHEL CORN PER ACRE

Expected Returns	
Grain @ 130 bu per acre x $2.75 per bu	$357.50
Expected Costs	
Fertilizer	$43.00
Seed	18.00
Chemicals	35.00
Buildings and equipment (DIRTI)	8.00
Machinery depreciation, maintenance, interest	20.00
Machinery operating expenses @ $10 per hour x 4 hours	40.00
Marketing and hauling @ $.20 per bu	26.00
Labor @ 5 hours x $4.25 per hour	21.25
Operating capital interest	14.00
Land charge	150.00
Total Costs Per Acre	$375.25
Net Return (loss) to Capital Per Acre	($-17.75)
Net Return to Capital and Labor Per Acre	$3.50

43

INCREASING MARKET LAMB PRODUCTION EFFICIENCY

Clarence V. Hulet,
S. K. Ercanbrack

INTRODUCTION

Lamb production efficiency should be measured in relation to the total pounds of lamb meat produced per pound of feed consumed by the entire flock. The quality and cost of that feed also must be considered. These assumptions imply that for greatest lamb-production efficiency--every female in the flock should begin production as early in life as possible and produce lambs as rapidly as management constraints permit. Lambs should be fed until efficient gains stop.

Four practices that can have a great impact on improving efficiency are: 1) breeding ewes to lamb at 1 year of age, 2) producing more twins, 3) improving rate and efficiency of gain, and 4) producing larger lambs.

PRODUCING A GOOD LAMB CROP AT 1 YEAR OF AGE

When a strike idles a manufacturing plant for a year it usually spells economic disaster. Similarly it is costly to maintain a ewe lamb for an extra year before she starts lamb production at 2 years of age. This 1-year delay is common practice in the range industry. Many years ago the fleece would about pay the year's maintenance cost, but this is no longer true. In 1981, it cost $50.92 to maintain a replacement yearling ewe for 1 year at the Sheep Station. The fleece produced by this ewe brought $15.85, reducing the net cost of maintenance to $35.07. On the other hand, those ewes that were successfully bred at 7 to 8 months of age on the range and produced an average market lamb, in addition to the fleece, showed a profit of $10.08. This is a total income difference of $45.15 per ewe.

The pioneers in the range sheep industry who are now breeding ewe lambs are to be congratulated for their progress. Under some native range conditions where the environment is severely restrictive, breeding to lamb at 1 year of age could not be recommended. However, under many other conditions one of the easiest and fastest ways to improve production and profit in the sheep enterprise is to breed ewe lambs to produce their first lamb crop by the time they are yearlings. In addition to the extra lamb crop, the total lifetime production of these selected ewes is greater than that when breeding is postponed. It is a commonly observed that ewes are easier to manage at first lambing at 1 year of age than at 2 years of age.

Table 1 shows what can happen to total lamb production when selection is combined with breeding ewe lambs. In the first group, ewes were bred as lambs and pregnancy tested; only pregnant ewe lambs were saved. The open lambs were sold as choice fat lambs for slaughter. The second group represents those under the common management practice of first breeding replacement ewes at 18 months of age so as to lamb at 2 years of age. Note the great advantage in lamb production each year of those bred (and selected for pregnancy) to lamb first at 1 year of age. Studies in New Zealand (Ch'ang, 1967) have shown that selection of ewes based on their ability to breed as ewe lambs will improve the twinning rate of a flock as rapidly as will selecting directly for twinning.

TABLE 1. LAMB PRODUCTION OF TARGHEE RANGE EWES BRED AND SELECTED ON PREGNANCY AS 1-YEAR OLDS VERSUS EWES LAMBING FOR THE FIRST TIME AS 2-YEAR OLDS (USDA, SEA-AR, DUBOIS, IDAHO)

Management practice	Age of ewe (yrs)	% lambs born[a]	% lambs weaned	Total lambs weaned (lb)
Lambed first as 1-year-olds	1	111	83	56
	2	143	115	84
	3 & over	158	134	107
Lambed first as 2-year-olds	1	0	0	0
	2	102	82	58
	3 & over	141	115	87

[a] Percentage of lambs born or weaned of ewes in the flock.

Breeding ewe lambs is not easy. A lot of planning and good management is required, but it pays good dividends.

Points for Successful Breeding of Ewe Lambs

Consider the breed. Research and practical experience have shown that certain breeds mature earlier and conceive more readily than do other breeds (table 2). We have been especially successful in breeding lambs of the new Polypay breed for production at 1 year of age. Crosses with as

TABLE 2. EFFECT OF BREED ON FERTILITY[1] AND PROLIFICACY[2] IN EWE LAMBS (1978)

Breed	% ewes		Prolificacy	
Rambouillet (R)	54		112	
Targhee (T)	42	45*	105	111*
Columbia (C)	34		115	
Polypay (Poly)	90		138	
Poly x R, T, C	88		134	
(Finn x T, C) x T, C	74	86*	122	141*
Finn x R, T, C	91		158	

1 Percentage lambing of ewes exposed to rams.
2 Percentage of lambs born to ewes lambing.
* Weighted means.

little as 12.5% Finnsheep breeding have high conception rates. Targhees, Columbias, and Rambouillets are slower maturing but usually can be managed for relatively successful early breeding with a combination of good nutrition, breeding at the right time of year, and pregnancy testing. Our studies and other research show that success in ewe lamb breeding will improve if managers select lambs for this ability.

Breed at the right time of year. Ewe lambs born early in the lambing season are more likely to breed as lambs. Studies by the University of Idaho and the U.S. Sheep Experiment Station, Dubois, show that the breeding season of ewe lambs is much shorter than that of mature ewes. November is the optimum breeding time for April-born lambs at Dubois (Dahmen and Hulet, 1974). October is a satisfactory breeding time for lambs born in January and February. September is less satisfactory. However, Polypay lambs have been bred successfully in August at Dubois. Ewe lambs can be bred a month to 6 weeks earlier in California, Texas, and other areas at lower latitudes.

Feed lambs to assure fast growth and development. Adequate nutrition is essential for lambs to reach the size and sexual development necessary for early breeding. The ewe lambs should gain at least 1/3 of a lb and preferably 4/10 of a lb per day after weaning at 140 days. During the grazing season, a palatable green-grass pasture, preferably containing some legumes, should be adequate. Protect pasture-managed lambs from stomach worms. If good pasture is not available, supplement grazing so that the ewe lambs will gain at the required rate per day. Beet tops, aftermath in grain fields, or alfalfa fields have been used successfully. Good quality range can be used with adequate supplementation with range cubes or other concentrates. For ewe lambs in confinement, feed alfalfa pellets at a rate of 4.0 to 4.5 lb per day. High-quality hay with 1/2 to 3/4 lb of grain per lamb per day can produce satisfactory gains.

Breed separate from mature ewes. Studies have shown that ewe lambs breed better when kept separate from mature ewes. Ewe lambs are shy and cannot compete well for the attention of a ram while an aggressive mature ewe is in heat. Furthermore, the lamb cannot compete as well for the available feed when fed with mature ewes.

Pregnancy test. Many relatively satisfactory pregnancy-testing devices are now available. The Scanopreg II and Pregmatic II have been tested at the Station and each was found to be quite accurate and efficient in ewe lambs. (Mention of a trademark or proprietary product does not constitute a guarantee or warranty of the product by the U.S. Department of Agriculture and does not imply its approval to the exclusion of other products that may also be suitable.) Accuracy is poorer in fat ewes. Ewe lambs can also be pregnancy tested using the palpation-rod technique. We recommend that ewe lambs be pregnancy tested from 60 to 70 days after the rams are removed. For greatest improvement, save only the pregnant ewe lambs and send the open lambs to market as fat lambs. Depending on breed and environmental conditions, the pregnant ewe lambs should produce from 80% to 120% weaned lambs.

PRODUCING MORE TWIN LAMBS

Studies have shown that ewes selected for lambing at 1 year of age will produce more twins than those not selected at that time. Crossbreeding with prolific breeds such as the Finnsheep can quickly increase both the number of lambs born and the weight of lambs weaned. Breeding of Suffolk rams to Finnsheep x native western ewes results in high twinning rates of growthy lambs with very satisfactory carcass quality. There has been some resistance to Finncross ewes in the sheep industry because of carcass quality. We believe that in many instances this is not warranted, particularly when Finncross ewes are mated to Suffolk rams.

The Polypay is a new, developing breed designed to fit the needs of those who want a prolific breed but object to the Finncross ewes. Polypay ewes in preliminary tests have shown outstanding lamb production under typical range conditions. Table 3 compares yearling and mature lamb production performance of Polypay ewes with that of Rambouillet, Targhee, and Columbia ewes. The Polypay yearlings weaned a 90% lamb crop; whereas lamb crops were 10% to 18% for the other range breeds. The mature Polypay ewes weaned a 164% crop, which was 42% to 62% percentage points better than the performance of the other breeds. We have over 50 Polypay ewes with a lifetime average production of over two lambs weaned per year, including the first year of age. This would equate to about 148 pounds of lamb weaned (140 days) per year. Five sheep equal one cow in animal units. This means that one beef cow would have to wean over 740 pounds of calf per year to equal the production of these ewes.

TABLE 3. COMPARATIVE LAMB PRODUCTION OF VARIOUS BREED CROSSES UNDER HERDED RANGE CONDITIONS AT THE U.S. SHEEP EXPERIMENT STATION (APRIL 1977)

Age at lambing and breed	Lambs born to ewes lambing	Lambs weaned of ewes bred
12 months	%	%
Rambouillet	106	18
Targhee	102	13
Columbia	104	10
Polypay	135	90
Mature		
Rambouillet	160	122
Targhee	160	110
Columbia	162	102
Polypay	207	164

The carcass quality of Polypay lambs appears excellent and the growth rate is very similar to that of the other range breeds. Polypay wether lambs on feed at Dubois reach market condition at an average of about 110 lb body weight. However, Polypay wool production, though of good quality, is about 2 lb per head per year lower than that of the Rambouillet or Targhee.

Dubois Polypay lambs gain at about the same rate as Rambouillet, Targhees, and Columbia lambs. Selection efforts are now being made to improve the growth rate of Polypay lambs and to extend the breeding season or eliminate the nonbreeding season so that ewes can be bred at any time of year and more frequently than once per year.

IMPROVING RATE AND EFFICIENCY OF GAIN

Efficiency of gain is related to the amount of gain made by lambs in relation to amount of feed consumed. When one improves efficiency of gain he reduces the amount of feed required per unit of gain. Rate of gain is highly correlated with efficiency of gain. This simply means that the faster a lamb grows the less feed required per pound of gain.

Our work has shown that young lambs gain much more efficiently than older lambs. This suggests, and research work demonstrates, that young lambs can be fed and marketed much more economically than older lambs.

A group of Polypay lambs were weaned at 31 days of age at the U.S. Sheep Station. They were fed diets consisting of about 60% cereal grains, 22% soybean meal, 15% alfalfa meal plus limestone, trace mineral salt, and vitamin complex. The average daily gain to 97.4 lb was .64 lb on an average of 3.55 lb of feed per pound of gain. (J. Doyle, personal communication).

Some lambs gain faster than do others and are more efficient. Is this trait heritable and, if so, how can one select to make improvements in both rate and efficiency of gain and what can the rate of that improvement be?

A preliminary five-year study at the Sheep Station showed that selection was ineffective when it was based on a 6-week test period. However, there was evidence that suggested that accuracy of selection would be improved if the feeding test period were extended over a longer period of time. Therefore, the test period was extended to 16 weeks. The test procedure was as follows:

- Spring-born lambs were weaned at about 80 days of age and placed on test. The test period was divided as follows:
- 4-week group-feeding period
- 6-week individual-feeding period
- 6-week group-feeding period
- The lambs were given all they would eat.
- The diet during the first 10 weeks was 37% barley and 63% alfalfa pellets.
- The diet during the final 6-week period was 100% alfalfa pellets.

The results of the study showed that accuracy of estimating genetic values increased over the 16-week feeding period (table 4).

TABLE 4. EFFECT OF TIME ON THE ACCURACY OF ESTIMATING GENETIC VALUES

Estimated heritabilities	Weeks on test						
	4	6	8	10	12	14	16
Body weight	.39	.38	.51	.54	.57	.65	.75
Gain per day	.22	.18	.38	.52	.61	.64	.71

Percentage superiority of selected lambs over control for rate of gain and efficiency of gain increased 3.5% and 2.9%, respectively, over the 4-year period (table 5).

TABLE 5. PERFORMANCE OF SELECTED LINE AND UNSELECTED CONTROLS DURING THE 16-WEEK TEST PERIODS OF 1978 THROUGH 1981

	Year			
	1978	1979	1980	1981
Rate of gain:				
Mean of selected line	.4528	.4698	.6216	.5914
Mean of controls	.3927	.4026	.5218	.4911
% superiority of selected line	15.3	15.7	19.1	20.4
Relative increase of selected line = 10.36%/year				
Efficiency of gain:				
Mean of selected line	287.2	245.9	326.3	289.9
Mean of controls	250.5	211.6	264.2	235.0
% superiority of selected line	14.7	16.2	23.5	23.4
Relative increase of selected line = 18.28%/year				

BIGGER MARKET LAMBS MEAN MORE EFFICIENT MEAT PRODUCTION

Can you imagine slaughtering beef cattle at 700 to 800 pounds? This would be comparable to the current practice of slaughtering lambs at 90 to 110 pounds. Heavier lamb carcasses are thought to be overly fat and wasty. This may have been true for the smaller breeds that were once popular. However, in the now-popular, larger, slower-maturing range breeds, lambs (especially ram lambs) are making efficient lean gains at weights well above 110 pounds.

Studies at the Sheep Station show that slaughter weights of Rambouillet, Targhee, and Columbia ram lambs may be increased to 140 pounds and ewe lambs to at least 125 pounds without significantly decreasing the percentage of lean meat in the carcass and without affecting tenderness or juiciness. Furthermore, slaughtering and processing costs per pound of meat at heavier slaughter weights are actually reduced, since it costs no more to slaughter, dress, and process larger lambs than smaller ones.

In cooperative work with the University of Idaho, taste-panel scores on overall acceptability and tenderness of cooked lamb racks were found to actually improve slightly as associated carcass weights increased from 46 pounds (95 pounds live) to 79 pounds (153 pound live) (Sauter et al., 1977). More important, actual consumer-acceptance tests, conducted in cooperation with Utah State University in both California and Utah retail markets and involving responses from over 338 customers, revealed that consumers liked cuts from heavier carcasses as well as those from lighter carcasses. Ninety-seven percent indicated that they would purchase the same cut again and the remaining 3% gave reasons for dislike that were in no way associated with carcass weight (Mendenhall and Ercanbrack, 1979). In further tests at Utah, cooking yields of roasts increased slightly, and tenderness also improved slightly as carcass weights increased to 70 pounds.

In the above tests consumers did not discriminate in the least against ram lamb carcasses as evaluated on the basis of tenderness, flavor, and juiciness scores. Nor were significant differences in cooking yields observed among roasts from ram, ewe, and wether lambs. Cuts from ram lambs also were found to be acceptably tender by criteria prescribed by the American Meat Science Association. The above findings are especially significant because ram lambs were found to be about 22% more efficient in feed conversion than were ewe lambs.

CONCLUSIONS

Lamb production efficiency can be markedly improved by proper management and breeding of ewe lambs, improving prolificacy in ewes, and marketing fast-growing, high-quality lambs at heavier weights.

ACKNOWLEDGMENT

The authors acknowledge cooperation of the University of Idaho, Moscow, Idaho.

REFERENCES

Ch'ang T. S. 1967. Indirect selection for higher fertility in the Romney ewe. Sheep Farming Ann., Massey University, Palmerston North, New Zealand. p 7.

Dahmen J. J. and C. V. Hulet. 1974. Boost profits by breeding ewe lambs. University of Idaho Current Information Series, No. 247.

Mendenhall, V. T. and S. K. Ercanbrack. 1979. Influence of carcass weight, sex, and breed on consumer acceptance of lamb. J. Food Sci. (In press.)

Santer, E. A., J. A. Jacobs and C. E. Robinson. 1977. Effect of carcass weights on taste panel score and shear values of lamb roasts. Univ. of Idaho Agr. Exp. Sta., Misc. Series 41:89.

44

TRIPLETS UNDER RANGE CONDITIONS? IT CAN BE DONE

Arthur Christensen

For at least 7000 years, man has tended sheep for food and fiber. In the U.S., we have seen the industry shrink to a point where we now eat less than 2 lb of lamb and mutton annually per person, and wool has been crowded out by synthetics. During the last year, lambs have sold for as little as $.40 per lb and wool has fallen under $.60. Is there a way out for the sheep industry? It is obvious that if we are to survive we must cut costs or increase production. Since most expenses are fixed, it would seem that our only hope is to produce more. It has been said that our methods have changed very little since biblical times.

I would like to share my views on increasing lamb production. In doing this, I shall review our 30 years of experience in raising twins and triplets under semirange conditions and discuss our most important production tool--a good set of records.

As a matter of introduction, allow me to present a short history and description of our operation. The Christensen family has raised sheep in the Dillon, Montana, area for over 60 years. I, my wife Margaret, and our children have been involved since 1950. We also raise cattle, grain, and hay. Our ranch varies in elevation from 5500 ft to 8400 ft. We are in a 12 to 14 inch rainfall area. In recent years we have run between 450 and 600 ewes in 35 sheep-tight pastures with no herder.

Our ewes are predominantly one-half Finn x one-half Targhee of Dubois, Idaho, breeding. Our main feed is native range for 9 months of the year. We shear in February and shed lamb in March and April. All of our animals are individually identified with ear tags. We have kept individual animal records since 1954. During the last 3 years our mature ewes have weaned a lamb crop of slightly over 200%. Our place is not very fancy--long on hay wire and a little short on paint.

Why did we get involved in this type of operation in the first place?

As a young man I was a charter member of the Montana Beef Performance Association. I worked my heart out with performance records and artificial insemination. I soon became disillusioned when I realized that a cow's production could be improved by no more than 20% or 30%. On the other hand, a ewe's production could easily be doubled. In 1954 my wife and I started keeping individual records on our Targhee ewes. We kept only twin ewe lambs from high producing dams. We built better facilities and improved our feed program. Our lambing percentage was raised from 120% to about 150%. After trying half a dozen breed combinations, we started with Finnsheep breeding in 1972. Over the last 10 years we have moved up into the 200% bracket. Our goal is to develop a ewe that, under proper management, can easily raise three 80 lb lambs in 5 months. Although we haven't done it, we believe that 1000 lambs from 400 ewes is possible. We know of several small operators who have produced a 235% lamb crop.

There is nothing wrong with an operation that raises a 1005 lamb crop, provided their costs are low enough. On the other hand those that have high land and labor costs, (that's most of us) must produce twins or more from each ewe each year to show a profit. Twins can be raised with only slightly more cost than a single. We have found that twins are even more profitable than triplets on most operations. Although we have raised some triplets for the last 20 years, it was not practical until 3 or 4 years ago. Despite many problems, it is possible to raise three lambs without excessive cost. There are at least a hundred management practices that contribute to high production. But if a ewe is to raise three or more lambs, the essential changes to be made are in genetics, feed, and management.

GENETICS

Ewes must be selected for their ability to raise litters. All of our ewe lamb replacements are of multiple birth and from ewes that have the potential to raise three or more lambs. Although we have some Targhee ewes that raise three and fewer lambs, the average ewe cannot compete with the Finn-cross.

What about the Finnsheep breed? The Finn is the ugly duckling of the sheep industry, but she is still the only ewe that can consistently contribute the genetic ability to exceed a 200% lamb crop. We have raised every combination from purebred to one-eighth crosses. The Finn ewes are very prolific (more so than any other common breed), are good mothers, and reach puberty earlier than most breeds. We have had two one-half Finn ewes that gave birth to sextuplets. I understand that purebreds have given birth to seven lambs. The rams are very aggressive and fertile.

On the negative side, purebred Finns have poor wool and poor conformation, are prone to have watery eyes, and are nervous. "They can jump the 4 ft fence that the Suffolk crawls under."

We find that half-and-half Finns have few undesirable traits and have a better chance of raising their lambs than purebred Finns.

FEED PROGRAM

On October 1, ewes are flushed on third crop alfalfa or grain stubble. They are kept on a high level of nutrition through breeding. November 15 mature ewes return to average mountain pasture. Replacement ewe lambs are kept on high-level rations, (alfalfa hay and No. 1 grain). A pregnant ewe lamb should never know a hungry day until she weans her first lamb. January 25, ewes are started on 1/3 lb of barley plus 1 to 2 lb of alfalfa hay. By shearing time, Feburary 10, ewes are eating No. 1 barley plus grazing and hay--depending on the weather. March 1, lambing starts. All ewes are fed a mixture of oats, rolled barley, and alfalfa pellets after lambing. Ewes with two or more lambs consume 7 lb to 10 lb per day. Mature ewes with single lambs are returned to pasture at 2 weeks of age. Twins, triplets, and quads are segregated into groups of no more than 10 ewes and their lambs. They are fed concentrate in self-feeders and choice alfalfa hay. April 15 twins return to pasture with a self-feeder, and by May 15 all sheep are turned out to pasture for the summer.

MANAGEMENT

Early shearing. Removing the fleece 3 weeks before lambing increases feed requirements and stimulates milk production. Ewes are cleaner, easier to handle, and require less space at lambing time. After lambing, sheared ewes seek shelter, which reduces the weather stress on their lambs. A few lambs are always bumped when ewes are separated for late shearing. Ewes should not be sheared in the winter in bad climate unless shed space is available. Cold rain causes more stress than low temperature or snow.

Lambing. Newborn lambs are placed in one of the 75 individual pens (jugs) in the lambing barn. These pens vary in size from 12 to 24 sq ft. Each contains feed and running water and is cleaned and rebedded every day. Triplets and quads are left in the lambing shed from 3 to 6 days, depending on the weather. A wood-burning stove keeps the temperature between 35° and 45°. Hot water and a warming box are available for chilled lambs. Extra, orphan lambs are kept on hand for grafting. A mature ewe that gives birth to a single is given a second lamb by the wet graft method. Portable stanchions are used to restrain the ewe. Ewes that lose their lambs are also given two grafts if they have sufficient milk. After trying about every known method of convincing a ewe that she should accept a lamb, our record is about 75%. "Nothing works all of the time." Those lambs that are not accepted by a ewe are raised artificially on milk replacer. However, our hand-raised lambs have not been profitable.

Gathering pens. Weather permitting, 5 or 6 ewes and their lambs are put outside in each of 20 small gathering pens after 2 to 5 days. No more than 25 ewes and 75 lambs are placed in a large gathering pen until lambs are at least 6 weeks old. One hundred ewes with twins is acceptable. These pens have plastic covered shelters.

We have made many management mistakes in our time, but turning triplets out too soon in too large a pasture has always led to trouble. On one occasion, we lost 75 head of lambs in a blizzard. By June 1, all ewes are bunched and rotated through fenced pastures until weaning time.

Labor. Our labor requirements are less than most operations from June 1 to March 1 because we do not herd. Our lambing operation is very intensive and requires a lot of labor. Ewes are checked every 20 to 30 min 24 hrs a day. This requires at least three full time lambers plus some extra help for feeding and pen cleaning. All outside pens are checked at least once a day. For years, nearly all of our lambers have been women. My wife Margaret is shed boss. During the last two seasons we have had student interns from Utah State University and we couldn't ask for better help. To many men, lambs are nothing more than stinking sheep. To most women, they are little babies that need to be loved and cared for. To be a successful shepherd you must love animals and enjoy the work. Our pens are set up to save labor, which includes self-feeders.

Health. I understand that nationwide 25% of all lambs born are lost before weaning. Keeping ewes and lambs alive and healthy is our biggest challenge. We try to do things right, but are still doing badly. We vaccinate ewes for Vibrio, Enterotoxemia, over-eating or sore mouth (between 2 and 3 weeks of age), and Enzootic abortion. The last experimental vaccine we used was not effective because we had about 60 ewes abort one or all of their lambs this year. We hope for better luck next lambing season with a new vaccine.

Mastitis is a major problem, both for us and nationwide. Unfortunately, we have no effective solutions nor do we know of a veterinarian with any. Animals can be successfully treated if mastitis is detected early enough.

No one has a good answer on what to do about nasal bots or progressive pneumonia, but we will eventually find some answers. Keeping sheep healthy is like wrestling an octopus, when one problem is solved, two takes its place. We don't have problems with foot-rot, worms, or lamb scours. We have an excellent veterinarian that we call on for most of our health problems. He also does a postmortem inspection of our losses from unknown causes. Our state research laboratory also works closely with us.

Now let's look to the future. I'm assuming that most of you are small operators or just getting started with sheep and don't really care whether the Christensen outfit raises two, three, or four lambs per ewe. You are concerned with what you can do to line up your ducks right. Dwight Halloway, former director of the lamb and wool project at Pipestone, Minnesota, recently said, "There is no livestock enterprise with such a wide gap between the industry production average and what is achievable in the future." The national average lamb crop is under 100%. We know of growers that are crowding 250%. The exciting thing about our story is not that we are doing so well but that we have done a fair job of raising our production in spite of our many shortcomings.

Most of us still have as many questions as we do answers about increasing lamb production. If the sheep industry is to prosper, we must find a better way. It's hard to argue with the man who says "Why raise two ewes when one can produce as much?"

45

STATUS AND USE OF LIVESTOCK-GUARDING DOGS IN NORTH AMERICA

J. S. Green, R. A. Woodruff,
Clarence V. Hulet

INTRODUCTION

The use of dogs to protect sheep and other livestock from predators is an ancient concept that can be traced back many centuries B.C. in Eurasia (Bordeaux, 1974). The Spanish conquistadors brought guarding dogs with them to the western hemisphere to help protect their flocks (Lyman, 1844). Darwin (1839) noted that guarding dogs were commonly seen with large flocks of sheep in South America. Despite this long history, the use of dogs to protect livestock in the United States is relatively new, with the exception of the Navajo Indians who have used dogs with their sheep and goats since the early 1700s (Black, 1981).

Guarding dogs continue to accompany and protect livestock in many Old World countries including Turkey (Nelson and Nelson, 1980), Italy (Breber, 1978), India (Bardoloi et al., 1979), and Yugoslavia (Coppinger and Coppinger, 1980a). Livestock-guarding dogs have primarily been used in the United States since the mid to late 1970s (Green and Woodruff, 1980), although isolated anecdotal accounts of successful dogs were reported earlier. Dogs are currently in use in the majority of states where significant numbers of sheep and goats are raised, with concentrations in Colorado, North and South Dakota, Oregon, Texas, and the New England states. There are no published figures on the number of guarding dogs currently in use, but the number is apparently growing steadily.

Only recently have researchers begun to ask pertinent questions regarding livestock-guarding dogs: which breeds are most suited to the task, how should dogs be raised and trained, and perhaps most important, can dogs effectively reduce the loss of livestock to predators (Coppinger and Coppinger, 1978; Linhart et al., 1979; Green and Woodruff, 1980).

Interest in guarding dogs is growing because of their effectiveness in reducing or eliminating predation on sheep. A survey showed that over 75% of the dogs were effective in reducing losses (Green and Woodruff, 1980). In addition, using a dog is relatively trouble-free, economically practical (Green et al., 1980), and environmentally and aesthetically appealing. Suitable guarding breeds are becoming more readily available, and no special skills or equipment are generally needed to raise a puppy to a successful adult (Arons, 1980; Coppinger and Coppinger, 1980b).

Research with livestock-guarding dogs in the United States is centered at two locations: Hampshire College's New England Farm Center (NEFC) in Amherst, Massachusetts, and the U.S. Sheep Experiment Station (USSES), USDA Science and Education Administration in Dubois, Idaho. Many cooperating sheep producers are assisting the research effort by rearing and working with dogs from the two research centers, and many independent sheep producers have purchased and raised their own guarding dogs. In addition, USDA has funded two cooperative research projects on guarding dogs--one at Colorado State University and the other at Brigham Young University.

Researchers at the NEFC are working with several breeds of dogs from Europe and Asia, including the Maremma (Italy), the Shar Planinetz (Yugoslavia), the Anatolian Shepherd Dog (Turkey), and various crosses of these breeds. Researchers at the USSES are working with the Hungarian Komondor, the Great Pyrenees, and the Akbash Dog.

Relatively few quantitative data have been compiled comparing livestock predation losses with and without the protection of a guarding dog. Probably 2 to 5 more years of study will be needed to provide accurate assessment of the effectiveness of guarding dogs.

DISCUSSION

What Is a Livestock Guarding Dog?

A successful livestock-guarding dog possesses several key characteristics: 1) it remains with or near the sheep continuously (or at least during times when the potential for predation is high), 2) it does not harm, chase, or harass the sheep, and 3) it is appropriately aggressive to potential predators (Coppinger and Coppinger, 1980b). Although a variety of dogs may possess the necessary characteristics, several specific Old World breeds appear to display the required traits most consistently.

There is no consensus as to which breed of dog is the best livestock guardian and, indeed, there may never be a "best" breed. Individuals have reported protection offered by a variety of breeds, some of which have traditionally never been guardians. Further research may identify which dogs are best for livestock protection. It is also conceivable that certain breeds or lines can be matched to specific guarding situations (pastures or open range).

Socialization and Training

Although attentiveness to sheep and aggressiveness to predators are traits that have been selected over the centuries, the guarding dog's inclination to remain with sheep is increased with early and continual socialization. It appears that placing a puppy (approximately 8 weeks of age) with several lambs enhances the formation of a bond to sheep. Such a bond is critical since the dog is later placed alone with sheep and away from human supervision. The guarding behaviors (aggressiveness and tendencies to patrol, scent-mark, and bark) appear to be largely instinctive but can only be used to an advantage when the dog remains near the sheep.

Generally, the guarding dog is not a pet, and it is important that this distinction be made at the outset. The dog must learn that its place is with the livestock and not at the farmhouse. The dog should be praised for correct behavior and reprimanded for unacceptable behavior. Consistency and repetition will help the guarding dog learn to perform its intended function.

Dog Age and Effective Guarding

Mature and effective guarding dogs are not generally available to most sheep producers. Most dogs are purchased as pups and must be reared under appropriate conditions until they become mature enough to repel predators. This level of maturity varies among individual dogs, and there is no predetermined age when an adolescent dog can be expected to become an effective guardian.

Several criteria indicate a dog's readiness to assume the guarding role. The following behaviors tend to increase in frequency as guarding maturity is reached: 1) male dogs (and sometimes females) use raised-leg urinations rather than squat urinations to scent-mark, 2) scent-marking (urination and defecation) becomes more deliberate, and marks are concentrated near the periphery of a pasture, 3) barking at novel stimuli becomes more frequent and direction oriented, 4) dogs become more interested in the sheep than in the handler, and 5) deliberate patrolling activities increase in frequency and duration. We have observed many of these behaviors in dogs as young as 4 1/2 months of age. However, before a young dog is placed where sheep predation losses are high, it should be capable of defending itself if confronted by several coyotes.

As a dog becomes more experienced, its behaviors (barking, scent-marking, patrolling) are usually more closely oriented to its guarding responsibilities. If coyotes are frequently near, the dog may mark and patrol more than if coyotes are rare. As a dog becomes more familiar with its area and the normal activities that occur there, random barking may occur less frequently. Successful guarding dogs have an appropriate mix of physical and behavioral maturity, combined with experience.

Daily Routine and Behavior

How should a new owner of a guarding dog expect the dog to behave during a 24-hour period? A guarding dog uses its senses and experience to know when and where to patrol and how to best keep predators away from sheep. Some people have mistakenly attempted to impose their own conceptions of the guarding routine on the dog. The dog should be given freedom to develop its guarding behaviors within the limitations imposed by each particular livestock operation.

Some people are surprised that a dog that appears to sleep most of the time can be a deterrent to predation. It is true that some guarding dogs, especially immature dogs, seem to spend a large part of their time sleeping. If the sheep are active (moving and feeding), the dog may also be active. However, dogs are not necessarily with the sheep constantly. The dog may sleep during the day while the sheep are feeding, or the dog may be away from the sheep investigating adjacent areas. With experience, the dog will learn when disturbances from predators are likely to occur (evening and early morning hours) and will be actively patrolling or alertly positioned at a selected location. A dog will often bed with the sheep, but is usually quickly aroused by any disturbance.

Dogs In Fenced Pastures

Over 80% of the people who raise sheep in the United States maintain their flocks in fenced pastures exclusively during some portion of the year (Gee and Magleby, 1976). This represents over 50% of the nation's sheep. It has been predicted that the greatest growth in the sheep industry will come from pastured flocks of sheep.

The majority of producers now using guarding dogs run their sheep in fenced pastures, which range in size from 10 to 500 acres (Green and Woodruff, 1980). Pastures, when compared to open rangeland, are particularly well-suited for using a guarding dog. The sheep are confined in a fenced area, and a dog can quickly establish territorial ties. A dog will generally patrol and scent-mark the fence line thus adding an additional element of "separateness" between the sheep within and the predatory canids without. Since the sheep are confined, they quickly become accustomed to the presence of the dog. Care and feeding of the dog are easier because the dog is in a predictable location from day to day.

Several factors must be considered to determine the number of dogs needed to achieve effective predator control. The performance of individual dogs will differ. Some experienced dogs may effectively patrol a large area containing hundreds of sheep, while younger dogs may not cover as much territory.

The topography and habitat features of the pasture also must be considered. Relatively flat, open areas can be adequately patrolled by one dog, but when brush, timber, ravines, and hills are in the pasture, several dogs may be required.

The behavior of the sheep is important in determining the number of dogs needed. For example, sheep that flock and form a cohesive unit, especially at night (a typical time of predation), can be more effectively protected than sheep that are continually scattered and bedded in a number of locations.

Under some circumstances, one experienced dog may be capable of adequately protecting several hundred sheep in a pasture of several hundred acres. However, if the terrain is rough and brush-covered, if the sheep remain scattered, or if predation pressure is severe, several dogs may be required. Each situation must be evaluated individually.

Dogs on Rangeland

Management practices differ between pasture and range operations and affect the overall concept of predator control by guard dogs. Pastures have fenced boundaries that provide a clearly defined, stationary territory for a dog to defend. There is little chance that the sheep will be lost if they scatter within a pasture, so a full-time shepherd is usually not needed. On the open range, however, fences are rarely encountered and a shepherd tends the flock, controls the grazing pattern, and provides some degree of protection from predators. A dog on the range must learn to identify the ever-changing area occupied by the sheep as a defendable territory. The dog must adapt to new areas as the shepherd implements the grazing plan, and since the dog remains unsupervised with the sheep much of the time, its behavior must not cause the flock to scatter.

Planning is critical to the successful use of a livestock-guarding dog on the range. Several months are required to socialize and prepare a pup for rangeland use. Preliminary information suggests that the first dog to be incorporated into a range flock should be between 7 and 10 months of age. Therefore, a producer would need to purchase a two-month-old pup 5 to 8 months prior to incorporating the dog into the flock. During this period, while gaining physical and behavioral maturity, the pup can be socialized to the stock it will guard as an adult.

An ideal time to place a dog with sheep is when the sheep are in a pasture or fenced area. Sheep producers who keep lambs in sheds can incorporate a dog into the flock shortly after the ewes have lambed and while the main flock is being formed. At this time the shepherd can get to know the dog and teach basic obedience commands such as "Come," "Sit," and "Stay." This period will also allow the shepherd time to observe how the dog and sheep interact prior to going on the range--and perhaps more importantly will allow the sheep to become accustomed to the dog.

Several factors, such as terrain and the type of sheep and dogs being used, make it difficult to generalize about how many dogs are needed to be effective on the range. Some range operators have reduced predation with a single dog, but others may require more than one dog to achieve similar results.

In most operations it would be advantageous to start with a single dog and add dogs as needed, preferably after the first dog is well established. Once the experienced dog develops an effective working pattern, it can often be used as a role model in training a second dog.

Herding dogs are an integral part of a range sheep operation. Herding dogs and guarding dogs can coexist. The guarding dog must be taught that its role is different from that of the herding dog. Immature guarding dogs may attempt to mimic the herding dog as it moves the sheep. This behavior cannot be tolerated. Juvenile guarding dogs can interfere with a herding dog when it is working and must often be restrained (tied or held) by the shepherd. As the guarding dog matures, it will learn that there are times when the herding dog is in charge (when the sheep are moved) but that it assumes the dominant position at all other times. Brief fights may result between the herding and guarding dogs while each dog learns its respective role.

Interactions of Guarding Dogs and Predator

The behaviors of guarding dogs discussed earlier (barking, patrolling, and scent-marking) are usually enough to repel many potential predators from the sheep. In other situations, when predators are more persistent and aggressive, physical encounters (fighting) may occur between a guarding dog and a predator. However, in a questionnaire sent to owners of working Komondor and Great Pyrenees dogs, only 3 out of 30 respondents claimed that their dog(s) had killed a predator (Green and Woodruff, 1980). Dog-predator encounters may be relatively uncommon, and to the casual observer it appears that "nothing ever happens (Coppinger and Coppinger, 1980b)."

Dogs are the principal predator of sheep in many areas. Some guarding dogs regularly repel strange dogs from the pasture, and all guarding dogs are potentially capable of similar behavior. If the dog owner routinely discourages all strange dogs from coming near the sheep, a guarding dog can be encouraged to do the same. As a guarding dog matures, its aggressive behavior to strange dogs should increase.

Problems with Using Guarding Dogs

The use of guarding dogs is not free of problems, especially on rangeland. If losses to predators are substantial, the sheepman may be more willing to spend time raising and training a dog. If losses are minimal a guarding dog may not be practical.

Some potential problems can be prevented by proper and early socialization of dogs to sheep. However, sometimes the most serious problem encountered is one of personal disillusionment with the guarding dog concept. Some livestock producers think that the purchase of a guard dog will immediately solve their predator problems. Unfortunately, this is rarely the case.

Livestock-guarding dogs mature slowly. Komondorok seem to reach behavioral maturity at 18 to 30 months of age while Great Pyrenees appear to mature somewhat earlier. During maturation, a dog experiences physiological and behavioral changes that may contribute to a form of emotional instability. The young dog may show strong desires for playful activities and seemingly irrational behavior. A puppy or adolescent dog should not be expected to match the performance of a mature, experienced guardian. When placed alone with livestock for the first time, a young dog will almost certainly make mistakes and may require an adjustment period of weeks or months, depending on the individual.

We have raised dogs side by side under similar conditions. Some become good livestock guardians, and others do not. After a certain point, it appears that no amount of proper training and early exposure will guarantee that every dog will become a good guardian. Instinctive ability must be present if a guarding dog is to be successful.

CONCLUSIONS

A growing number of guarding dogs are being used to combat predation. They appear to be particularly effective in fenced-pasture conditions; there is growing interest in the use of guarding dogs. However, dogs are not the solution to all predator problems. Considerable time, effort, and good fortune are required to bring a puppy to maturity and with no assurance that the dog will become an effective flock guardian.

The coyote is an adaptable predator, and experience has shown that a variety of control methods are necessary to minimize coyote predation. Livestock-guarding dogs are another tool available to livestock producers for protection from predators. In some situations, use of a guarding dog may be ineffective, whereas in others a dog may be all that is necessary to stop predation. Between these two extremes, dogs may be used to supplement electric fencing, trapping, aerial gunning, or other forms of control.

ACKNOWLEDGMENT

The authors acknowledge cooperation of the University of Idaho, Moscow, Idaho.

REFERENCES

Arons, C. 1980. Raising livestock guarding dogs. Sheep Canada 5(3):5.

Bardoloi, R.K., S.S. Tamhan and K.K. Pant. 1979. Sheep rearing in northeastern Indian Arunachal Pradesh. World Rev. Animal Prod. 15:23.

Black, H. 1981. Navajo sheep and goat guarding dogs: A new world solution to the coyote problem. Rangelands 3:235.

Bordeaux, E.S. 1974. Messengers from Ancient Civilizations. Academy Books, San Diego.

Breber, P. 1978. Il Cane de Pastore Maremmano-Abruzzese. Editoriale Olimpia.

Coppinger, L. and R. Coppinger. 1980[a]. So firm a friendship. Nat. History 89:12.

Coppinger, R. and L. Coppinger. 1978. Livestock guarding dogs for U.S. agriculture. Hampshire College, Amherst, Mass.

Coppinger, R. and L. Coppinger. 1980[b]. Livestock-guarding dogs, an old world solution to an age-old problem. Country Journal 7:68.

Darwin, C. 1839. The Voyage of the Beagle. P.F. Collier and Son Corporation.

Gee, C.K. and R.S. Magleby. 1976. Characteristics of sheep production in the western United States. USDA Econ. Res. Ser. Ag. Econ. Rep. No. 345.

Green, J.S. and R.A. Woodruff. 1980. Is predator control going to the dogs? Rangelands 2:187.

Green, J.S., T.T. Tueller and R.A. Woodruff. 1980. Livestock guarding dogs: Economics and predator control. Rangelands 2:247.

Linhart, S.B., R.T. Sterner, T.C. Carrigan and D.R. Henne. 1979. Komondor guard dogs reduce sheep losses to coyotes: A preliminary evaluation. J. Range Manage. 32:-238.

Lyman, J.H. 1844. Shepherd dogs. The Amer. Agriculturists 3:241.

Nelson, D. and J.N. Nelson. 1980. The livestock guarding dogs of Turkey. National Wool Grower 70:12.

Part 10

HEALTH, DISEASE, AND PARASITES OF SHEEP AND GOATS

46

INFECTIOUS LIVESTOCK DISEASES: THEIR WORLDWIDE TOLL

Harry C. Mussman

By good fortune, plus nearly a century's worth of cooperation between stockmen and government, the United States has managed to wipe out or keep out the most serious of the world's livestock diseases.

Contagious bovine pleuropneumonia, introduced in the mid-1800s, was the first major disease we battled together and eradicated. Hog cholera was the last to be wiped out--just four years ago. In between these two efforts were nine outbreaks of foot-and-mouth disease, which has not been back since the last remains were buried in 1929.

These diseases and others still bring major losses and reduce livestock productivity in much of the world. This is particularly true in the developing nations of Africa, Asia, and Latin America. The popularly held notion that Third World livestock are infected with all of the most dreaded diseases is not true, but many of these countries do live with at least a few of them, often superimposed one on the other. Losses due to animal disease in some developing countries are estimated at 30% to 40% annually--twice as great as losses recorded by most industrialized nations.

Looking at meat and dairy productivity, developing nations produce one-fifth the beef and veal per animal that developed countries do, one-half the pork, one-half the eggs, and one-eighth the milk. North America and Europe are roughly four times more efficient in mutton, lamb, and goat meat production than is South America, and 25% more efficient than Africa.

The so-called Third World is hungry already, and is projected to host 90% of the global growth in human population expected by the year 2000. Obviously, greater and more reliable sources of protein must be found and developed, or many of these people will not survive. A diseased animal that may not survive can hardly be counted a productive resource. A virus or other disease agent can wipe out entire herds in a very short time--or, in a chronic state, can leave them debilitated, more expensive to feed, and capable of producing far less milk or meat.

Of all the factors involved with production, animal diseases have by far the most dramatic impact. Unfortunately, disease control is not improving rapidly enough in the Third World. And many of the most feared diseases--because of the tremendous economic toll they can take--are appearing in locations where they never existed before or reappearing in places where they had been eradicated years ago.

Some of these diseases are listed next.

Foot-and-Mouth Disease. Foot-and-mouth disease (FMD) is the most feared of all the animal diseases worldwide because of its ease and speed of spread and the susceptibility of all clovenhooved animals. It has long been established in livestock populations of the Mideast causing abortions, deaths, weight loss, mastitis, and lowered milk production. Europe has managed to keep these Asian strains at bay in Turkey, however.

Prior to World War II, FMD swept through Europe every 6 to 8 years, severely reducing livestock productivity. Finally in the 1950s, better quality vaccine was introduced and stemmed the spread of the epizootics. Outbreaks that occurred in the early 1960s, and again in 1973, were halted by teams from the United Nations working with the affected countries and regions. Then, in March 1981, FMD was confirmed in France, in Brittany, and near Cherbourg. Within a few days it broke out on the Isle of Jersey and the Isle of Wight in Great Britain. Just prior to that positive diagnosis came in from the south of France, from Portugal, and from Spain--outbreaks that most likely originated in Spain. Austria, free of FMD for six years, was infected during the winter of 1981 from imported Asian buffalo meat.

After 10 years of freedom from the disease and despite a vigilant quarantine program, Denmark discovered an outbreak of FMD on the Isle of Fyn just last March. The origin is unknown, but suspicions are that it came from one of the Eastern bloc countries where severe FMD outbreaks were reportedly occurring just prior to Denmark's infection. The loss Denmark has suffered from those outbreaks is estimated at almost a quarter of a billion dollars. Most of the loss has come from the cessation of fresh pork sales to the U.S., Canada, and Japan, although sixteen other countries were also compelled to embargo Danish meats and animals. The cost of eradication alone was also significant.

U.S. policy prohibits imports of fresh, frozen, or chilled meat products, or swine, or ruminants for a full year from any country that experiences an FMD outbreak. We had to refuse United Kingdom exports for a year following their 1981 outbreaks. We also refused farm products from across the Canadian border in 1952 following an outbreak there--which also curtailed big game hunting for U.S. hunters up north. (Certified "clean" animals, however, may be brought in from FMD-infected countries through the Harry S. Truman Animal Import Center on Fleming Key, Florida, after three months of high-security quarantine and testing at the importer's expense; or they may be shipped directly to an approved zoo.)

Some call this "politics," aimed at protecting the home meat industry. U.S. meat producers do gain a competitive edge from such an embargo, certainly, but that is not the aim. The productivity of more than 200 million fully susceptible cattle, swine, sheep, and goats is at stake. The collective worth of that stock is estimated at $24 billion --if they stay healthy. Healthy animals with their high-quality protein will keep America healthy, help alleviate worldwide hunger, and help keep the U.S. trade balance healthy.

If FMD were to invade the U.S. and enter a major market, it could spread to over a dozen states within 24 hours. By the end of the first year, direct losses would be an estimated $3.6 billion--indirect losses to allied industries and curtailed exports could climb to $10 billion.

Rinderpest and Peste de Petits Ruminants. Another serious, almost 100% fatal among cattle and resurging as a serious problem in Africa, is rinderpest. As with FMD, the United States (by law) cannot import cattle or other ruminants from rinderpest-infected countries (unless, as with FMD, they are brought in through the Truman Center on Fleming Key or are destined for an approved zoo.)

Rinderpest was essentially eliminated from Africa in 1975, following a decade of intensive and widespread vaccination. A team of international cooperators trudged from West to East Africa, vaccinating some 80 million head of cattle at a cost of $30 million. Unfortunately, though, the people became complacent once the disease had been suppressed. As the livestock populations were built up again, the young were not all vaccinated and were allowed contact with remaining carriers and wild animals.

So, the incidence is climbing again, especially in West Africa, and the virus is spreading east. Egypt is affected now, and the Arab Gulf States are reporting such frequent outbreaks that it appears the disease has become enzootic.

Peste de petits ruminants is a virus quite similar to rinderpest and readily infects goats and sheep in West Africa. It is actually classified as a strain of rinderpest that has lost its ability to infect cattle under natural conditions.

Rift Valley Fever. Another disease that has moved up from Kenya through the Sudan into the Sinai is Rift Valley fever. RVF primarily affects sheep but may also affect goats. It is also a serious human health problem. An epidemic in Egypt a few years ago killed nearly 600 people, along with a large number of livestock. The disease can spread rather easily. A human body incubating the virus could carry it anywhere in the world or a mosquito could carry it while hitchhiking on an airplane.

Israel has been vaccinating its animals against RVF for self-protection as well as to help curb further spread of the disease around the Eastern Mediterranean Basin.

Because of grave concern over Rift Valley fever, the U.S. Department of Agriculture has developed diagnostic capability of its own and is making preliminary vaccine studies. Work on possible domestic insect vectors is also planned.

African Swine Fever. Over the past 20 years, swine-producing nations have faced a rising risk of African swine fever infection. ASF has been established for many years in wart hog, bush hog, and giant forest pig populations in tropical Africa, south of the equator. Other animals may not have the same tolerance to the virus: in 1909 Europeans tried bringing domestic hogs into Africa, and the hogs were dead from the virus in a matter of a few days.

The first appearance of ASF in Europe was in Portugal in 1957 where the disease was at first mistaken for classic hog cholera and finally recognized as ASf. More than 16,000 pigs became infected before the virus was recognized as ASF and finally stamped out. In 1959, the disease was found in Spain, and in 1960 reappeared in Portugal. France was invaded by ASF three times in the 1960s and 1970s, though the French managed to eradicate it each time at an early stage--a tribute to their veterinary surveillance system.

Italy became infected in 1967 through the Rome airport. More than 100,000 swine were slaughtered before the disease was wiped out. Finally, about four years ago, Malta and Sardinia became infected. In Malta, the entire swine population was slaughtered to achieve eradication--at a cost of some $50 million (U.S.). Italy is still working on eradication of the disease on Sardinia.

The Western Hemisphere managed to avert an invasion of the dread ASF until 1971 when the virus hit Cuba. Approximately half a million pigs died or were destroyed before the disease was eradicated. Then, another Cuban outbreak occurred in 1980 and was successfully eradicated in about six weeks. ASF appeared in Brazil 5 years ago and is still there today. Positive diagnosis came from the Dominican Republic in 1978 and in Haiti early in 1979.

The U.S., Canada, and Mexico are particularly and acutely concerned about African swine fever in the Western Hemisphere. All three countries are free of the disease but are working together in the Dominican Republic toward eradication. The program used a hard strategy: they completely depopulated the country's swine farms after which they restocked with healthy, imported pigs. Not a trace of the virus has reappeared since the restocking was completed. The Haitian effort repeated the model program (and first of its kind in the West) completed in the adjacent Dominican Republic last year. Equal success is anticipated in Haiti.

Success is crucial to the three big cooperators. Mexico's Yucatan Peninsula is just a short hop across the

Caribbean from the Dominican Republic; Canada has trade interests all over Latin America and the Caribbean; and the threat of ASF lies right at the doorstep to the United States. The U.S. strictly prohibits swine imports from ASF-infected countries, but the virus could enter through contaminated food waste from ships or planes, through farm soil still clinging to a returning traveler's shoe, or as a result of illegal immigration. The investment in such an eradication program is sizable. Direct costs for the Dominican Republic--shared by the UN's Food and Agriculture Organization, the Inter-American Development Bank, and the United States--amounted to approximately $20 million. The bill for Haiti should run about the same.

The cost of chronically battling a disease like ASF can be far higher, however. Spain, for example, has spent roughly $315 million fighting it over the past 20 years. In addition, exports from Spain (along with Portugal and Brazil) have been restricted because of ASF-infected status.

Heartwater Disease. Latin America was jolted last year by the diagnosis of heartwater disease in a goat on Guadaloupe, an island in the eastern Caribbean. This was its first appearance in the Western Hemisphere and the implications are serious. We are cooperating with other countries in the region to make certain that it does not spread.

Heartwater is native to southern Africa and Madagascar and may be a major cause of animal deaths reported in West Africa. We have no simple and reliable means of diagnosing the disease; a blood test, especially, is badly needed.

Contagious Bovine Pleuropneumonia. Contagious bovine pleuropneumonia (CBPP) recently broke out in France for the first time in years. The disease may have been smouldering in the Basque region in Spain for some time. CBPP was the first major exotic disease to invade U.S. shores back in the 1800s. Cattle ranching had become so productive after the Civil War that a surplus of more than a quarter million head was available for export to England. When pleuropneumonia showed up in the animals, however, Britain closed off the market. That was the spark that set off the first animal disease eradication program in the United States.

Goat Diseases. In the United States evidence is accumulating of two possibly serious diseases of goats: contagious caprine pleuropneumonia and caprine arthritis encephalitis. CCPP is well established in much of the developing world--the regions that depend heavily on goats. The true impact of the disease is clouded as yet, largely because of common and widespread disagreement among diagnosticians. More research is needed to develop definitive diagnostic tests and effective vaccines.

Caprine arthritis encephalitis was only recently recognized in the United States and may exist yet undiscovered in other countries also. Research may give us some answers to this relatively new entity.

Trypanosomiasis. A full third of Africa--taking in most of the wide band of savannah and forest across the middle of the continent--is plagued by infestations of the tsetse fly. A blood parasite of animals as well as people, the tsetse fly transmits trypanosomiasis, or animal sleeping sickness, from animal to animal. If the tsetse-infested areas could be reclaimed for cattle production, it is estimated that annual meat production in Africa could double.

Bacterial Disease, Intestinal Parasites, and Others. The more dramatic virus diseases, such as FMD, ASF, rinderpest, bovine pleuropneumonia, and external parasites that transmit East Coast fever and trypanosomiasis, are the most highly visible of the livestock plagues worldwide.

Much more research remains to be done on bacterial diseases and intestinal parasites, particularly with small ruminants. Also more research is needed to understand the incidence and significance of the clostridial diseases--tetanus, anthrax, blackleg, and malignant edema--and paratuberculosis, and brucellosis. In the U.S., eleven states are free of cattle brucellosis now and another 25 have an extremely low incidence. Eradication of the disease requires only time. The damage done by different strains of the disease worldwide, however, is not entirely known. Since all strains can infect people as well as animals, the public health problem is always important. Anaplasmosis and babesiosis also are economically significant blood-parasite diseases in many tropical areas of the world, but the actual extent of the damage they do is not fully appreciated, much less known.

Another virus disease--one less dramatic in its toll, but gaining more and more visibility--is bluetongue. Bluetongue affects cattle, sheep, goats, and wild ruminants. It is particularly damaging to sheep, with a mortality rate running as high as 50% among affected animals; in cattle and goats, bluetongue lowers reproductive ability. Currently, U.S. cattle cannot be exported directly to Europe because of bluetongue in the South and southwestern states. Canada has a competitive edge over the United States in the European livestock trade--with the exception of the Holstein dairy breed--because the small gnat that carries the disease does not appear that far north.

Like caprine arthritis encephalitis in the United States, the possibility of diseases yet undiscovered still remains--diseases that may have been smouldering since the first goat herds on record were tended some 10,000 years B.C.

THE NEED FOR STRONGER CONTROLS

Proper vaccines, a solid border inspection and quarantine program, and accurate diagnosis could take care of many of the most severe disease problems in the world.

Vaccine

The alarming resurgence of rinderpest in Africa shows what can happen when a successful vaccination program breaks down. On the other hand, a laboratory opened in Mali just 5 years ago has been researching, tracing, and producing vaccine for several diseases rampant in West African livestock --rinderpest and bovine pleuropneumonia. The Central Veterinary Laboratory (CVL), as it is called, is also coordinating regional vaccination programs, and CVL veterinarians and technicians are being welcomed now by Malian herdsmen as they drive their cattle back and forth for the grazing season. The incidence of disease appears to be declining, at least locally.

Quarantine

Contributing to the rinderpest problem in Africa was the failure to keep sick animals from the healthy. Quite simply, quarantine can prevent the spread of many infectious diseases. For example, double fencing, as practiced in some parts of Africa, effectively protects domestic swine from African swine fever.

Border Inspection

Strict import controls over meat, feed, and animal products and a tight border guard are essential to keeping disease out. Years ago when most travel was by boat, a virus disease could probably not survive a transocean trip. Today a few hours on a jet plane are little threat to virus survival.

Customs and Agriculture inspectors at all 82 U.S. ports-of-entry inspect as much luggage and as many carry-on bags and travelers as possible. The goal of 100% inspection is a difficult one to achieve, however. International travel rose 8% in the United States last year, and 6% worldwide. Also more and more cargo is containerized every year.

Diagnosis

Research must proceed and develop simpler and more effective means of diagnosing heartwater disease, contagious caprine pleuropneumonia, and other diseases. Clinical signs of certain diseases are changing--and we must keep up with them. Genetic research must continue so that animal science can continue to breed for resistance to disease. Biological, chemical, ecological, and other means of controlling insect vectors are unfolding through field and laboratory trials.

CONCLUSION

Each country or region of the world will have its own formula for prioritizing the danger that different animal diseases present. Generally speaking, the most dangerous diseases are those that are spread not only by animal-to-animal contact (contagious diseases) but whose virus, mycoplasma, bacteria, or rickettsia may be transmitted by live vectors, inanimate fomites, or meat scraps. The most dangerous diseases do significant economic damage to producers and exporters and carry a human health hazard, such as brucellosis or Rift Valley fever.

Most important before the animal disease situation in the Third World can improve, veterinary services will have to be strengthened. Developing countries have nearly 50% of the world's total livestock population but less than 20% of the world's veterinary forces. A number of developing nations, in fact, have little or no organized veterinary structure at all. Certain nations in Africa, in fact, have only one or two veterinarians working the entire country; and in others they are often young, with little or no field experience. The expertise of technicians, if it can be found, is usually limited--gained through in-service training. Veterinary technicians have no reliable means of communication or transportation--yet national animal health policy decisions are based on their input. Thus, the global data on disease prevalence, productivity losses, and effectiveness of control measures on which many veterinary directorates rely is woefully deficient.

It becomes clear why diseases such as contagious bovine pleuropneumonia, foot-and-mouth, and rinderpest still exist and spread. With existing reliable diagnostic tests for both, as well as effective vaccines, eradication of CBPP and rinderpest would require little more than organization and finance, backed up by the government's strong commitment. With heavier investment in livestock health programs--and greater cooperation among nations--the great toll that diseases take can be curbed. The people <u>can</u> be fed.

47

IMPACT OF ANIMAL DISEASES IN WORLD TRADE

Harry C. Mussman

IMPACT OF ANIMAL DISEASES ON WORLD TRADE

Animal diseases are an important factor inhibiting world trade and hampering the free movement of both live animals and animal-derived products. The foot-and-mouth disease outbreak in Denmark in early 1982 provides a recent example of the impact of animal disease on trade. At one point, Denmark's export trade in meat and dairy exports was suffering badly--$7 million was lost each week because of the outbreak. The U.S.--as well as several other countries--placed Denmark on the list of countries from which animals and animal products cannot be imported until free status is regained. Overall costs of the outbreak will run in excess of one billion dollars.

Another example closer to home is our bluetongue situation. In 1980, the European Economic Community banned animal imports from the U.S. because this cattle and sheep disease was found in a portion of our country.

U.S. COMPETITIVE EXPORT POSITION

Despite the bluetongue situation and other domestic diseases such as brucellosis, tuberculosis, and leukosis, the U.S. export position remains competitive. In 1981, our animal and animal-product exports had a market value of $3.24 billion; and, in the same year, we had a $307 million trade surplus in these commodities, the first since 1977.

A U.S. animal export health certificate enjoys high credibility. One reason is that we have eradicated 12 major animal diseases that still plague many other countries of the world--diseases such as foot-and-mouth, rinderpest, hog cholera, and contagious bovine pleuropneumonia. Only a half-dozen or so other countries can claim freedom from these diseases. Largely because of our strict animal import health requirements and procedures, we are successful in keeping animal diseases out of our country.

IMPORT PROCEDURES

Livestock destined for this country must first be examined, tested, and certified by government veterinary officials as being healthy and meeting U.S. requirements in the country of origin. The foreign exporter must obtain health certification papers from his government and, in a majority of cases, an import permit in advance from us. At the U.S. port of entry the livestock must be examined again, this time by our veterinarians. This examination sometimes includes further testing and port-of-entry isolation, depending upon the kind of animal and the country it came from.

Although we are strict in our import controls, we try to meet the needs of American importers, particularly U.S. livestock breeders needing new bloodlines and exotic breeds of cattle. We recently opened a specialized import-quarantine facility--the Harry S. Truman Animal Import Center, at Fleming Key, Florida. Imported cattle are carefully tested and held there in quarantine for 3 months. They are mingled with a select "sentinel" group of susceptible U.S. cattle and swine to make sure they will present no disease threat to U.S. livestock.

Because some swine diseases, such as African swine fever, could devastate our swine herds, we do not accept swine imports from most of the world. Imports of sheep and goats are also severely limited because of the threat of scrapie, which has an incubation period of up to four years or more.

Restrictions on horse imports are generally less stringent than those required for swine, cattle, sheep, and goat. Still, incoming horses are tested for such diseases as dourine, glanders, equine infectious anemia (EIA), equine piroplasmosis, and contagious equine metritis (CEM).

Because of Venezuelan equine encephalomyelitis, horses from all Western Hemisphere countries--except Canada and Mexico--are quarantined for at least a week. Horses from countries known to have African horse sickness are quarantined for two months.

Contagious equine metritis is a recently identified disease of breeding horses. It has been found in parts of Europe, Japan, and Australia. As a result, horses cannot be freely imported from these countries.

Stallions and mares can be imported only after extensive treatment and negative culturing of the genitalia, both in the country of origin and again in the U.S. while under quarantine.

PROCEDURES FOR IMPORTS FROM CANADA AND MEXICO

The entry procedures for animals from Mexico and Canada are generally less strict than those for animals from overseas nations because Mexico and Canada have animal disease situations relatively similar to ours. The exception is hog cholera in Mexico, and for that reason we do not import swine from that country. In the case of other animals, however, entry quarantine and advance import permits are not required for either Mexico or Canada.

The U.S. has 16 crossing points on the Mexican border and 43 on the Canadian border where APHIS veterinarians examine and process animal imports. Entry procedures vary. For example, cattle from Mexico must be dipped in a pesticide solution as a precaution against cattle fever ticks and scabies.

ANIMAL SEMEN, EMBRYOS

Since animal semen is as much a potential disease threat as live animals, it must be subjected to strict standards for collection, handling, and shipping. These standards are spelled out in agreements between USDA and the foreign countries involved. While the technology for testing animal semen is well researched and established, the same cannot be said for embryos, the newest practical method for exporting animals. More research is needed to determine the diseases to which embryos are immune.

Our restrictions on animal imports may seem extreme to some. But we have a major responsibility for maintaining the health of our livestock--for example, a $9 billion swine industry, a $35 billion cattle industry, and a $10 billion poultry industry. Considering what is at stake, our restrictions are reasonable.

EXPORT PROCEDURES

Just as we insist that only healthy livestock be imported into this country, we have an obligation to see that only healthy animals are exported to other countries.

We ensure the health status of our exported animals in two ways. First, we establish our own health rules for exports; second, we cooperate fully in meeting the import rules of receiving nations.

All animals we export are subjected to special testing and certification requirements to indicate freedom from certain diseases found in the United States: bluetongue for cattle, sheep, and goats; brucellosis for cattle and goats; anaplasmosis for cattle; equine infectious anemia for horses; and pseudorabies for swine.

Our own export rules are necessary because we do have some health problems, i.e., EEC bluetongue of concern to foreign importers. We want to avoid damaging our position in the world market by making sure the animals and animal products we export are free of disease.

The animal health requirements of foreign countries vary, reflecting the particular animal disease problems and danger they face in their part of the world. Testing and certification are performed by private veterinarians accredited by us (usually at the shipper's expense) who normally conduct their tests on the farm or cattle ranch. Adult dairy and breeding cattle and goats must be tested and found free of brucellosis and tuberculosis within specified time limits before shipment. An animal health certificate is endorsed by the APHIS area veterinarian after the private accredited veterinarian completes his testing. The endorsement certifies that the private veterinarian is qualified to conduct the examination and tests.

Once the certificate is endorsed, livestock can move to a port of embarkation. There they must rest for 5 hours at a USDA-approved facility while the animals and paperwork are given a final check by an APHIS veterinarian. If the animals are healthy, if they are properly identified, and if the health certificate is in order, they are loaded on a ship or aircraft under APHIS supervision for export.

COMPLEXITY OF FOREIGN REQUIREMENTS

The health requirements imposed by foreign governments can be quite complex, so whenever possible we work out agreements with these countries. We do all we can to negotiate standard requirements, but we have over 150 agreements with some 70 foreign governments, and they can be changed on short notice. It is nearly impossible for an exporter or an examining veterinarian to keep track of the different rules for every overseas livestock shipment. To avoid delay and frustration, the exporter and the accredited veterinarian are urged to contact the APHIS Veterinary Services office in their state before attempting the process livestock or animal products.

Each APHIS Veterinary Services office keeps a current file of foreign animal import health requirements. Each area office of APHIS Veterinary Services also has a veterinarian assigned to work with exporters and their veterinarians. His job is to check the export health tests and certifications and place the final endorsements on the health papers before the livestock can leave this country.

CANADIAN, MEXICAN EXPORTS

As with our imports, our exports to Canada and Mexico are handled more simply and quickly than those to overseas nations. Exported livestock do not need an APHIS veterinary examination at the port of export. Once the health tests, certification, and APHIS endorsements are completed, the animals move directly to the border where they are examined by the Canadian or Mexican officials. Canada and Mexico are by far our biggest customers for exported livestock (and poultry as well), so it is important to devote some special attention to health matters on shipments across our borders.

Exports to Mexico move with few special problems. However, Canada has some requirements that exceed our own export rules. The Canadians are particularly concerned about chemical residues in cattle shipped to slaughter, so feed additives and antibiotics should be withdrawn from livestock within the recommended time limits. This is the responsibility of the exporter.

CLOSE COORDINATION REQUIRED

Successful U.S. exports, particularly those to overseas nations, require close coordination between the exporter, the private veterinarian, APHIS officials, the forwarder, the broker, the insurance underwriter, and the carrier. Among the most common causes of costly delay is the failure to conduct all required tests and failure to allow enough time for completion of the tests at the diagnostic laboratory.

Even if a plane or ship is waiting, animals cannot move to the port of embarkation unless APHIS endorses the health papers. Therefore, exporters should be aware of all the requirements and plan to allow sufficient time for testing when making plans to ship animals to another country.

INTERNATIONAL INVOLVEMENT

The U.S., along with its major partners, is actively involved in international organizations such as the Food and Agriculture Organization (FAO), the Office of International Epizootics (OIE), and the General Agreement on Tariff and Trade (GATT) in dealing with animal diseases worldwide and taking steps to assure the expeditious movement of healthy livestock and livestock products. We are actively involved in international health programs because the more we can do to reduce diseases worldwide, the more freely our own animals will move in international markets--an advantage of the U.S. exporter.

CONCLUSION

APHIS does all it can to help the stockman with his exports. We safeguard his markets by making sure no diseased animal gets out; and we make sure no diseased animal gets in to infect his livestock. We are against the unduly restrictive animal health import requirement that functions as a nontariff barrier and, more often than not, serves to protect a country's livestock industry--more from foreign competition than from foreign animal diseases.

The U.S. strongly supports "free trade" and endeavors, whenever possible, to make it a dominant principle in world trade, which includes trade in live animals and animal products. This is compatible with our free-enterprise tradition. If we adhere to this principle and keep our disease defenses strong, our livestock should remain the healthiest in the world. And export opportunities for U.S. stockmen should expand as world trade in animals expands.

48

BIOLOGY AND CONTROL OF INSECT PESTS OF SHEEP AND GOATS

R. O. Drummond

Although the number of sheep and goats (Angora goats, dairy goats, and meat goats) has been declining in the U.S. in recent years, they remain important and generate considerable income for their owners in many areas. Sheep and goat owners continually battle a variety of pests (lice, keds, flies, bots, ticks, and mites) that bite, sting, irritate, annoy, injure, transmit diseases to, live in tissues of, and suck blood from sheep and goats. Insect pests cause annual losses in wool, meat, and milk amounting to tens of millions of dollars, despite the use of insecticides to control these pests on sheep and goats. Only increased awareness of the pests' biology, life history, and local abundance and proper use of safe and effective chemicals and other control methods can help in reducing losses and costs of treatment, while controlling pests without danger to the animal, the applicator, the consumer, or the environment. This article contains information on the biology of a number of pests of sheep and goats, describes accepted techniques for their control, and lists precautions for the safe use of insecticides on sheep and goats.

Dairy goat owners must be extremely careful when selecting and applying insecticides to their goats. Certain insecticides can be applied to nonlactating goats only, because application to lactating dairy goats would create illegal residues of insecticides in milk and milk products. Other insecticides, when applied as recommended, can be used directly on lactating dairy goats without creating residues.

EXTERNAL PARASITES

Lice

The most common insect pests of sheep and goats are biting lice and sucking lice. All lice have a similar life cycle in that all stages are found on the host, females attach eggs (called nits) to wool or hair, and immature forms (called nymphs) hatch from eggs and molt one or more times before they become adults.

Biting lice. Sheep and goats are infested with four species of biting lice: sheep are infested with the sheep biting louse, Bovicola ovis (Schrank); goats are infested with the so-called hairy goat louse, Bovicola crassipes (Rudow); the Angora goat biting louse, Bovicola limbatus (Gervais), and the goat biting louse, Bovicola caprae (Gurlt). Biting lice live off skin scales, debris, wool, hair, and other matter on the surface of the animal's body. Often large infestations cause sheep and goats to itch and rub, which causes a loss of wool and hair.

Sucking lice. Sheep and goats are infested with four species of sucking lice: sheep are infested with the so-called sheep sucking body louse, Linognathus ovillus (Neumann), and the so-called sheep sucking foot louse, Linognathus pedalis (Osborn); goats are infested with the goat sucking louse, Linognathus stenopsis (Burmeister), and Linognathus africanus (Kellogg and Paine). All these species suck blood from sheep and goats, and massive infestations can cause anemia, loss of condition, decreased efficiency of feed conversion, a scurfy appearance, and loss of wool or hair. Louse infestations may lead to the death of the animal.

In general, lice on sheep and goats can be controlled by application of effective insecticides to the body of the animal. Sheep and goats may be dipped, sprinkled, or dusted thoroughly for louse control. Sheep and goats are often treated after shearing when it is easier to achieve thorough treatment.

Flies

Sheep and goats are attacked by bloodsucking flies (commonly called "biting flies") such as keds, horn flies, biting gnats, and mosquitoes; nonbiting flies such as fleece worms, blow flies, and screwworm flies; and nonfeeding flies such as sheep bot flies.

Biting flies. The sheep ked, Melophagus ovinus (L.), a wingless, bloodsucking fly that is often called the sheep "tick," but is not a tick at all, is found on most sheep and rarely on goats. The adult female cements a small, whitish pupa to the wool, and in about 2 or 3 weeks the adult fly emerges. (The maggot stage of the sheep ked is passed inside the female.) A sheepskin defect, called cockle, is the result of feeding of the adult fly. Keds can be controlled with the same insecticides that are used to control sheep and goat lice; thorough treatment is necessary for complete control.

Other biting flies such as the horn fly, Haematobia irritans (L.), the stable fly, Stomoxys calcitrans (L.), black flies, mosquitoes, and other species are sometimes found on sheep, but are very difficult to control. Some control can be provided by sprays, dips, or mist sprays of insecticides applied directly to sheep and goats. A biting gnat, Culicoides variipennis (Coquillett), which attacks sheep and transmits bluetongue, a virus disease, is also difficult to control. The larvae breed in sewage- and manure-polluted water.

Nonbiting flies. Sheep and goats are attacked by the larvae of several blow flies including Cochliomyia macellaria (F.), the secondary screwworm fly, Phormia regina (Meigen), the black blow fly, Phaenicia sericata (Meigen), and other species. These flies lay eggs on dead and decaying flesh, and the maggots (called fleece worms) do not destroy living flesh as does the maggot of the screwworm fly, Cochliomyia hominivorax (Coquerel). The screwworm fly has been eradicated from the U.S. and is currrently the subject of a highly successful eradication campaign in Mexico. Although screwworms are eradicated from the U.S., producers of sheep and goats along the Mexico border should routinely examine their animals for wounds. If the wounds contain larvae, specimens should be sent for identification to the Screwworm Eradication Program, P.O. Box 969, Mission, Texas 78572. All wounds should be treated thoroughly with a spray, dust, aerosol, or smear of insecticide to kill maggots and protect the wound from reinfestation. It may be necessary to thoroughly dip or spray sheep or goats that have large areas of their skin infested wth many fleece worms.

Nonfeeding flies. Sheep and goats are infested with larvae of the sheep nose bot fly, Oestrus ovis. The nonfeeding flies infest sheep pens, and female flies fly around the heads of sheep. Larvae hatch inside the female fly, and the female lays these tiny living larvae in and around the nostrils of sheep. The young larvae live on the mucous surfaces of the nasal passages, and as they grow larger they move toward the sinuses until the grown maggots are found in nasal sinuses. When the larvae (called bots) are fully grown, they exit from the sheep via the nostrils and fall to the ground where they pupate, and in several weeks (depending upon temperature), they become flies. Heavy infestations of bots in nasal passages and sinuses of sheep can create problems of excess mucous or blood from nostrils. On occasion a bot may enter the brain of an animal and cause the animal's death. Heavy and continuous attack by flies causes sheep to "fight" the flies, feed irregularly, and as a result to gain weight slowly.

Unfortunately, there are no treatments now registered for the control of sheep bots in sheep and goats. A number of animal systemic insecticides that control cattle grubs in cattle and horse bots in horses have been tested as oral, dermal, or injectable treatments for bots in sheep, and many were found to kill all of the larvae. None are now on the market for use in sheep. The anthelmintics available for sheep are not effective against sheep nose bot larvae.

Ticks

Sheep and goats occasionally are attacked by ticks, but these attacks are rare and losses caused by these 8-legged relatives of spiders are minimal.

Soft ticks. Sheep and goats are infested by only one species of "soft ticks," a title given to a large group of ticks because of the wrinkled, leathery texture of their "skin." This species is the spinose ear tick, Otobius megnini (Duges), which lives deep in the ears of sheep and goats. Adults of this species are free living (they do not feed) and are found in protected places such as cracks and crevices of buildings, fences, under salt troughs, etc. Females are bred and lay eggs in these protected places. Tiny 6-legged larvae hatch from eggs, seek sheep and goats, attach deep in the animal's ears, feed for a short period, and molt to the spiny-appearing nymphs that may feed for several months. This species is found on sheep and goats in most states but is commmon in the southwestern states.

Spinose ear ticks are usually controlled by the thorough treatment of ears of sheep and goats with insecticides as dusts, low-pressure sprays, aerosols, and smears. Effective treatments will provide adequate control for a month or longer. Attempts to control the adults in the environment are generally unsuccessful.

Hard ticks. Most ticks in the U.S. have a hard covering on all or part of the back and thus are called "hard" ticks. These ticks have 2 types of life cycle. One is the "1-host" cycle in which larvae (or seed ticks) attach to a host, engorge on blood, and molt to the next stage (the nymph), which engorges and then molts to the adult male or female. These adults mate on the host; the females fill with blood, detach, drop to the ground, find a secluded spot, and lay eggs from which larvae will hatch. All the molting and engorging takes place on a single host. The other cycle is the "3-host" cycle. Larvae of a 3-host tick feed on a host and, when fully fed, drop off the host and molt to nymphs on the ground. The nymphs find another host on which they engorge. Fully engorged nymphs drop off the host and molt on the ground to the adult stage. The adults find a third host, and mate; then the females engorge, drop off, and lay eggs.

Each region of the U.S. has a particular group of hard ticks that may attack sheep and goats. The Pacific Coast tick, Dermacentor occidentalis (Marx), is found on the Pacific Coast west of the coastal mountains. The winter tick, D. albipictus (Pacard), is generally found throughout the northern tier of states and as far south as Texas. The Rocky Mountain wood tick, D. andersoni (Stiles), is generally distributed in the northern Rocky Mountain States. The American dog tick, D. variabilis (Say), is found generally distributed over the eastern half of the U.S. The blacklegged tick, Ixodes scapularis (Say), is found in the southcentral U.S. The lone star tick, Ambylomma americanum (L.), is found throughout the southeastern one-third of the U.S. The Gulf Coast tick, Ambylomma maculatum (Koch), is limited to south Atlantic and Gulf Coast States, although large populations are found in eastern Oklahoma and surrounding states.

Because of the variety of tick species, their variable life cycles, and their seasonal abundance, the sheep and goat raiser must know which species is attacking his animals so that he can apply the most effective treatments at the proper time of the year. Generally tick control (when needed) consists of treating infested sheep and goats thoroughly with the same insecticides applied as dips or sprays for the control of lice.

In some situations, ticks may be controlled by the application of approved insecticides to the ground to kill larvae, newly molted forms (of 3-host species), and engorged females.

Mites

Sheep and goats are infested with several species of itch, mange, or scab mites. These very tiny species live on or in the skin and can cause intense irritation, itching, loss of wool or hair, thickening of skin, and considerable discomfort to infested animals. One species, the common scab mite, Psoroptes ovis (Hering), has been eradicated from sheep in the U.S. Other species of mange, itch, or scab mites on sheep and goats can be controlled by the thorough application of insecticides by whole-body spray or dip.

USE OF INSECTICIDES ON SHEEP AND GOATS

Treatments

Insecticides remain our first line of defense against insect pests of sheep and goats. Properly used, approved insecticides can kill the pests and thus prevent or reduce losses due to the infestations of flies, lice, ticks, mites, and others. It is necessary to seek the advice and recommendations of local officials such as county agents, and agricultural advisors who have current informtion on kinds of pests in their area, correct techniques of application, recommended times of treatment, and a variety of other facts about local pests, their biology, and their control.

Restrictions have been placed on the use of insecticides for the control of insect pests of milking goats in order to avoid residues of insecticides in milk. The nonlactating goat herd may be treated with a larger variety of insecticides. Nevertheless, it is important for the goat herder to observe the withdrawal period after treatment of nonlactating goats before they freshen in order to avoid insecticide in milk.

Precautions

The insecticides that can be used to kill insect pests of sheep and goats also can be toxic to the animals and the humans who apply them. In addition, these insecticides can create illegal residues in milk and tissues of treated animals and can be destructive to the environment if not used and handled safely and correctly. The following are a few precautions to follow when choosing and applying insecticides for the control of insect pests of sheep and lactating and nonlactating goats.

- Use only those insecticides recommended by a recognized authority, usually a government official such as an agricultural agent or advisor, and approved for use on sheep and lactating or nonlactating goats.
- Use a formulation of the insecticide that is approved and especially designed for use on sheep and goats. In particular, in dipping vats use only those formulations designed specifically for dipping vats.
- Follow the label directions exactly. The label contains all the information on dilution, time of retreatment of animals, antidotes for poisoning, methods of disposing of unused insecticide, intervals between treatment and freshening, and other important facts.
- Avoid treating sheep and goats in cold, stormy weather, and avoid treating stressed, overheated, or sick animals.
- Be sure that spraying equipment is clean and working properly and especially that it provides sufficient agitation to allow for thorough mixing of insecticides.
- Be aware of safe practices when mixing and applying. Wear protective clothing and do not smoke, drink, or eat while applying insecticides. Do not contaminate feed or feed and water troughs.
- Learn to recognize signs of insecticide poisoning in livestock (and humans) to avoid delay in instituting antidotal measures.
- Store all insecticides in original containers. Do not store insecticides with food or where they can be reached by children, animals, or unauthorized persons.

49

THE INTERNAL PARASITES OF SHEEP AND GOATS

Thomas R. Thedford

Internal parasites of small ruminants (hair and wool sheep; meat, milk, and hair goats) are a primary limiting factor to the successful production of these animals. Thus a primary goal of this paper is to provide a better understanding of the internal parasite problem and to help reduce related production losses. Better control of internal parasites can lead to improved weight gains, milk production, reproductive rate, and fleece or hair production.

Internal parasites are a problem to sheep and goat producers all around the world. Death and production losses run into the millions of dollars annually. Such losses are difficult to document because they include decreases in gain and feed conversion as well as unreached potential of milk production and fleece and hair weight.

My discussion deals specifically with:

- How the parasites affect sheep and goats.
- General symptoms of internal parasitic infections.
- The various types of internal parasites.
- The basic life cycles of these internal parasites.
- Where the adult and/or diagnostic form of the parasites are located in the host.
- Recognition of specific symptoms of each parasitic type.

RECOGNIZING SYMPTOMS

The first step in recognizing a parasite problem is that of identifying the symptoms. The symptoms of parasitism are:

- Emaciation or extreme thinness. This can be due to loss of nutrients from the host's digestive tract or from chronic loss of blood.
- Anemia. Blood loss is a common symptom caused by bloodsucking roundworms of the stomach and upper intestine.

- Diarrhea. The animal's feces may be bloody or dark black (from partially digested blood), or it may be of normal color, but very watery. Black or red coloration can be caused by toxic materials produced by the worm, or bleeding at the attachment site while the worm feeds or after it leaves.
- Icterus or jaundice. Yellowing of the mucous membranes can be caused by blockage of the bile ducts by liver flukes or by toxic reactions that might destroy red blood cells.
- Bottle jaw. Submandibular edema or swelling and accumulation of fluid under the skin below the jaw (hence "bottle jaw") is a common symptom of parasites, especially those that suck blood.

DESCRIPTION OF PARASTIES

Each worm and its life cycle are discussed next as a basis for understanding the causes of certain symptoms and the methods of prevention and treatment.

Roundworms of Stomach and Intestines

Primarily bloodsuckers, roundworms include those of the species Cooperia (Coopers worm), Haemonchus (barber pole worm), Nematodurus (thread-necked worm), Chabertia, and Ostertagia (brown stomach worm).

Life Cycle. These parasites have what is called a direct life cycle, which indicates that they have no living intermediate host and the infection passes directly from animal to animal (figure 1). This life cycle usually requires 21 to 36 days to complete.

The eggs are passed from the host in the feces, which falls to the ground where the eggs hatch into first-stage larvae.

While living on the ground, the larvae develop into 2nd-stage larvae and then into 3rd-stage larvae. These 3rd-stage larvae are capable of infecting the host if they can enter either by ingestion or penetration. This development process through the first 3 larval stages requires 8-15 days under most conditions.

To get into position to be ingested by a suitable host, the infective 3rd-stage larvae are able to move up grass or other plant material by wiggling through moisture droplets such as rain or dew on the grass. The sheep or goat swallows the infective 3rd-stage larvae with the grass while grazing.

Within the host, the infective 3rd-stage larvae changes into a 4th-stage larvae and does one of two things:

- It becomes an adult, starts feeding, breeds, and starts laying eggs, thus completing the life cycle, or...
- It encysts in the wall of the stomach or intestines of the host as an inactive, non-parasitic form. The larvae remain in this inactive form until a stimulus is received that makes them leave the wall of the gut and develop into a parasitic adult. These adults complete the life cycle by feeding, breeding, and laying eggs.

Several things can trigger this movement from the encysted to the active form including: death of many adult parasites (for example, after a deworming procedure), the birth process in the host, and the start of milk production by the host.

The rapid change of these 4th-stage larvae into adults can produce a severe parasitic debilitation or death of the host in about 2 to 3 weeks after deworming.

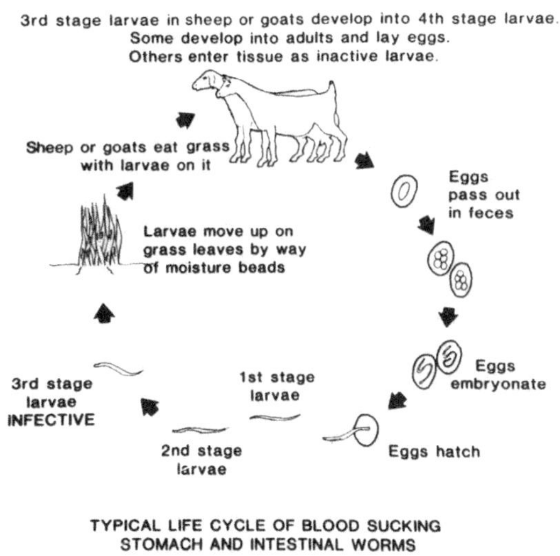

Figure 1. Typical life cycle of blood sucking stomach and intestinal worms

<u>Symptoms</u>. Acute infection is characterized by anemia, occasional bottle jaw and diarrhea, and sometimes sudden

deaths. Chronic infection results in poor condition, poor performance, anemia, diarrhea, bottle jaw, and death after a week or two.

Whipworms

Whipworms do not usually cause serious problems, unless there are heavy infections. Diarrhea and straining are the most common symptoms. Sometimes a prolapse of the rectum may follow the diarrhea when very high numbers of these parasites occupy the ceacum. These parasites primarily feed on the contents of the gut. Whipworms have a direct life cycle that is similar to that of the stomach and intestinal worms. The main differences are that they are slower (up to 3 months) to develop from the larvae into an adult, and the eggs and larvae are very resistant to environmental conditions.

Flukes or flatworms. The liver fluke provides an example of the fluke and their effects on ruminants. Liver flukes cause severe tissue damage by their migration process; bacteria may invade the tissues damaged by this migration process and cause infections such as Black disease. As the fluke feeds, the animal may lose some blood.

Figure 2 shows the indirect life cycle of the liver fluke. An indirect cycle requires that the life stages of the parasite must go through a different host (in this case a snail) before the parasite becomes infective for sheep and goats.

The fluke's eggs are passed in the feces, and if they fall into water will hatch into a miracidium or a free-living form in the water. The miracidia, while free in the water, will penetrate the body of certain types of snails. The snail is a secondary host to the liver fluke and is necessary to complete the life cycle.

While in the snail the miracidia divides asexually and eventually produces many cercaria. These cercaria leave the snail and again are free in the water.

When the cercaria come in contact with vegetation they attach and become encysted. These encysted forms are called metacercaria.

The sheep or goat swallows the metacercaria with the grass eaten while grazing near wet spots or edges of ponds.

The encysted metacercaria breaks away from the grass during the digestion process. It then penetrates the walls of the gut and migrates through the animal until it reaches the final site for adult development, such as in the liver.

Here the larval form develops into an adult, starts to lay eggs, and completes the life cycle of about 4 months. Adult flukes can live for years in sheep and goats.

Symptoms. Liver fluke symptoms are loss of weight and poor body condition accompanied by paleness of the mucous membranes (anemia), submandibular edema (bottle jaw) and an enlarged and painful abdomen. Sudden death is also seen when this parasite infects sheep and goats.

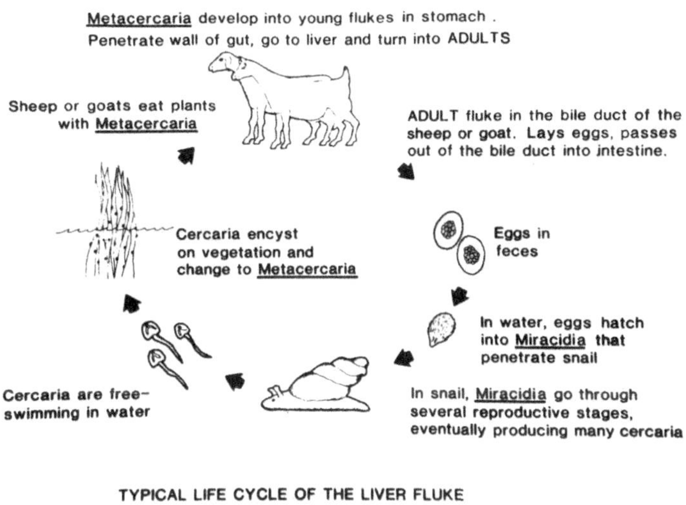

Figure 2. Typical life cycle of the liver fluke

Fluke infection is usually diagnosed when the adult fluke is found in the liver (bile ducts) of a dead sheep or goat. Also, special diagnostic procedures are used involving sedimentation and microscopic examination of feces.

Tapeworms

Tapeworms also have an intermediate host in their life cycle but are not considered a major problem. They rarely cause any disease and then only in the very young animals that have a heavy infection. These parasites feed on the gut contents of the host.

The life cycle of the tapeworm is seen in figure 3.

The eggs are passed from the host in a white, mucoid segment (proglottid) that clings to the fecal pellets. These segments are actually egg containers that break away from the adult tapeworm and are easily visible on the fecal pellets. After falling to the ground with the fecal pellets, the egg bearing segments (or proglottids) break open, releasing many eggs. The eggs are swallowed by a soil mite as it feeds and they hatch into larvae while in the mite. The mite lives on or near plant roots. When sheep or goats pull up plants, or graze forages very closely, they swallow the mites containing the larvae.

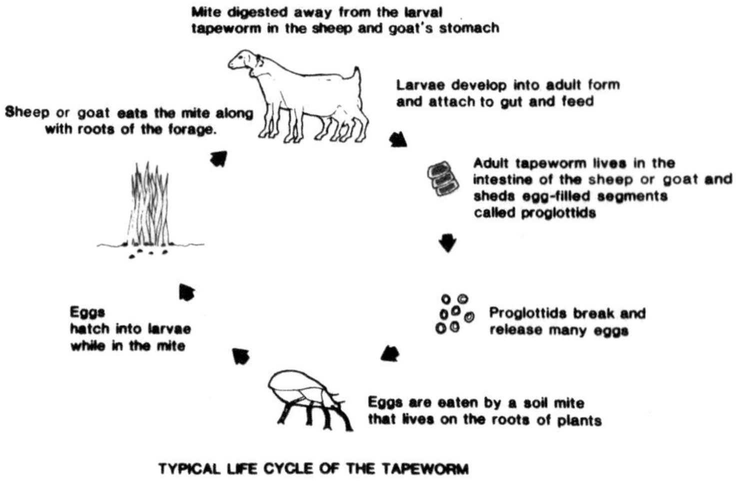

Figure 3. Typical life cycle of the tapeworm

In the digestion process, the larval tapeworm separates from the mite while in the rumen of the sheep or goat. It then passes into the small intestine and develops into an adult tapeworm. Fertilization takes place, proglottids develop and are passed with the feces, completing the life cycle.

Lungworms

The most common lungworm of sheep and goats has a direct life cycle similar to the illustration in figure 2. Other lungworms of small ruminants have an intermediate host (snail or slug) but are not discussed here. The adult lives in the lung of the host and the eggs are laid in the lung and coughed up in the mucous.

The mucous containing the eggs is then swallowed and the eggs develop and hatch while in the gut. The live 1st-stage larvae are passed in the feces.

The larvae develop into infective 3rd-stage larvae while on the ground where they attach to plant material that may be eaten by the sheep or goat.

After the plant material is eaten by the sheep or goat, the larvae migrates through the lymph system to the lungs where the adult form develops, and the life cycle is completed. (The lymph drainage system is similar to the blood circulation system. It is the system that connects lymph nodes throughout the body.)

The symptoms of lungworm infection include progressive loss of condition; severe wet sounding cough; rapid, shallow breathing; and fever if a pneumonia has developed.

Coccidia

Coccidia are an intestinal protozoa that cause blood loss and cell damage leading to bacterial infection.

The life cycle of coccidia, figure 4, is very complex. One infective stage of the coccidia enters the host, develops asexually into many infective forms, and causes much tissue damage. After many multiplications, the asexual forms become male and female. One of each unites in a

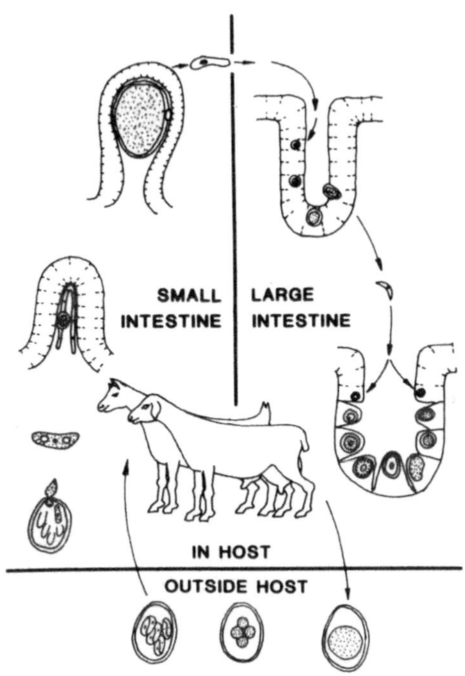

Figure 4. Typical life cycle of the coccidia

sexual change and develops into a zygote that turns into the infective stage. The zygotes's infective stage then ruptures from the cell, causes more cell damage, and passes with the feces as an oocyst. When the host swallows the oocyst, the life cycle is complete.

Almost all coccidia are host specific. This means that sheep and goats only have sheep and goat coccida. They do not get them from birds or other animals.

Symptoms. The coccidia of small ruminants involve the small intestine. They cause a profuse diarrhea that may or may not contain blood. Straining occurs with the diarrhea and usually many kids or lambs are sick at the same time. The death loss may be very high and many of the young are commonly found dead before symptoms have time to develop.

REFERENCES

George, J. R. Parasitology for Veterinarians (2nd Ed.). W. B. Saunders Co., Philadelphia, London, and Toronto.

Hall, H. T. B. Diseases and parasites of livestock in the tropics. Intermediate Tropical Agriculture Series. Longman group, N.Y. and London.

Howard, J. L. (Ed.). Current Veterinary Therapy - Food Animal Practice. W. B. Saunders Co., Philadelphia, London, and Toronto.

50

IDENTIFICATION, TREATMENT, AND PREVENTION OF INTERNAL PARASITES OF SHEEP AND GOATS

Thomas R. Thedford

In addition to an understanding of the life cycles and symptoms of internal parasite infection (see "Internal Parasites of Sheep and Goats," this book), the producer requires some knowledge of the equipment and procedures involved in diagnosing these diseases. Although the producer probably will not be making such diagnoses due to the cost of equipment and technology, an understanding of these procedures will help in applying the therapeutic and preventive measures described in this paper.

Many procedures are used to diagnose parasitism in the laboratory; however, the technique of fecal flotation (a process of floating parasite eggs out of feces) is most commonly used. The equipment and technique for the process are outlined below.

Specifically, my paper will:

- Describe the equipment necessary to diagnose parasitism by fecal flotation.
- Explain the technique of fecal flotation and the identification of parasite eggs usually seen on flotations.
- Describe the drugs used to treat internal parasite infection and describe their effectiveness.
- Describe the procedures effective in reducing exposure and keeping infections of internal parasites at a manageable level.

EQUIPMENT AND TECHNIQUE

The equipment and procedures necessary to diagnose parasite infection from fecal samples include:

- Microscope--that will magnify up to at least 400 X; also a light source -- either electric or the sun.
- Flotation liquid--consisting of a super saturated solution of salt or sugar. Either of these solutions will be satisfactory. The solution is made by adding sugar or salt to a specific amount of liquid until it will leave some of the

solid on the bottom of the container. No more of the sugar or salt will go into solution. The sugar solution cannot be stored for extended times because it will mold and be unusable.

- Small containers--one for mixing the feces with the flotation liquid; another "flotation container" to hold the feces and flotation fluid so that the eggs can float to the top.
- Stirring stick--the fecal pellets and the flotation liquid are placed in the first container. Stir the fecal pellets until they are completely broken up. A piece of gauze may be used to strain the larger particles of the feces from the fluid mixture. When throughly mixed, pour the fluid through the gauze into the flotation container from the mixing container. Fill the flotation container completely.
- Glass cover slip--placed on top of the flotation container and left for 7 minutes. The container should be completely full so that the liquid touches the bottom of the cover slip. This allows the eggs to float up and stick to the bottom of the cover.
- Place a glass slide on a flat surface, carefully lift the cover slip straight up and lower it onto the glass slide. You are now ready to look at the material under the microscope. To examine the entire slide, begin viewing at the top and corner edge of the slide and scan across the slide. Then drop down 1 field width and move the slide the opposite direction. Continue this process until all eggs have been differentiated and identified.

DIFFERENTIATION OF PARASITE EGGS

Strongyle eggs are oblong with rounded ends; they are smooth and contain a developmental stage called a morula (figure 1).

In differentiation of common parasite eggs seen on fecal flotation, strongyle-type eggs of the following parasites may be found. (It is almost impossible to differentiate among these parasites by examination of their eggs): *Trichostrongylus* sp., *Ostertagia* sp., *Haemonchus* sp., *Cooperia* sp., *Bunostomum* sp., *Oesophagostomum* sp., *Chabertia* sp. The eggs of *Strongyloides* sp. and *Nematodirus* sp. are sometimes confused with Strongyle types.

Trichuris eggs or whipworms (figure 2) are "double operculated." That means they have a trap-door-like opening on each end. They are more elongated than the strongyle-type egg.

Moniezia eggs or tapeworm eggs (figure 3) have a thick shell; they are square-shaped and contain a pear-shaped embryo.

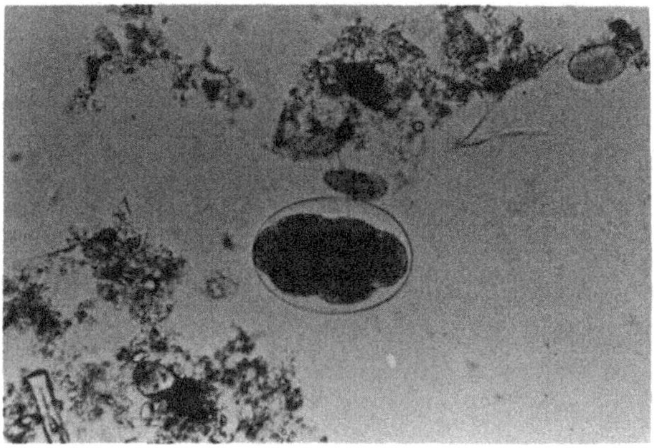

Figure 1. Strongyle egg

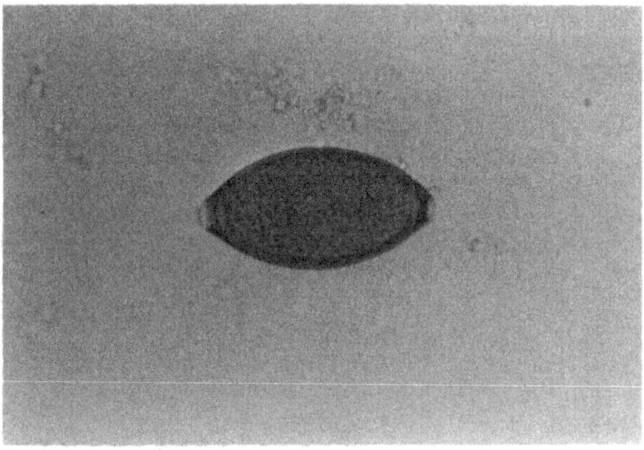

Figure 2. Trichuris egg

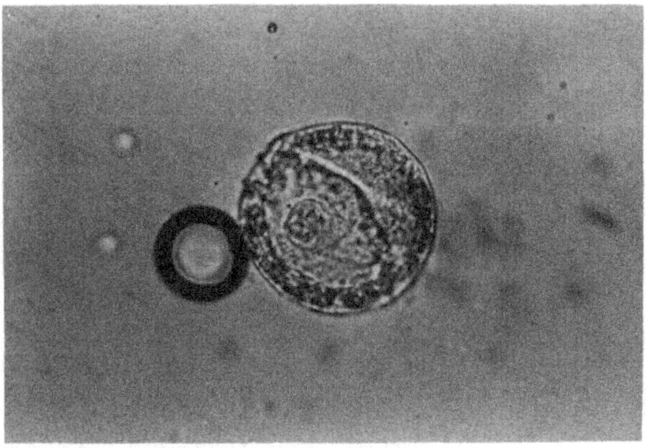

Figure 3. Moniezia egg

Coccidia oocysts (figure 4) are usually seen in ruminant feces; they are smooth, small, and do not shine under a microscope. In animals showing symptoms, large numbers are usually necessary to indicate infection. Figure 5 shows the relative sizes of coccidia and strongyle eggs.

Trematode or fluke eggs (figure 6) are larger than most other eggs and have only one operculum. Fluke eggs, which usually will not float, must be washed and centrifuged to locate. Diagnosis is usually made by finding the adult trematode on necropsy (postmortem exam). Note: the flotation techniques and photos described here will not identify problem protozoan parasites. Consult with a veterinarian or a diagnostic laboratory for diagnosis of these problems.

ANTHELMINTICS OR DEWORMERS

Due to the international character of this school, I am discussing many anthelmintics that do not have FDA approval for use on sheep and goats in the U.S. Many of these products, however, do have approved usage for other species and information is available on dosage for other countries. Practicing veterinarians may prescribe products unapproved for a species providing they have a knowledge of the farm and the animals on that farm and are willing to accept the responsibility of the use of the product on those animals. This is a very common circumstance and explains my discussion here. It is generally recommended that these products not be used on milk animals or on animals going to slaughter in 60 days or less. However, no data is available on withdrawal dates on these unapproved usages.

Dewormers are provided in several dosage forms: pills, liquids, paste, injectable, and feed additives. The dosage forms of each product are equally effective as long as the correct dose is administered to the animal. The generic types, their effectiveness, and the parasites they affect are outlined below and in table 1. Listing of these drugs is not intended as a recommendation. Always consult with a veterinarian and thoroughly read the manufacturers' recommendations. Discussion here is limited to generic drug names only to minimize risk of misinformation. These names appear on the container of all anthelmintics along with a trade name that may vary depending on the company, the country, and the way the drug is marketed. Dosage rates are given in milligrams per kilogram of body weight, unless otherwise noted.

Benzimidozoles

The benzimidazoles are the largest class of dewormers and are generally very good anthelmintics. They are easy to administer, relatively safe, and come in powder, paste liquid, bolus (pill), and feed additive form.

TABLE 1. DRUGS FOR CONTROLLING INTERNAL PARASITES (ANTHELMINTICS)

Drug	Roundworms	Larvae	Whipworms	Tapeworms	Lungworms	Flukes	Coccidia	Precautions (Always check label for withdrawal information.)
*Albendazole	5-10 mg/kg			5-10 mg/kg		10-20 mg/kg		Do not use 1st third of pregnancy--75 mg/kg fatal.
*Oxfendazole	5 mg/kg	5 mg/kg	5 mg/kg	5 mg/kg				Do not use in pregnant animals--safe to triple dose otherwise.
*Cambendazole	10-15 mg/kg	10-15 mg/kg		20-25 mg/kg	40 mg/kg			Do not use in 1st third of pregnancy, do not overdose.
*Fenbendazole	5 mg/kg	5 mg/kg	5-10 mg/kg	5-10 mg/kg				Safe in pregnant animals.
Thiabendazole	44-66 mg/kg							Resistance common. Safe in pregnant animals.
*Oxibendazole	5-10 mg/kg	5-10 mg/kg						Safe in pregnant animals.
Mebendazole	13.5 mg/kg			13.5 mg/kg				Safe in pregnant animals.
Levamisole	8 mg/kg				8 mg/kg			Do not overdose or use on milking goats. Safe in pregnant animals.
*Haloxon	50 mg/kg							May cause posterior paralysis.
Phenothiazine	12.5 gm/ 11 to 27 kg 25 gm over 27kg 1 gm/hd/da							Do not use last third of pregnancy--only fairly effective. Do not use on lactating does. Do not overdose or use on debilitated animals. Reduces egg production and hatchability.
*Morantel	10 mg/kg							Safe in pregnant animals.
*Amprolium							10-14 mg/kg	Give for 5 days to 21 days. Long term use may cause thiamine (B_1) deficiency.
							50 mg/kg	One time drench.
*Pyrantel	25 mg/kg							Safe in pregnant animals.
*Monensin							15 gm/ton	Fairly toxic. Feed throughout feeding period.
Sulfa drugs (dimidine,							200 mg/kg	Reduce dosage by 1/2 on subsequent days. Treat for 3-5 days.
guanidine, methazine,							or 1 1/2 gr/#	Reduce dosage by 1/2 on subsequent days. Treat for 3-5 days.
quinoxaline)								Make sure drinking water intake is normal.
*Lasalocid							20 gm/T	In feed.
							90 gm/T	In salt.
*Decoquinate							0.5 mg/kg	In feed for 28 days.
*Avermectins	200 µg/kg	200 µg/kg						Also effective against external parasites. Dosage in micrograms.
*Febantel	5/mg/kg							
Nitrofurazone							7-10 mg/kg	Prescription drug.

*Drugs not approved for sheep and goats in the U.S.

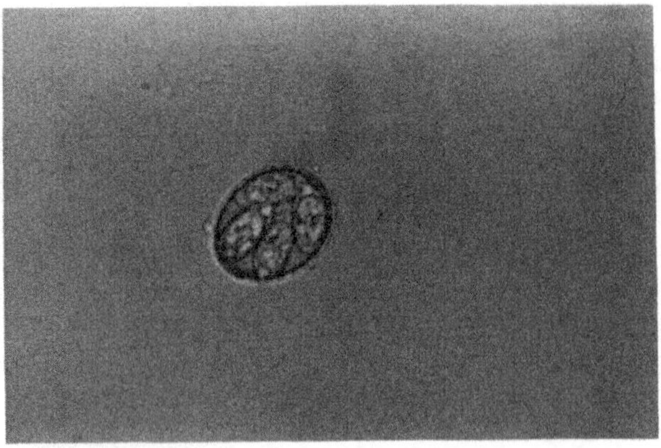

Figure 4. Coccidia oocyst

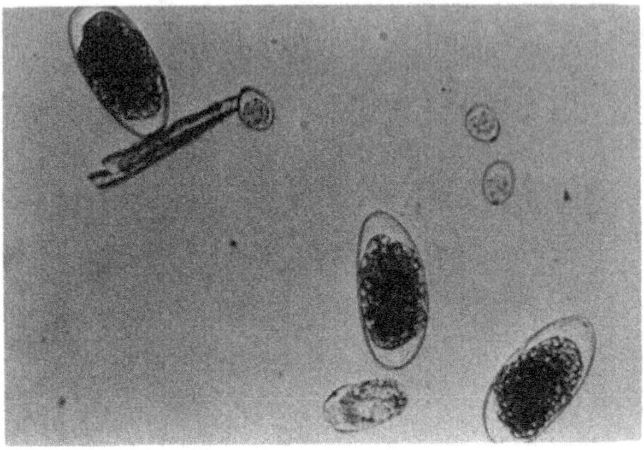

Figure 5. Relative sizes of coccidia

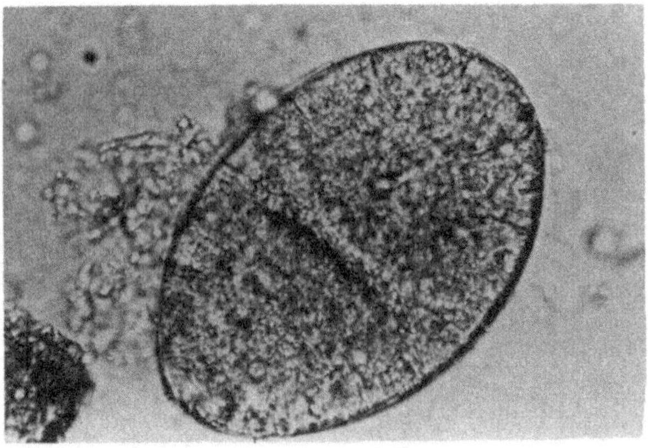

Figure 6. Fluke egg

Thiabendazole. Thiabendazole is the oldest, most common dewormer of this group; it is very safe (FDA approved) for adult sheep and goats. It is effective against most stomach and intestinal roundworm parasites, but many of these parasites have developed resistance. The dosage range is 44 to 66 mg/kg or higher. Do not use this drug for 30 days before slaughter. Milk should not be used for human consumption for 96 hours after the last dose was given.

Albendazole. Although not approved for sheep and goats in the U.S., albendazole is widely used in other parts of the world. The dose of 5 to 10 mg/kg is very effective against all adult stomach and intestinal roundworms and tapeworms. A dose of 10 to 20 mg/kg is equally effective against liver flukes. Albendazole is not as safe as some of the other benzimidazoles. Its use during the first one-third of pregnancy may cause deformed lambs and kids. A dose of 75 mg/kg is fatal.

Cambendazole. Cambendazole is not approved for use on sheep and goats in the U.S. A dose of 10 to 15 mg/kg is very effective against larvae and adult stomach and intestinal roundworms. At 20 to 25 mg/kg it is effective against tapeworms, and at 40 mg/kg it will kill lungworms. The drug is not safe above these dosages and should not be used during the first one-third of pregnancy.

Fenbendazole. Fenbendazole is not approved for use on sheep and goats in the U.S. It is very effective, however, against all stages of stomach and intestinal roundworms at a dosage of 5 mg/kg. When the dose is increased to 10 mg/kg it is equally effective against whipworms and tapeworms. It is very safe and can be used on pregnant sheep and goats.

Oxfendazole. Oxfendazole is not approved for use on sheep and goats in the U.S. At a dosage of 5 mg/kg, it is excellent against all stages of stomach and intestinal roundworms, whipworms, and tapeworms. It is fairly safe at levels up to triple dosage but should not be used on pregnant animals at any time.

Oxibendazole. Oxibendazole is not approved for use on sheep and goats in the U.S. At a dosage of 5 to 10 mg/kg, it is 90% to 100% effective against all stages of stomach and intestinal roundworms. It is safe and can be used on pregnant animals.

Mebendazole. Mebendazole is not approved for use on sheep and goats in the U.S. A dosage of 13.5 mg/kg (6 mg/#/) is very effective in removing stomach and intestinal roundworms and tapeworms. It is safe and can be used on pregnant animals.

Other Dewormers

Although these dewormers are grouped here for discussion, their actions and spectrum have little in common.

Levamisole. Levamisole is FDA approved for use in sheep and is commonly used in goats. The dosage of 8 mg/kg is very effective for removal of adult forms of stomach and intestinal roundworms and lungworms. It is not as safe as the benzimidazoles and should never be overdosed. Do not use within three days of slaughter. It is safe for pregnant animals. It is supplied as a powder, liquid, bolus (pill), and is marketed in an injectable form for cattle. If used in the injectable form, the instructions for cattle can serve as a guide. (This is an unapproved usage.)

Pyrantel. Pyrantel is not approved for usage in sheep and goats in the U.S. It is effective and safe at 25 mg/kg for the removal of adult stomach and intestinal roundworms. It may be used on pregnant animals. This drug is sold in the U.S. as a horse dewormer and as a feed additive to deworm swine.

Morantel. Morantel is sold as a feed additive and very similar to pyrantel. It is not approved for usage in sheep and goats. Used as a feed additive at a dose of 10 mg/kg, it is effective against adult stomach and intestinal roundworms. It is safe and can be used on pregnant animals.

Phenothiazine. Phenothiazine is one of the oldest dewormers on the market and is FDA approved. It is only 75% or so effective on adult stomach and intestinal roundworms. The dosage is 12.5 gm for 11 to 27 kg sheep and goats and 25 gm for those over 27 kg. It should never be given to pregnant animals during the last one-third of pregancy, nor to goats being milked, because it causes the milk to turn red. The urine also turns red and will stain the fibers of the wool-producing breeds of sheep and of hair goats. Do not use within 4 days of slaughter. This drug can cause a photosensitivity to animals when overdosed and to man when skin contact occurs.

Phenothiazine is effectively used at a continuously low level when mixed with salt or mineral. At 1 gm/head/day, it is effective in reducing parasite egg production and decreasing the hatchability of eggs produced. When used extensively, it will assist in decreasing pasture contamination by parasite eggs. It also will reduce fly populations by killing fly larvae that breed in manure.

Avermectin. Avermectin is a new type of anthelmintic that is scheduled for approval for use in sheep and goats in late 1982. A dosage of 200 ug (microgram) per kg is highly effective against all stages of stomach and intestinal roundworms. It also is effective against many external parasites including lice, mange mites, and nose bots. It is to be marketed as both an injectable and oral product. Read the instructions carefully, as this is a very potent compound.

Anticoccidia Drugs

The anticoccidia drugs are amprolium, monensin, lasalocid, nitrofurazone, decoquinate and the sulfa drugs.

Amprolium. Amprolium is not approved for usage in sheep and goats in the U.S. It has been widely used in other countries and by veterinarians at a dose of 10 to 14 mg/kg in drinking water for 5 to 21 days . A one-time drench of 50 mg/kg also has been used. The lower dose in the drinking water is likely to be more effective due to the life cycle of the parasite. Long-term use of amprolium can cause a thiamine (Vitamin B_1) deficiency.

Monensin. Monensin is not approved for use in sheep and goats in the U.S. It has been found to be effective in controlling coccidia at a rate of 15 gm/ton of feed during the feeding period, primarily for feedlot lambs. Extreme care should be taken when using this product because monensin is more toxic to sheep than it is to cattle. It is very toxic to horses.

Lasalocid. Lasalocid is used as a chicken coccidiostat but is reported to be effective when fed to sheep and goats at a dosage of 20 gm/ton in feed or 90 gm/ton in loose salt.

Nitrofurazone. Nitrofurazone is a reasonably good coccidiostat in lambs and kids at a dose of 7 to 10 mg/kg for one or two doses. It is a prescription product.

Decoquinate. Decoquinate is effective as a preventative and treatment when fed as a feed additive for at least 28 days at a level of 0.5 mg/kg (23 mg/100 lb) body weight. It is not approved for use in sheep and goats in the U.S.

Enteric sulfas. Enteric sulfas are good coccidiostats and are approved by F.D.A. The dosage of 200 mg/kg for the first day and 100 mg/kg for 3 to 5 additional days should be strictly adhered to. Ensure normal water intake because the sulfa drugs will precipitate out and plug the kidneys.

PREVENTIVE MEASURES FOR PARASITE PROBLEMS

Because it is not practical and virtually impossible to maintain parasite-free animals and facilities, some exposure to parasites is necessary as sheep and goats will develop some resistance to them after exposure. There is no set program that will work for all producers in all places. However, some common sense rules do apply to most situations and to most producers.

From the time kids and lambs start to nibble feed (3 weeks old) until about 1 year old they are very susceptible to parasites. Their exposure to them can be lessened by:

- Creep feeding.
- Grazing only on cultivated crops.
- Grazing rapidly across areas used only once per year. Desert or mountain grazing should be completed in less than 3 weeks.

Keep young animals in good physical and nutritional balance. Well fed lambs and kids are usually less susceptible to the effects of parasites.

Deworm ewes and does as soon as possible after birth of their offspring. Egg production of most parasites increases massively when lactation begins. Deworming at this time reduces contamination of pastures.

Deworm all animals in the herd at the same time before changing pastures; afterward allow 24 to 48 hours before moving the herd to a new pasture. This delay will help reduce egg contamination of the new pasture.

Take advantage of the egg-reducing effect of low-level phenothiazine in salt or mineral, if this treatment is practical and economical. Low-level phenothiazine will reduce egg output and hatchability only--it does not kill adult worms. This treatement must be in conjunction with a good deworming program.

When deworming, make sure all animals are dewormed and that the drug goes into the animal. It is poor economy to injure an animal trying to set a world record for number dewormed in one hour. Always coordinate the deworming with pasture rotations so that wormy animals are not moving onto clean pastures.

Many products that are nontoxic to adult animals may be very toxic to the young. Never deworm kids and lambs less than 6 weeks old. Most prepatient periods are at least 3 weeks, and kids and lambs are unlikely to pick up parasites until after they start eating forage, which is usually at about 3 weeks of age. This is why most veterinarians do not recommend a deworming program for kids and lambs less than 6 weeks of age. An exception to this rule would be an infection of *Strongyloides* and possibly coccidia.

DOSAGE CALCULATION

To calculate the dose of a specific dewormer use the following formula and example.

- To find body weight in kilograms: divide weight (lb) of an animal by 2.25 to calculate animal weight in kilograms (kg).

  wt (lb) ÷ 2.25 = kg of body weight

- To find total dose per animal in milligrams: multiply kg of body weight by recommended drug dosage (mg/kg). This gives the total recommended drug dose for that animal.

Bodyweight (kg) X recommended dosage (mg/kg) = total mg of drug required

- To find concentration of the drug: now check the label on the dewormer container; it will give you the concentration of the drug for that specific dewormer, probably in milligrams of active ingredient per unit of volume, i.e., "contains 10 mg per ml of active ingredient."
- To find dose for that animal: divide the total recommended dose by the concentration of the dewormer to calculate the recommended amount of dewormer for that animal.

Total recommended dose ÷ concentration on label = dose in ml or ounces. For example: a sheep weighs 112.5 lb and we intend to give it a drug with a recommended dosage of 10 mg/kg. The drug concentration is 100 mg per milliliter, as shown on the container label of a particular dewormer product, therefore:

112.5 lb ÷ 2.25 = 50 kg of body weight
50 kg X 10 mg/kg = 500 mg of recommended drug
Concentration noted on drug container = 100 mg/ml
500 mg ÷ 100 mg/ml concentrate = 5 ml dosage of that particular product for that specific animal.

REFERENCES

George, J. R. Parasitology for Veterinarians (2nd Ed.). W. B. Saunders Co. Philadelphia, London and Toronto.

Hall, H. T. B. Diseases and parasites of livestock in the tropics. Intermediate Tropical Agriculture Series Longman group. New York and London.

Howard, J. L. (Ed.). Current Veterinary Therapy - Food Animal Practice. W. B. Saunders Co. Philadelphia, London and Toronto.

The Merck Veterinary Manual. 5th addition. Merck and Co., Inc., Rahway, N.J.

Jensen and Swift. Diseases of Sheep (2nd Ed.). Lea and Febiger, Philadelphia.

51
RAM EPIDIDYMITIS

Thomas R. Thedford

Ram epididymitis is a highly infectious, easily transmitted sheep disease. About 90% of the cases are caused by Brucella ovis, an organism in the same family that causes Bangs disease or contagious abortion in other animals and undulant fever in man. B. ovis, however, has not been found in man or isolated from other animals, and sheep are not commonly infected with other Brucella sp., with the possible exception of B. melitensis.

Ram epididymitis was first seen in Australia in 1942 and in 1955 was found in California in the U.S. Since that time it has spread eastward with the first case reported in Oklahoma in 1976.

The incidence of B. ovis infection within a state has varied greatly, with percentage of infected rams in a flock sometimes reaching 70% or higher. The number of flocks infected varies depending on the control program and level of awareness of producers.

SYMPTOMS

Symptoms are swelling or enlargement of the epididymis,progressing to hardness, abscesses, and adhesions of the testicle to its tunic. These symptoms may cause total sterility or reduced fertility in a ram. As a result, there may be no lamb crop or a reduced lamb crop with extended lambing time.

Brucella ovis has been indicated as a cause of abortion in New Zealand; however, this result has not been documented in the U.S., although the organism may be involved in early fetal wastage.

Figure 1 shows the normal ram testicle. The sperm cells develop in the testicle proper, requiring 40 to 45 days. They then start their 21 to 28 day passage through the epididymis. At maturity, the sperm pass from the epididymis through the vas deferens and are ejaculated from the penis to fertilize the ova.

The epididymis is a long, coiled, single tube about 150 feet long. Inflamation and infection anywhere along this tract can cause adhesions and blockage. The blockage can cause the epididymis to rupture and cause development of abscesses and tumors called seminomas. These cause the knotty feeling of the epididymis when palpated.

TRANSMISSION

Pathogenesis, or the method of transmission, may occur when the ram swallows material contaminated by either urine or semen. For example, the ram may eat contaminated feed or lick or sniff the reproductive organs of other rams or of ewes that have just been bred by an infected ram. The organism may be transmitted if it gets into the mucous membrane of the nose or eye or on the prepuce. Infected rams can transmit the organism to clean rams by mounting them.

The ewe transmits the organism only if she is bred by an infected ram and then bred again by a clean ram within a few hours. She usually doesn't develop the disease.

Epididymitis spreads rapidly by natural means--in one herd all of the clean rams pastured with an infected ram were infected within about 4 months. In an experimental study in which rams were infected by prepucial inoculation, all rams showed white blood cells in their semen in 2 to 6 weeks; the organism was isolated in 50% of the rams in 4 weeks and in 100% of the rams within 5 weeks after inoculation. Shedding of the organism occurs at this time. Lesions can usually be felt by 4 months. Thus, rams that appear normal when palpated can, in fact, be infectious.

In other experimental cases, antibodies in the blood stream have been detected on the 9th day after inoculation and have persisted for 7 months. At that time the antibody titer began a gradual decline. Many such animals are infectious for a long time after the blood titer returns to negative. They may also develop lesions.

DIAGNOSIS

Diagnosis of ram epididymitis should be based on: palpation, blood test (complement fixation test, Enzyme Linked Immunosorbent Assay (ELISA) Test, or Indirect Haemagglutination Assay), semen evaluation, and semen culture.

PALPATION

Palpation (examination by feeling with your hands) can be done by placing the ram in a sitting position or by feeling his testicles while he is standing. Figure 2 shows the proper way to grasp the testicles with one hand while feeling with the other. The testicle on the left is obviously larger; such swelling suggests further checking.

In figure 3, the pointer shows the top of the epididymis. All the tissue below the pointer is epididymis.

In figure 4, the darker pointer (above) is showing the uppermost limit of the normal epididymis and the lighterpointer (below) indicates the infected one. Although only one testicle may show lesions, both could be infected.

Figure 5 shows testicles with the capsule or tunic partially removed. The testicle on the left shows an indentation between the epididymis and the testicle proper. This is normal. On the infected testicle on the right, the complete epididymis is enlarged, hard and knotty, and the tunic has adhered to it. This infected testicle is smaller than is normal in this case.

Figure 6 shows both testicles inside the tunic. Both are enlarged and have hard, knotty epididymides. Figure 7 shows these same testicles with the tunics removed. Note the adhesions--especially on the left testicle. Figure 8 is a closeup of the adhesion and knots on an infected epididymis.

SEROLOGY

Palpation will not always reveal infection, thus you may chose to use one of several available blood tests for epididymitis. The complement fixation test at the USDA lab in Ames, Iowa is the official test and is reasonably accurate. (The ELISA test has been used at Colorado State University with good results, and the Indirect Haemagglutination Assay test is used in Australia.) Regardless of which test is used, the important finding is that blood tests usually show positive indications before lesions are apparent in the testicles. Early identification and removal of infected animals from the herd reduces the spread of infection.

SEMEN EVALUATION

Semen evaluation is very important in the diagnosis. Early in the infection, a high number of white blood cells are found in the semen. This is not diagnostic but gives an indication of possible infection. The semen evaluation also gives prognostic information and assists in identifying rams that have reduced fertility. Most infected rams are not sterile, only less fertile, and are a source of infection to other rams.

Semen culture is the most specific test. If the organism can be recovered, infection exists. However, a ram may be infected even if <u>no organism</u> is recovered from the semen--he simply may <u>not have</u> released organisms at the time of the test. Infected rams will not show all symptoms at all times nor will all tests be positive at all times. To control this disease it is important to use all four parts of the diagnostic scheme. Herd history also is valuable in establishing a diagnosis.

EFFECTS OF EPIDIDYMITIS

The effect of ram epididymitis on reproduction was demonstrated in one test using infected rams that produced a lamb crop average of .77 lambs per fertile ewe and required 6.8 services per ewe for breeding. Normal rams produced 1.55 lambs per fertile ewe and required only 1.7 services per ewe.

Treatment is of no value after lesions have developed; the damage done to the epididymis cannot be repaired. These rams should be sent to slaughter to reduce the possible spread to clean rams. Their removal will give the uninfected rams more opportunity to increase the lamb crop. If the ram maintains some fertility and is very valuable, an isolated herd may be maintained but at great risk since the infection could still spread to the main flock.

PREVENTION

To prevent introduction of infected males into the herd, animals should not be bought from infected herds. Ask about health problems and palpate all rams purchased. All new ram additions should be isolated for 30 to 45 days before using them or turning them into the ram pen. Each year have all rams palpated and semen checked by your veterinarian before turning in with the ewes. Blood tests and cultures should be done on suspect animals.

Any ram palpated and showing symptoms or that has a poor semen evaluation should be removed from the breeding herd. If the ram is valuable, do not cull him until 2 or more tests have been run or he until has failed 2 or 3 semen evaluations. All possibly infected rams should be removed from the breeding herd and kept in isolation.

Vaccination is practiced in many commercial herds and, when incorporated into a program of palpation and semen evaluation, has been found to reduce the incidence of new infections. This is probably a sound program for the large commercial breeder. New ram additions should be retested 30 to 45 days before being added to the breeding flock. This second test helps to pick up those rams that could have been incubating the disease or in the very early stages at the first test.

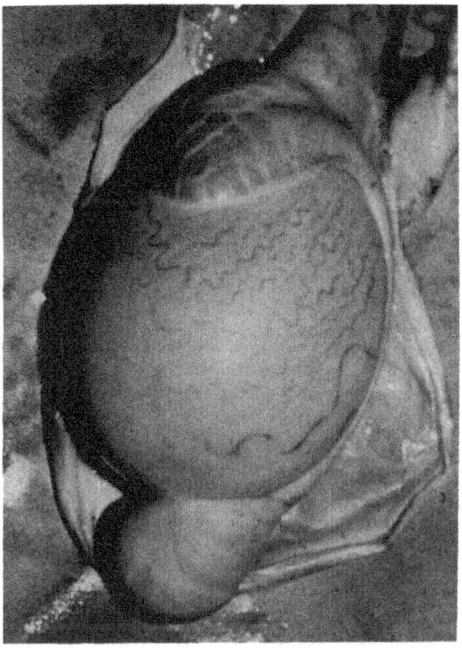

Figure 1. Normal ram testicle

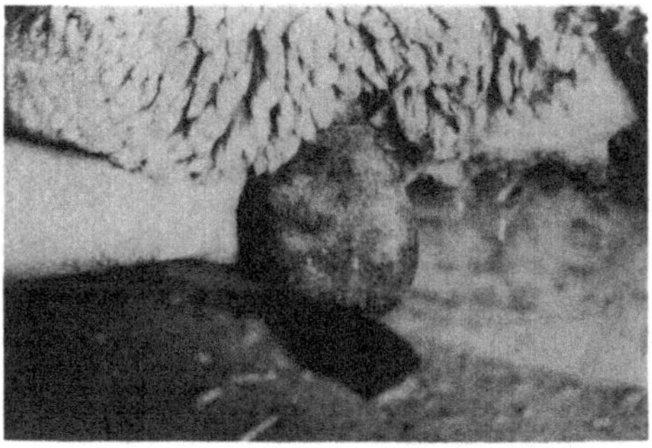

Figure 2. Proper palpation technique

Figure 3. Tissue below pointer is epididymis.

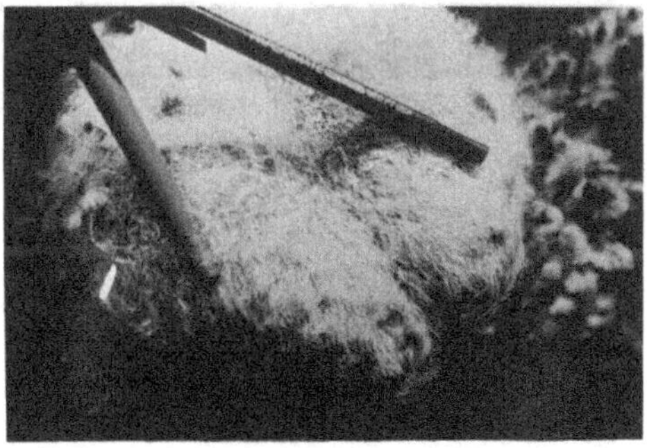

Figure 4. Dark pointer shows top of normal epididymis and light pointer identifies normal.

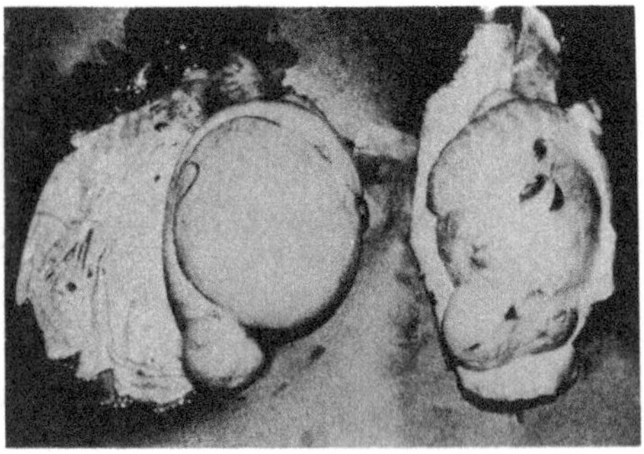

Figure 5. Left--normal epididymis Right--infected epididymis

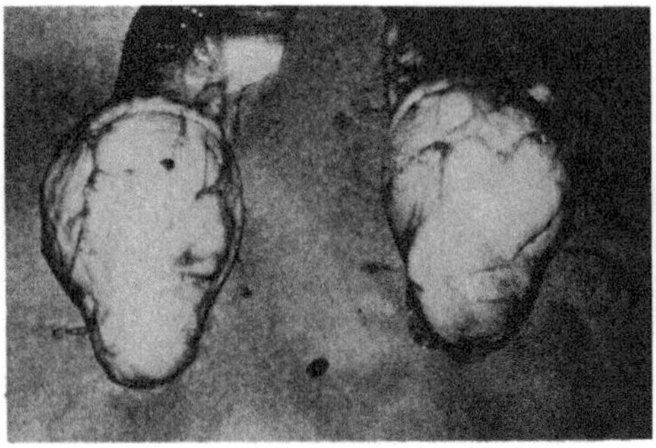

Figure 6. Left--testicle inside tunic, both infected.

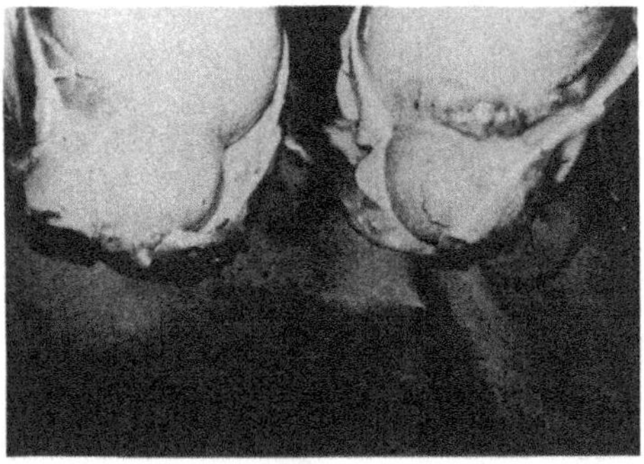

Figure 7. Tunics removed showing infected epididymides.

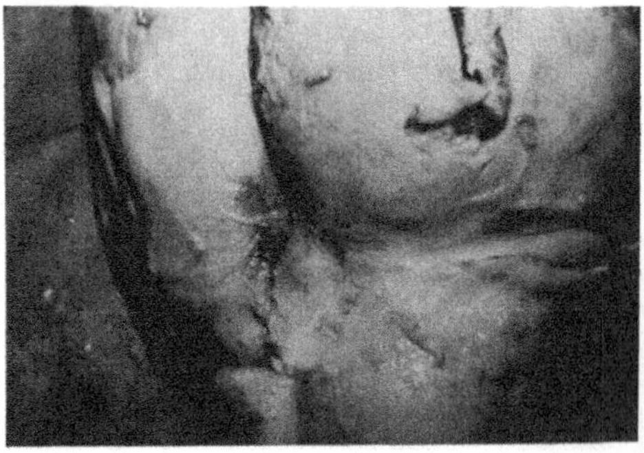

Figure 8. Close-up of infected epididymis.

52
BLUETONGUE: A DISEASE OF RUMINANTS

Thomas R. Thedford

Bluetongue was first diagnosed in sheep in Texas in 1948. The disease had been recognized in South Africa for over 100 years, but the virus that caused the disease was not isolated and identified in the U.S. until 1954 when the disease spread to California. The disease was first observed in cattle in the U.S. in 1959.

There are 21 types of bluetongue widely spread across the world. In the U.S. only 4 types currently infect our livestock: types 10, 11, 13, and 17. All are fairly mild and cause few deaths.

CAUSE AND TRANSMISSION

The orbivirus that causes bluetongue is tranmitted primarily by the bite of the gnat, Culicoides Variipennis. Other names for this vector are: biting gnat, sand fly, biting midge, and no-see-um. The gnat responsible for transmission is about the size of a straight-pen head. The virus reproduces within the infected gnat so that it becomes more capable of transmission. The gnat remains infected the rest of its life.

Requirements for transmission of this disease are: 1) a vector (such as the gnat); 2) readily available susceptible hosts and, 3) a virus reservoir (carrier animals). The disease normally is not transmitted from animal to animal except through the vector (Culicoides varripennis). However, it has been reported that infected bulls can transmit the disease through semen and infect calves in the uterus.

In the southern U.S., moderate temperatures and high populations of gnats increase the incidence of the disease. Wet seasons cause puddles of water to accumulate that offer breeding places for the gnats. Bluetongue is also commonly seen in southwestern U.S.

Bluetongue virus has been known to undergo antigenic drift, which causes strains of low virulence to become highly virulent, especially in cattle. Thus, bluetongue could evolve into a more severe clinical illness than presently observed.

Culicoides gnats prefer to feed on cattle rather than sheep. Thus, where cattle are numerous, they become more important as a potential carrier of the virus; infected cattle may remain carriers for 3 years or longer. Usually less than 5% of the cattle show symptoms. Reproductive losses (abortion and malformed fetuses) may reach 10% or higher. The virus type and susceptibility of the flock determine the rate of infection and death losses in sheep.

SYMPTOMS

In ruminants, bluetongue symptoms are:

- Lameness that disappears with exercise.
- Fever of 106°F/41.1°C or higher.
- Ulcerative lesions of the tongue, dental pad, and mucous membranes of the mouth and lips.
- Excessive salivation or drooling.
- Burned or cracked appearance to the muzzle.
- Sudden drop in milk production in cattle.
- Cracking and sloughing of the skin around the nose, feet, teats, and vulva.
- Copious pus-like material plugging the nose, necessitating oral breathing.
- Laminitis or reddening of the coronary band at the junction of the hoof and the skin of the foot that may progress into severe lameness.
- Swelling of the mouth, face, and sometimes ears, especially in sheep.
- Death, when it occurs, usually from a terminal pneumonia.
- Wool loss in recovered sheep caused by high fever.
- Reproductive failure, which may be fairly high. Abortion may occur if the dam becomes infected very early in gestation. If the infection occurs later, malformed fetuses develop and are born near term. These losses may reach 30%.

Other ruminants are infected with bluetongue virus; goats, however, may have a positive blood test, but show no clinical signs. Deer and pronghorn antelope, on the other hand, are severely affected and have a high death loss from bluetongue. The symptoms in deer are identical to those of epizootic hemorrhagic disease (EHD).

TREATMENT AND PREVENTION

Bluetongue is a viral disease, thus there is no specific treatment available at this time. Antibiotics are used to reduce the possibility of a bacterial pneumonia that usually causes death in the infected animal. Prevention and control of bluetongue are very difficult. Once the disease is introduced and established it is almost impossible to eradicate.

Possible Control Measures

Vaccination. International Mineral Corporation is currently working toward clearance for use in cattle and sheep of promising quadrivalent, modified-live-virus vaccine developed by Texas A&M Agriculture Research Station at San Angelo. A killed-virus vaccine is in the development stage in California, and a monovalent vaccine is on the market now that protects only for type 10, probably the least prevalent type in the U.S. This vaccine causes some malformation of fetuses when used on pregnant sheep.

Management. Eliminate shallow mud puddles that are rich in organic matter. Leaking water tanks, shallow muddy swamps, or edges of ponds that have manure in or around them are all breeding places for gnats. If practical, movement of stock to higher and drier ground during the vector season will help. Housing susceptible animals at night to keep them away from gnats will work but often is not practical. Elimination of carrier animals helps to reduce the source of the virus. Sometimes lambs or calves who were infected before birth carry many viral particles in their body, test negative, and remain lifetime carriers. If practical, sell all offspring for slaughter during the year that an outbreak occurs. This reduces carriers and the source of the virus.

Biological. Some subspecies and certain individual gnats are not capable of harboring the virus. Promising work is under way to determine if the genetic inability to transmit the virus could be bred into gnats.

Significant economic damage is inflicted by bluetongue because of the interstate and international restraint on the exportation of cattle and sheep. This applies to live animals, semen, and embryos. Other economic losses stem from a decrease in reproduction, growth, wool production, and efficient feed conversion.

REFERENCES

Committee on Bluetongue. Proceedings of the U.S. Animal Health Association each year. U.S.A.H.A. Richmond, VA.

Goltz, J. 1978. Bluetongue in cattle: a review. Canadian Veterinary Journal 19:95.

Howard, J. L. (Ed.). Current Veterinary Therapy - Food Animal Practice. W. B. Saunders Co. Tronto, London, and Philadelphia.

Luedke, Jochim and Jones. 1977. Bluetongue in cattle: Effects of Culicoids variipennis transmitted bluetongue virus on pregnant heifers and their calves. American Journal of Veterinary Research 38:1687.

53
A HEALTH PROGRAM FOR DAIRY GOATS

Samuel B. Guss

UNDERSTANDING THE HERD HEALTH PROGRAM:

Too often, dairy goat owners conceive herd health programs to be "do it yourself" veterinary programs. At the very outset, it is important that there be a good understanding between the goat herdsman and the veterinarian. Establishing a good relationship is the most important requirement for success. However, owners may have had unhappy experiences with veterinarians who seemed indifferent, incompetent, and less than compassionate when called to render veterinary assistance. Of course, the other side of the picture, from the veterinarian's viewpoint, may be an owner who does many things that make it virtually impossible to maintain a healthy, highly productive herd. Armed with a wide array of drugs, biologics, and antibiotics, some owners depend upon a goat book or the experience of other owners to diagnose and manage disease problems; they are naively oblivious of the factors that foster disease.

The availability of improved drugs, antibiotics, and biologics has sometimes given owners the false impression that these are the primary ingredients of a good health program. Actually, it is far more important to recognize and understand the disease entities and their sources. Both newly assembled herds and long-established herds require the same essential basis for a successful herd health program: (1) a thorough examination of the animals in the herd, (2) establishment of a good record keeping system, and (3) a free, open owner-veterinarian relationship.

Most owners of dairy goats are well-educated people who have full-time jobs. They are deeply interested in the health and welfare of their animals, and they are willing to sacrifice time and energy if it will help their animals. Thus a frank discussion should take place between owner and veterinarian before a herd health program is begun. A large number of topics for discussion will emerge and should be explored and resolved at the outset. The veterinarian and the goat owner should decide on a suitable time and place for this discussion so that it can take place leisurely, informally and openly, without rancor or outside interference. There should be "before the fact" mutual understanding that the veterinarian must be paid for his time, but if the time for discussion is selected well ahead, that charge can be minimal.

There are a few formidable obstacles that must be recognized and eliminated before real progress can be made. The veterinarian may have had unhappy, frustrating, and disappointing experiences working with dairy goat owners and their animals, including the owner and the herd under consideration. The veterinarian may feel that his veterinary education and clinical experience leaves him unqualified to provide the kind of service he would like to be able to provide.

The owner of the herd may have his own set of reservations about developing an open, frank relationship with a veterinarian such as bad memories of an unsatisfactory past experience, perhaps involving unsuccessful use of veterinary service (including fees charged and payment difficulties).

Answering and discussion of these questions requires time. Spend at least a few hours with your veterinarian. Prepare a detailed history of the herd and go over it openly and honestly with the veterinarian. Ask and answer questions to the best of your ability--honestly and frankly.

HISTORY OF THE HERD

What was the source of the animals of breeding age in the herd? What, in the past, has been the source of sickness, lowered production, and death? What has been done in the past to expose or prevent exposure to disease? What about medicine costs and production loss?

The reproductive history is important. What is the source of insemination--a single herd sire or artificial insemination? Are the does naturally bred to outside sires? What is the source of semen, if bred artificially?

Regarding conception: What is the average length of breeding season? Are drugs used to enhance heat detection or cycling? How many services per pregnancy? Is there a pregnancy examination or laboratory confirmation? What is the incidence of known abortions and dystocia? What is the number of live fetuses born per doe?

Questions on parturition: What is the frequency of parturition, the age at first parturition, and the interval between parturition? What is the number of kids per doe per year? How much milk does each doe produce each year? Are there or have there been metabolic and infectious diseases?

Finally, what is currently being done to define, prevent, and control both diseases and parasites in the herd?

A CLOSED HERD

Most veterinarians who have had food-production-animal experience recognize the value and primary importance of establishing a closed herd. What is a closed herd? It is a group of animals maintained with no association or minimal association with other animals of the same species (or even with other species).

It is essential that dairy goats be kept as closely as possible in a closed-herd situation. Some aspects demanding consideration are listed below:

Source of Replacement Animals

All herd replacement animals should be born and reared in the herd. A whole list of dairy goat diseases can be introduced into herds by acquisition of normal-appearing carrier animals. Any new replacement animals acquired should be kept in isolation without contact with the resident herd for not less than 30 days. During this quarantine period, the new animals can be subjected to an immunization routine established for the herd.

Fecal examination and treatment is essential to minimize exposure of the resident herd to introduced new internal parasites or drug resistant parasites. At least two initial applications of an external parasitacide chemical should be applied during the quarantine period.

Reproductive Disease

Ideally, risk of introducing diseases related to reproduction is minimized by introducing virgin animals into the herd. Young males should be examined carefully for libido, masculinity, and testicle conformation and size. Collection and examination of semen for volume, motility, spermatozoa concentration, and viability should at least be considered when young herd sires are purchased.

A good economical and satisfactory method of evaluating young virgin future herd sires is to allow them to copulate with at least two or three virgin cycling females showing normal heat cycles. Closely observe the females for recurrence of heat, genital discharge, and pregnancy. This need not be a major enterprise, nor one entailing great risk.

The new male and the virgin females to be bred can be immunized with one of the multivalent abortion disease vaccines available during the quarantine period and at least 15 days before any breedings are attempted.

Breeding herd females to on-trial sires via natural service is to be totally avoided, as it could be a major source of disease introduction into the herd.

Use of the Herd Sire to Breed Outside Females

This practice can be a major break in the wall intended to prevent introduction of outside disease. Unfortunately, most reproductive pathogens are readily infective to males during natural service, and it is possible that they can become established in the male, making him a carrier of reproduction disease for the remainder of his life.

Artificial Insemination

Prevention of disease spread was one of the major reasons for the development and use of artificial insemination for dairy cattle. Bulls were used subject to prior negative test for diseases known to be spread via copulation: brucellosis, Johnes' disease, and tuberculosis. Semen was treated with antibiotics known to be effective against vibriosis.

A major development against the spread of cattle disease was the initiation of the young sire evaluation program. In this program, young sires are purchased as calves, reared away from cows, immunized, carefully examined physically, and their semen evaluated. As soon as it is possible to collect and evaluate their semen, a large sample number of females is artificially inseminated. The young sires are not introduced into regular service and their semen is not used commercially until their daughters have been evaluated.

This program, which also helps AI cattle services meet export requirements for semen, has been a major advancement against disease spread. Unfortunately, much of the dairy goat frozen semen offered for AI today carries no assurance for the purchaser against spread of major goat disease entities. No significant effort has been made by regulatory veterinary agencies against spread of disease via AI in cattle or dairy goats.

Fortunately, as a result of foreign countries' purchases of frozen semen and their government veterinary requirements regarding health status of animal sources of semen, great progress has been made in prevention of spread of disease, both within this country and abroad.

DETERMINING THE HEALTH STATUS OF THE HERD

A close examination of the animals and available history should uncover the possible presence of endemic disease. Immunization procedures can be set up to minimize, control, or possibly, eradicate those diseases. An increasing number of drugs and biologics are available for this purpose and their judicious use can be highly effective.

IMMUNIZATION PROGRAM: TO PROTECT ADULTS AND THE VERY YOUNG AGAINST DISEASE

Initial and booster immunization against enterotoxemia, tetanus, and other contagious or infectious diseases endemic within the herd or the area should be established where justified.

Internal and external parasite control, both from the standpoint of strategic medication, use of indicated drugs and evaluation of management demands initial and ongoing surveillance by professional examination.

Herd parasite control or elimination can only be accomplished by thorough understanding and cooperation in development of treatment and management procedures indicated to be effective. Following the initiation of any changes, evaluation by means of simple laboratory means will provide ongoing surveillance for status and progress.

Lest owners and veterinarians who read this think that the above is a difficult or impossible effort, it should be recognized that any attempts toward the goals set for the herd are positive and worthwhile. What ultimately evolves as the health program for the individual situation may be minimal and very simple, or complicated and very difficult, or somewhere in between, depending entirely upon the relationship between the herd owner and the veterinarian.

Who will do the work? When the proper owner-veterinarian relationship is established, the veterinarian can be retained and paid for what he knows, and the owner can be trained and depended upon to administer the herd health program.

The ultimate program devised should be a good working relationship that is satisfactory, both from a health and production standpoint and an economic standpoint. A well-thought-out herd health program developed cooperatively by a veterinarian and his client should be mutually satisfying and profitable for both. This is the only justification for its existence.

It is extremely difficult for anyone to outline a herd health program that will satisfy needs for all herd situations in all areas of the country. There is no reason why owners and veterinarians cannot develop ideal programs if they determine to go about it openly, honestly, and cooperatively. The number of contagious and infectious diseases and their recent spread certainly justifies it.

A SUGGESTED DAIRY GOAT HERD HEALTH PROGRAM FOR NORTH AMERICA

The following suggestions were assembled for a "typical" dairy goat herd in North America. In other areas of the world, certain diseases of importance in this suggested herd health program may not be endemic, and others not considered in this program may be present. Both of those situations make it imperative that herd health programs be set up by the herd manager and the veterinarian, using all of the resource people available: nutritionists, agronomists (for soil management, forage establishment, and utilization), and all animal agricultural resource people, including regulatory veterinary medical officials.

SEASONAL HEALTH CONSIDERATIONS

Certain herd health recommendations are best considered on a seasonal basis, and these are listed first. Where certain herd disease problems are found, procedures for their control and/or elimination can be added to the routine herd health program below.

Fall and Early Winter

- Clean the barn thoroughly before the onset of bad weather.
- Clean and disinfect kidding pens and housing for baby kids. Keep these clean, dry, and free of bedding until needed.
- Give all animals two thorough cold water sprayings of Coumaphos (Co-Ral) at the rate of 1-1 1/2 lbs of 25% wettable powder per 100 gallons of water for external parasite control. A thorough spraying must be done. Dipping is preferable to spraying and should be done where practical.
- Spot apply insecticide that has been found to be highly effective on cattle and sheep when applied at the onset of the winter season. Where these products are approved for food producing animals, they greatly simplify external parasite control in sheep and goats. They merit trial and evaluation, but label directions should be carefully followed.
- Check and trim feet.

Winter

- Check housing to be sure that there is no moisture condensation anywhere inside.
- Be sure bedding remains dry and that there are no drafts which will unnecessarily chill animals.
- Check watering facilities for cleanliness and proper function. Goats will drink much more water if it is at least 15° above freezing, and milk production will benefit greatly.
- Remove horns and scurs from animals that are using them. Again, check the feet and trim hooves, if necessary.

Spring

- Check fencing and gates in lots and pastures. Two weed-chopper electric fence wires placed on the inside of woven wire or other permanent fencing will prevent fence damage and goat injury. Goats are more difficult to confine

within conventional fencing than any other domestic animals, except dogs and cats. New high tensile strength smooth steel wire fencing developed in New Zealand should be considered. Three to seven wire fence will simplify restraining the animals and prevent injuries sustained by goats when they attempt to escape through ordinary fencing. In addition, the new electric fence, properly installed, will keep out predators - dogs, coyotes, coons, etc.
- Remove all sources of possible injury from lots and pastures.
- Check and trim the feet, if necessary.

Summer

- Provide dry shade, cool fresh water, and trace mineral salt for all animals. Only loose salt will enable animals to consume adequate amounts during hot weather.
- Check bucks for physical condition and vigor. Feed them additional grain (dairy feed) to be sure they are gaining in condition before the onset of the breeding season. Physically examine all bucks for genital abnormalities or injuries.
- Check the herd for internal parasite burden during the height of the summer and treat, if necessary.
- Check and trim feet, if necessary.

HEALTH RECOMMENDATIONS FOR CLASSES OF ANIMALS

Dry Does

- All does bred for 70 to 110 days should be examined for pregnancy before drying them off. After the does are dried off, examine their udders carefully.
- Does that have had clinical evidence of mastitis, or older animals that have large pendulous udders, may benefit from administration of one-quarter dose of a dry cow mammary infusion in each udder half. Where mastitis is a health problem, routine treatment of all glands, when they are dried off, may be justified.
- Strict cleanliness and antiseptic teat end preparation are necessary for safe dry treatment. Dip teats for at least three days after drying them off. Examine udders carefully during the dry period; strip them out and treat them again if they show any abnormal signs.

- Dry does must gain in condition for the last month before kidding. However, calcium intake must also be controlled for the last month before kidding. Restrict the feeding of alfalfa and feed a grain mix containing no added calcium supplementation if any alfalfa forage has been fed.
- Administer a dose of selenium-alphatocopherol 60 days before kidding. The dose may need to be repeated again 15 days before kidding.
- Administer a booster dose of C. perfringens CD toxoid-bacterin and a booster dose of tetanus toxoid not less than 21 days before kidding. Blackleg bacterin or other clostridial bacterin may be administered where indicated at this time.

Kidding Does

- Make sure the does have adequate exercise right up to kidding.
- Confine in maternity pens when kidding is imminent. Both the doe and the maternity pens should be clean at kidding time.
- Wash the udder and soiled hind parts of the doe at the end of kidding.
- Assist the doe in cleaning and drying kids.
- Dip navel cords of kids in tincture of iodine immediately. Remove kids and hand feed at least 2 to 3 ounces of colostrum as soon as possible.
- Remove placenta and discharges as they are expelled by the doe; retained placentas should be removed after 24 hours.
- Examine does that show anxiety with straining after kidding.

Baby Kids

- Keep a small supply of antibiotic bactericidal scours medicine available for use if a kid develops diarrhea.
- Examine all kids carefully for navel infection. In some herds, it may be advisable to re-treat all navels a second time with tincture of iodine.
- Check all kids for congenital abnormalities (atresia ani, genital hypoplasia, inter sex, cleft palate, etc).
- Disbud kids at 3 days to 2 weeks of age. Castrate surplus bucks, remove wattles and extra teats at the same time.

SPECIAL SANITARY MEASURES FOR REARING KIDS

Many diseases of young goats can be prevented, controlled, or at least minimized when necessary sanitary measures are understood:

1. Keep kids separate from adult stock until they have reached at least six months of age or have been bred at least two months. Early separation of kids from their dams is important (after they have been cleaned and dried and have had their first feeding of colostrum). Does can be trained to let down their milk for the person who milks them by hand or machine. Letting kids nurse for even one day often makes it difficult to train first kidders to cooperate in a managed milking program, so the new kids can be moved immediately.
2. A basic common-sense hygiene program for any dairy goat herd, large or small, demands separate, clean quarters for the annual new generation of young animals. Several diseases (coccidiosis, caseous lymphadenitis, internal parasitism, paratuberculosis) are passed to successive generations of baby and weaned animals, when adult and young animals are housed or pastured together.
3. Dip navel cords in 2% tincture of iodine as soon as possible after birth and put baby kids into clean, disinfected pens before they are one-day old. This effectively eliminates two pathways of infection from disease-producing organisms in their environment.
4. Baby kid pens should be cleaned by scraping and scrubbing or using high pressure detergent steam spray after kids are removed at weaning time. Allow the pens to remain empty until needed for baby kids again. Chlorhexadine (Nolvasan) solution, chlorine bleach solutions, or iodophor dairy sanitizer solutions are adequate and safe disinfectant agents.

WEANED KIDS

- Provide 9 to 12 square feet of pen space for weaned kids. Inedible material--shavings, sawdust, sand--is preferable for bedding. All grain, forage, and water should be made available through keyhold fence line feeders outside the bedded or yard area.
- Keep weaned kids in clean, dry, well-drained outside lots without any growing forage. From

the standpoint of internal parasite control, these are better than are permanent pasture lots.

- Feed green chopped forages from fields not pastured. This may be the best approach to feeding green food to both young and milking stock. However, in large herds, annual pastures (sudan grass, cow peas, millet) may be the only practical sanitary pasture for young animals.
- Provide a large area where first kidders and dry does can be given daily exercise and have at least one or two hours of browse; nothing is better for bringing them into the best condition for kidding. Our Pennsylvania hillsides of birch, maple, ash, brush with shrubs, grapevines, and honeysuckle are a veritable paradise for goats of all ages during the period 2 or 3 months before kidding.

SPECIFIC DISEASE PREVENTION MEASURES

Coccidiosis

Amprolium given to kids at the rate of 5 mgm per kilogram (2.2 lbs) body weight for 5 days is effective. It is available in bolet or water soluble form.

Sulfonamides (sulfaguanadine, sulfamerazine, sulfamethazine) or long acting sulfas (sulfadimethoxine) are effective coccidiostats. The systemic drugs (sulfamerazine, sulfamethazine or sulfadimethoxine) have the additional advantage of overcoming coccidial forms (as in toxoplasmosis) that may be found outside the digestive tract. Medication with one of those drugs for 4 days per week for two weeks has been highly effective. The dose: 50 mgm/lb.

Monensin, a new feed lot drug for promoting growth and feed efficiency of feeder calves, when fed to young goats from weaning to breeding at the low rate of 10 to 20 PPM in feed, recently has been reported to provide excellent control for coccidiosis in lambs and is highly effective for goats.

(Caution - If there are horses on the farm, be sure they do not have access to feed containing monensin. The drug appears to be highly toxic to equidae.)

Tetanus and Enterotoxemia

Administer *C. perfringens CD* toxoid-bacterin and the first dose of tetanus toxoid at 3 to 4 weeks of age. Repeat both in two weeks.

Young does should be revaccinated again at least two weeks before kidding the first time or no later than one year of age. All bucks in the herd should be revaccinated when adult females are vaccinated.

Because tetanus protection may not last for more than 3 to 6 months following administration of tetanus toxoid, deep wounds justify the additional protection conferred by 150-300 IV of tetanus antitoxin.

Abscesses

Vaccination programs against Corynebacterium ovis or C. pyogenes herd infections (abscesses) should be implemented where these problems are identified. The first injection of autogenous bacterin should be given at one month of age, followed by another in one month.

In most abscess-problem herds, a repeat vaccination of all animals at 4 to 6 month intervals may be helpful.

Weaned Kids

- Check for internal parasites one month after weaning.
- Administer contagious ecthyma vaccine at least two weeks after weaning and not less than three weeks before the first show.
- Segregate buck kids from doe kids at three months of age or earlier.
- Recheck polled kids for genital abnormalities. Trim feet before turning kids out.

The Milking Herd

- Feed animals grain, individually, according to their needs. (Prebreeding recommendation). Be sure does gain in weight for at least two to three weeks before breeding.
- Record all heats noticed, length of heat, intervals between heats, and all breeding dates.
- Breed doe kids at 7 to 10 months of age or as soon as they are large enough to breed.
- Check all animals for parasites or routinely deworm the herd one month before breeding.

Preventing Mastitis

- Avoid milking injury and develop a good milking routine.
- Wash and dry udders before milking.
- Use a strip plate.
- Dip teats routinely in a nonirritant teat dip after milking.
- Check milk with the California Mastitis Test (CMT) every two weeks or when there is any reason to suspect mastitis (swelling, fever, tenderness or abnormal milk).

Bucks

- Administer all biologic injections and all parasite treatments to bucks at the same time these are given to the other animals in the herd.
- Provide abundant exercise and keep their feet trimmed.
- Administer selenium and alphatocopherol at the same time dry does receive it and again several weeks before the beginning of the breeding season.
- Be sure that bucks are in good flesh at the beginning of the breeding season.
- Examine all bucks for genital abnormalities or genital injuries before the beginning of the breeding season.

INTERNAL PARASITE CONTROL: A RECOMMENDED STRATEGY

Goats are extremely susceptible to infection by internal parasites. It is imperative that conditions that foster build-up of internal parasite burdens be recognized and eliminated.

Typically, blood-sucking worms inhabit the mucosa (lining) of the gastrointestinal tract (stomach and intestines). They are small enough, in many species, to be difficult to see with the naked eye. However, they feed on the host's blood and are often present in numbers large enough to produce severe anemia and damage to the gastrointestinal tract mucosa.

Adult female worms may produce thousands of ova (eggs) daily during warm, wet seasons. The ova are passed in the feces, and under warm, damp, shaded conditions, they hatch within a few hours. The developing and hatched larvae (microscopic baby worms) are highly resistant to freezing and excessive moisture. However, drying and exposure to sunlight quickly destroy both the newly hatched larvae and those that have developed to the infective stage, a stage they usually attain in a moist, shaded environment when they are protected from dessication (drying out) by the sun.

Feeders and waterers must be designed so that they provide conditions that will prevent fecal contamination of feed and water and also hinder successful development of larvae to their infective stage. Infective larvae, when ingested (eaten) in feed or water, rapidly develop to maturity. When conditions in the environment are best for worm multiplication, they complete their life cycle and begin to fertilize and produce ova. It is essential, to understand the conditions necessary for rapid multiplication of blood-sucking worm burdens.

Feeding Facilities

Feeders that provide grain or forage to goats should be designed to prevent fecal contamination. Goats should not be able to deposit feces into them nor get their feet into them.

Ideally, feeding facilities should be constructed and placed outside of the area in which goats live and move about; for example, fence line feeders with which goats are able to put their heads through a horizontal space in the boundary fence to feed out of a bunk. The widened space in the horizontal rail fence should enable them to put their heads through easily, but prevent them from getting their feet into the feed bunk.

So-called "keyhole" feeders have been widely used. They have one disadvantage--they confine goats that are feeding and enable "boss" animals to attack feeding animals and drive them away from the feeders, sometimes injuring or severly intimidating timid animals.

Water can be provided with the same arrangement, either by a shaded water trough or lick waterer, in summer, or an electrically heated waterer in winter.

A roof or shed arrangement should protect feeding and drinking animals from exposure to rain or snow and hot sunshine. Goats rarely consume enough feed and water unless they are exposed to the aforementioned conditions.

Worm Burden Diagnosis and Surveillance

Intelligent, efficient control of internal worm burdens demands that routine microscopic fecal examinations be conducted when worm ova are being produced. A competent necropsy (postmortem) examination and a thorough search for worms should be conducted each time an animal dies. By means of microscopic fecal examination and necropsy, the herd veterinarian can ascertain the species of worms present, their relative number, and whether or not treatment with a certain drug is indicated.

STRATEGIC INTERNAL PARASITE CONTROL

Parturition Medication

In the last two weeks of pregnancy and a short period following birth of the kids (parturition), dormant worms suddenly begin feeding and reproducing creating large numbers of fertile ova. Thus, it becomes judicious to administer deworming medication to pregnant does in the last ten days of pregnancy and through the first week following kidding. Medication selected must be highly efficient and relatively nontoxic for the welfare of the doe and her fetuses. Thiabendazole or one of the other benzimidazole anthelmintic drugs would be preferable over organic phosphate drugs--used strictly according to label directions. Either paste form or palatable granular form of these drugs is preferable to other forms that might be administered only with great difficulty to certain "medicine wise" animals.

Baby Kid Medication

Depending upon the sanitation level in their environment, baby kids should be checked for evidence of coccidiosis before they reach one month of age or certainly within two weeks before or after weaning. A diagnosis of coccidiosis is justified if large numbers of coccidia oocysts are present on microscopic fecal examination or characteristic small necrotic lesions are found on necropsy.

Coccidiosis is very common in dairy goats. Therefore, diagnosis and decision to institute medication for control or treatment for coccidiosis should be considered when kids are kept in an unsanitary, crowded environment and when fecal examination (finding large numbers of oocysts) or necropsy reveal the presence of clinical coccidiosis. Finding a few oocysts upon microscopic fecal examination may not justify medication and it may enable the young animals to actually profit by minimal exposure and coccidiosis resistance they might develop from it.

Undoubtedly, much coccidiosis medication is wasted or used uneconomically when a fecal examination of healthy vigorous kids reveals only a few oocysts.

When justified, medication against coccidiosis is worthwhile and necessary to prevent anemia, stunted growth, and(or) losses of young kids. The usual time for medication closely parallels weaning. Routine administration of sulfonamide or amprolium drugs is recommended to produce treatment levels of the drug used for at least three days each week, for 2 or 3 weeks.

The use of long acting (72 hour theraputic level) sulfonamide drugs for this course of treatment may provide additional protection against toxoplasmosis. Monensin (Rumensin - Lilly) fed in the grain ration at the rate of 15 to 20 PPM has been found highly effective against coccidia fed for periods of at least six weeks.

Pasture Deworming Medication

Goats allowed to browse green pasture often experience severe internal parasitism from a rapid rise in worm burdens at the onset of warm wet weather toward the height of the pasture season. This is one reasonably good indication for routine deworming treatment. Use of levamisole may be indicated for all animals in the herd if there has been any suspicion or evidence of worm resistance to one of the benzimidazole drugs.

If goats have access to green pasture at this time, they should be moved to fresh pasture immediately following treatment and the first pasture mowed or used for cattle or horses to remove the grass. Cattle and horses do not share internal parasites with goats and sheep. Using pasture at the proper time will furnish more feed for two or three species than it would have produced for any one of those species.

Prebreeding Deworming

In dairy goat herds where there has been evidence of clinical worm burdens, deworming the female animals to be bred in late summer before breeding may have two advantages: it is a recognized fact that gain in condition and reduction of even mild anemia results in production of more ova (eggs) at breeding, with a corresponding increase in the number of kids born into the herd.

Preferably, this deworming should be done about 3 weeks before breeding and repeated again one month following breeding, if necessary.

It is not advisable to deworm dairy goat does during the first 6 weeks of pregnancy with any kind of deworming drugs. During the last 50 days of pregnancy, they should not be dewormed with organic phosphate drugs.

Deworming Bucks

Herd sires and buck kids should be given deworming treatment any time female animals in the herd are given treatment. Male goats are often kept in unsanitary pens and pastures and neglected when medications of any kind are administered to the female members of the herd. Sometimes, there are tragic results from this neglect.

REFERENCES

Gall C. (Ed.) Goat Production. Academic Press, New York.

Guss, S. B. Management and Diseases of Dairy Goats. Dairy Goat Journal Publishing Co., Arizona.

Merck Veterinary Manual, 5th Ed. Merck and Co., Inc. New Jersey.

Part 11

GENETICS, SELECTION, AND REPRODUCTION OF GOATS

54

DHI AND AI AS AIDS TO DAIRY GOAT MANAGEMENT

Jack D. Stout

"When performance is measured, performance improves; when performance is measured and reported, the rate of performance accelerates." The original author of that quote has been lost through the years but the principle has never been more true for dairy goat breeders. Dairy Herd Improvement (DHI) records and Buck Summaries are now providing complete performance data on their animals. Artificial insemination (AI) provides the opportunity for acceleration of performance by allowing the good bucks to be used more extensively. At no previous time has there been a greater potential for the improvement of dairy goats.

The need for improvement is as great as the potential. Cost studies in Arkansas, Oklahoma, and 10 Northeast states indicate that producers must receive $25.00 per cwt on 1700 to 1800 lb production to break even if all costs are figured. Having a sufficient market for all fluid milk produced is a problem for most producers. Marketing the surplus through other sources at a reduced value brings the blend price of the total production down drastically. In the long run, improved efficiency--more milk from fewer does--must occur if the dairy goat industry is to remain profitable.

A look at the profile of Texas dairy goat herds (table 1), will give some insight as to why I feel DHI and AI must be used more extensively in dairy goat management. Although this is a Texas survey, the distribution of animals would be similar on a national basis but with herd size reduced to an average of 10 does. However, using this data, an average herd size of does with 32% of the farms having two or more breeds makes AI a natural. The male to female ratio of 1:3.2 seems to make AI a must.

AI must become as successful and as popular in the breeding of dairy goats as it is in the dairy cattle industry. AI is the prerequisite to the dairy goat reaching its maximum potential for: 1) genetic progress, 2) youth projects, 3) urban producers, and 4) cost efficiency.

DHI records must be a part of the same "ballgame." A herd without production and genetic records is similar to a world traveler charting his progress and making future plans without a map. DHI record would put the "man" in management decisions.

TABLE 1. PROFILE OF TEXAS DAIRY GOAT HERDS*

Breeds per farm
68% of herd owners had only 1 breed
25% of herd owners had 2 breeds
7% of herd owners had 3 or 4 breeds

Herd size
Herd size surveyed - 20 per farm
25% had 5 or fewer animals
42% had 10 or fewer animals
24% had over 20 animals
8% had over 40 animals

Age distribution
Kids made up 27% of population
Yearlings - 26%
Adults - 47%

Male - female ratio
One breeding age male for every 3.2 females

*515 herd owners surveyed--32% return.

With the development of the group testing program or even the use of owner-sampler mail-in system there now seems every reason for producers to use DHI records. Group testing permits breeders of an area to assume the responsibility of the DHIA supervisor and to alternate testing of herds within the group. Thus, the cost of official DHI records has been reduced to an acceptable level for the small producer. By visiting other herds of the group, the producer observes different management systems. DHI has grown phenomenally since group testing was adopted.

A look at the information available through DHI records should easily convince goat breeders of their usefulness in herd management (Appendixes 1, 2, 3). DHI data usually can be divided into four areas: volume of milk produced, profitability of production, breeding and reproductive summaries, and genetic evaluation.

The volume of production is covered most thoroughly because of NCDHIP rules and the use of level of milk production as the basis for all other phases of evaluation. Sample-day milk weights and butterfat percentages are summarized in various ways to provide:

- Lactation to date: the LTD is the total accumulation of lb of milk and fat produced from freshening to current sample day.
- Projected 305-2x-ME: LTD standardized to 305 days; mature equivalent (ME) corrected for age at freshening, breed, season of freshening.

- Difference from herdmates: the number of lb of milk or fat the current 305-2x-ME record is above or below the average of herdmates adjusted for genetic level of herdmates.
- Persistence: a percentage figure that tells how well the doe is holding up in production during the lactation.
- Summit milk: average of the two high sample-day weights for first and other lactations.
- Sample day summary: average of each sample day for previous 12 months including number of animals, percentage in milk, daily average milk, rolling herd average (RHA).
- Lactation average: average 305-2x-ME lactations by age group.

The production data can be used in many ways in formulating management decisions. Summit milk, as an example, will furnish insight into several areas of herd management such as feeding program, milking management, growing of doelings, or dry-doe care. For each 1 lb change in summit milk, the RHA will change by 215 lb in the same direction. Table 2 shows the expected herd averages at various levels of summit milk. Calculations for this table were made assuming 25% replacement rate of 1st lactation animals and standard 305-day lactations.

TABLE 2. EXPECTED HERD AVERAGES* AT VARIOUS LEVELS OF SUMMIT MILK

Expected herd average	Level of summit milk		
	1st lact. does	Other lact. does	All lact. does
lb	lb	lb	lb
800	2.2	3.8	3.4
1000	3.1	4.7	4.3
1200	4.0	5.6	5.2
1400	4.9	6.5	6.2
1600	5.9	7.5	7.1
1800	6.8	8.4	8.0
2000	7.8	9.3	8.9
2200	8.8	10.3	9.9

*Assuming normal distribution of ages (25% 1st lact).

Some limitations of summit milk flow rate are:
- Genetic ability.
- Improper body condition at freshening.
- Mastitis and other diseases.
- Metabolic disorders (ketosis, milk fever).

- Improper handling of doe at freshening or in early lactation.
- Incomplete milking.
- Inadequate energy in early lactation.
- Inadequate length of dry period.
- Individual doe care at time of freshening.

Length of lactation seems to be a problem with Oklahoma herds. Most breeders will assume the short lactations are a result of drying-off at times of low demand for milk. However, I feel genetics and management are the real culprits. Most breeders would add the extra labor to complete a 305-day lactation if their does were projected at high levels. Similar problems must occur on a national scale also. Forty-three percent of all lactations reported to USDA are less than 275 days in length. Table 3 shows the influence of percentage of days in milk on various production factors.

TABLE 3. DAIRY GOATS: INFLUENCE OF PERCENTAGE OF DAYS IN MILK (DIM)

	<60%	61-70%	71-80%	>81%
Number herds	6	5	6	3
Number does	50	91	116	21
Average DIM	42%	65%	74%	86%
Lactation length	153	237	270	314
Lb milk/doe	865	1270	1474	2274
Lb milk/day	5.7	5.4	5.5	7.2
Lb fat/doe	31	52	52	74
Value roduction/doe	$132	$179	$204	$358
Milk price/cwt	$ 15.76	$ 14.03	$ 13.83	$ 15.74
Feed cost/doe	$122	$104	$136	$142
Feed cost/cwt milk	$ 14.01	$ 8.18	$ 9.22	$ 6.24
Income over feed cost	$ 10	$ 65	$ 68	$216

Pounds of milk per day are fairly constant on lactations less than 270 days in length. This is why I think length of lactation is a genetic factor. Milking a doe 45 more days at 5.0 lb/day would only increase lactation average by 225 lb.

The profitability of production through DHI records accounts only for feed costs. Each breeder provides their particular milk price and butterfat differential plus the prices for grain and forage. At current Oklahoma feed prices and average value received for goat milk, it requires approximately 1000 lb of milk to cover feed costs. Budgets calculated by OSU Agricultural Economics Department indicates a loss of over $50.00 per doe when all costs are figured at $17.00 milk and 1800 lb RHA. Our producers sorely need a cheese milk market for their product to improve blend price. An increase of $3.00 in value or 300 lb in production is needed to break even. The increasing of milk production per doe and reducing herd size would be the most logical and practical with our current demand for goat milk. Culling the low 20% of does based on milk production would raise herd average by 250 lb in Oklahoma herds having over 15 does.

The breeding and reproductive summary of DHI records are not as useful to goat breeders as they are in dairy cow management because of the seasonal breeding and small herd size. However, if complete breeding records are provided, summaries or projections will include: 1) days open, 2) number of services per conception, 3) due date, 4) gestation length, and 5) projected freshening interval.

The genetic evaluation is probably the most valuable portion of DHI records for goat breeders. The potential benefits also are increasing with the development of the Buck Summary. The Mid-States Dairy Records processing Center, Ames, Iowa, has been offering the following genetic evaluation for several years:

- Difference from Herdmates (Diff. H.M.)--the number of lb of milk fat that the current 305-2x-ME record is above (+) or below (-) the average of herdmates with adjustment for genetic level of herdmates.
- Estimated Producing Ability (EPA)--the EPA is the best estimate of the female's ability to produce under the conditions of her previous environment.
- Estimated Average Transmitting Ability (EATA)--the EATA is the best estimate of the female's ability to transmit to her offspring. It is based on the production of her paternal sister, dam, maternal sisters, and daughters, as well as her own production.
- Predicted Difference (PD)--the PD is an estimate of the sire's ability to transmit production superiority to his offspring.
- Pedigree Estimate of Breeding Values (PEBV)--the PEBV is the combination of the sire PD and Dam EATA, an estimate of the inherited production level from the parents.

The use of these genetic evaluations will depend on the goal to be accomplished. The Diff. H.M. and EPA should be used when making culling decisions. Each is based only on the doe's own performance. The EPA uses past performance weighted for number of records to predict future production. The Diff. H.M. on the monthly Individual Lactation Report is based on the current lactation. On the Herd Ranking Summary it is the average Diff. H.M. for all lactations. A doe will produce approximately half of her lactation in the first 100 days, so culling decisions could be made before daily production drops below a profitable level. The EATA should be used when selecting dams of replacement animals. Bucks should only be kept out of does having the top EATA.

The PEBV was developed for ease of ranking offspring on inherited capacity of milk production. A doeling receives half of its genetic value from each parent figured at Buck's PD + Dam's EATA. The PEBV serves as an estimate of expected performance. Of course, with small numbers there will be considerable variation. However, the principle is sound. Table 4 shows the results of ranking 1980 doelings of the five largest DHI goat herds in Oklahoma into high, medium, and low groups based on PEBV. The Diff. H.M. and 305-2x-ME are based on their 1st lactation in 1982. The weighted Rolling Herd Averages for the five herds in 1982 was 1788 lb milk. There was a 99 lb difference in PEBV between the high and medium groups and 35 lb between the medium and low groups. Difference from Herdmates was +381, +186 and -46 for the three groups, respectively. The 305-2x-ME production varied from 2243 lb for the high group to 1596 lb for the low group.

TABLE 4. DOELINGS OF FIVE OKLAHOMA HERDS STRATIFIED BY PEBV

Group	PEBV lb	1st lact Diff. H.M. lb	1st lact 305-2x-ME lb
High	+ 90	+ 381	2243
Medium	- 9	+ 186	2013
Low	- 46	- 179	1596

The genetic tools are available to build genetic progress. Dam EATA, Buck PD and Doeling PEBV, when used properly, will improve the production performance of dairy goats. Yet, Grossman, in his study of approximately 100,000 records reports "essentially no increase in average milk production over the last decade." The lack of improvement is the result of not using proven bucks on a high percentage of the does. AI to proven sires must become the accepted method of breeding in goats as it has in dairy cattle. The top PD bucks must be allowed to produce more progeny to assure breed improvement.

Figure 1 shows the production distribution of 10 does and the distribution of production of offspring of those does when bred to bucks of varying genetic transmitting levels. The use of bucks with increasingly + PD levels in each succeeding generation will continue the herd progress.

Figure 2 shows the production distribution on an Oklahoma herd that used the best bucks available for three generations. Herd size was maintained relatively constant and older does were culled as genetically better replacements were available. In year one, the 305-2x-ME herd average was 1800 lb with a 54% to 46% distribution above and below the

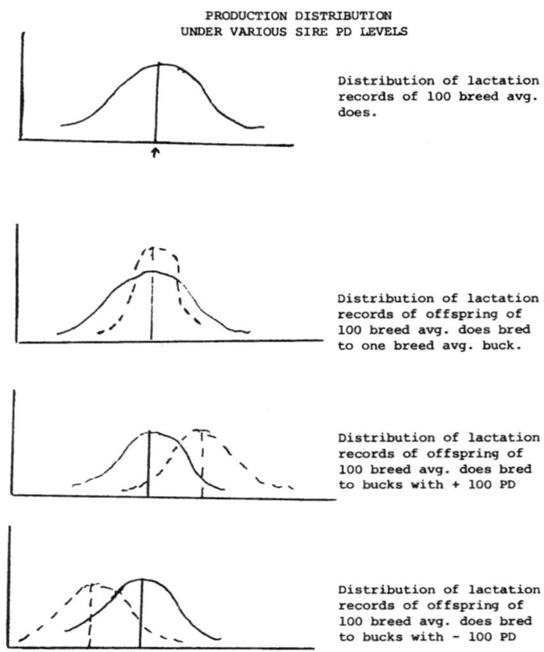

Figure 1.

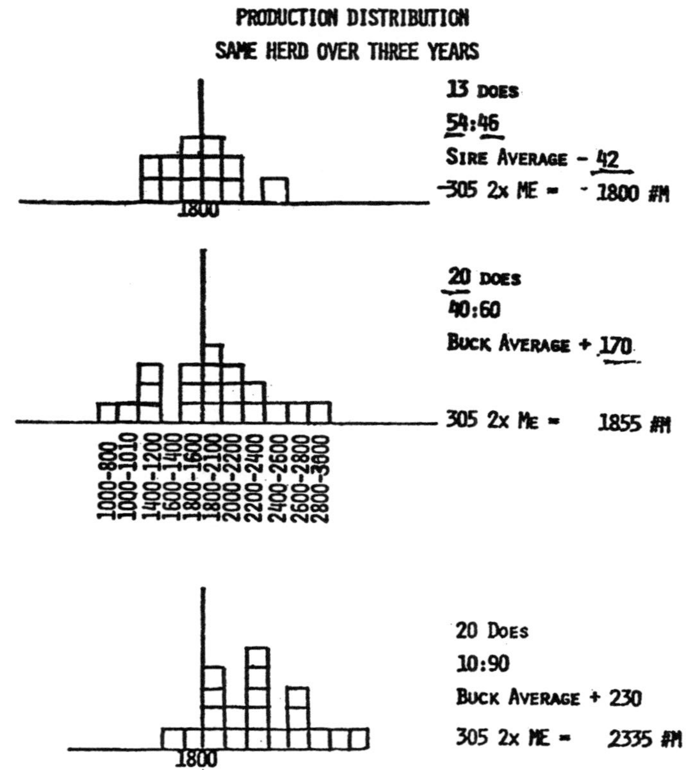

Figure 2.

average. In the second generation, 40% of the does were under 1800 lb while 60% were above. In the third generation only 10% of the does had ME lactations less than 1800 lb. During the three generations, the buck average breeding value increased from -42 lb to +230 lb.

The data presented is quite thorough in support of DHI and AI in promoting genetic progress of dairy goats. It would also aid in the general promotion of dairy goats as youth projects and for urban producers. AI would remove the need for maintaining a buck. Thus numbers of animals, space requirements, and labor would be reduced. Our 4-H agents report that elimination of the buck (with his offensive odor) would greatly stimulate dairy goats as youth projects. The urban family, even in some restricted areas, could also maintain a few does for family milk and meat. DHI records would provide the information necessary to make dairy goat projects a complete educational experience in teaching young people management, feeding, caring, breeding, genetics, and the economic results of their decisions.

447 John A. Dairyman Midtown Oklahoma 73

HERD CODE	BREED	TEST INTERVAL LENGTH	FROM	THRU SAMPLE DATE	DATE RECEIVED	DATE MAILED	ASOC	SUPR-VISOR	PLM OR PROTEIN	OPTIONS REQUESTED	TYPE OF RECORD	LISTING ORDER CMP	BN	HERD	STR.	SAMPLE DATE	PAGE
73-45-9063	RC	32	5-30	6-30-82	7-14	7-17	95	MS	NO		OFFICIAL DHIR	X		X		6-30-82	1 OF

DHIA-200

SAMPLE DAY AND LACTATION REPORT

COMPUTER NUMBER	IDENTIFICATION NUMBER	BREED	SAMPLE DAY DATA: MILK POUNDS	FAT %	FAT POUNDS	INCOME OV FEED COST	GRAIN MIX FED	GRAIN MIX NEED	STRING	BARN NAME OR NUMBER	LACTATION TO DATE: LACT NO.	DAYS DRY	FRESHENING DATE	AGE	DAYS	MILK	FAT %	FAT POUNDS	COND AFF REC	INCOME OV FEED COST	PERSIST-ENCY %	305-2X-M E MILK	305-2X-M E FAT	DIFFERENCE FROM HERDMATES MILK	DIFFERENCE FROM HERDMATES FAT	DUE DATE	ACTION NEEDED OR SERVICE SIRE
9	233803	A	13.1	2.5	.3	2.43	0	4	1	LAURA	5	31	3-08	5-09	115	1490	3.6	53		273	105%	2900	103	+573	+1		B 8-30
45	194410	A	4.6	3.1	.1	.76	0	3	1	DUSTY	7	79	2-01	8-08	150	960	3.9	37		66	99%	1520	59	-742	-27		B 7-26
48	283903	A	6.2	2.5	.2	1.08	0	3	1	LADY	4	49	1-30	4-07	152	970	3.8	37		110	105%	1540	59	-453	-20		B 7-24
59	304436	S	10.5	2.8	.3	1.91	0	3	1	LAYLA	3	42	2-21	3-09	130	1260	3.6	45		223	108%	2360	84	+154	-1		B 8-15
64	246460	N	6.4	3.7	.2	1.11	0	3	1	APPLE	3	9	1-27	4-11	155	1330	4.3	57		227	100%	2060	88	+104	+11		B 7-21
76	333772	A	10.7	2.3	.2	1.97	0	3	1	CATHY	2	54	2-02	2-10	149	1510	2.8	43		210	105%	3010	85	+818	+0		B 7-27
86	373835	L	6.7	3.7	.2	1.20	0	3	1	CORSGE	2	31	2-03	1-10	148	950	5.6	53		139	105%	2140	119	-94	+35		B 7-28
88	366843	A	6.2	3.0	.2	1.09	0	3	1	NYMPH	2	60	2-26	2-00	125	710	3.7	26		119	109%	1790	65	-462	-21		B 8-20
89	370309	L	5.8	3.3	.2	1.04	0	2	1	KEEMUN	2	75	1-23	1-11	159	970	3.8	37		85	104%	2030	77	+75	+0		B 7-17
91	372868	A	8.8	2.6	.2	1.61	0	3	1	LEE	2	45	3-01	1-11	122	1000	3.6	36		181	107%	2630	94	+273	+5		B 8-23
93	360545	A	9.4	2.4	.2	1.73	0	3	1	MAYTME	2	35	2-22	2-08	129	1160	4.1	47		227	105%	2670	107	+459	+23		B 8-16
95	375275	S	11.1	3.2	.4	2.03	0	3	1	MYLA	2	41	3-23	2-00	100	1130	3.8	43		210	104%	3480	131	+1173	+44		B 9-14
116	414523	A	5.6	2.4	.1	.98	0	2	1	NETTE	1		4-09	1-01	83	420	3.6	15		76	110%	1770	63	-831	-32		B 10-01
120	427840	N	5.3	2.6	.1	.90	0	3	1	PIE	1		5-12	1-02	50	240	3.3	8		43		1590	52	-1023	-44		B 11-03
129	417252	L	9.6	3.8	.4	1.76	0	3	1	GAY	1		3-30	1-01	93	880	4.0	35		168	104%	3310	132	+980	+44		B 9-21
130	R 26	C	6.9	3.5	.2	1.22	0	3	1	PAMLRA	1		6-03	1-02	28	160	3.8	6		34							B 11-25
138	428190	S	11.0	2.6	.3	2.04	0	3	1	MYLENA	1		3-19	0-11	104	1040	3.4	35		200	108%	3550	120	+1229	+32		B 9-10
149	369994	A	11.5	2.5	.3	2.13	0	3	1	PATTI	2	60	4-10	2-01	82	960	4.0	38		173	103%	3540	139	+1061	+49		B 10-02
152	372867	A	8.3	2.6	.2	1.47	0	3	1	DOLLY	2	22	4-10	2-01	82	720	3.5	25		109	102%	2650	92	+110	-1		B 10-02
156	370315	L	7.3	3.7	.3	1.32	0	3	1	CREMORA	1		2-02	1-10	149	1050	3.8	40		185	104%	2350	89	+127	+4		B 7-27
157	374729	A	6.4	2.3	.1	1.13	0	3	1	BLOKPNT	1		2-01	1-10	150	980	3.2	31		170	104%	2190	69	-39	-17		B 7-26
158	334253	A	6.9	3.5	.2	1.18	0	4	1	OMEGA	1		2-24	2-11	127	990	3.4	34		180	100%	2240	77	+13	-8		B 8-18
168	375596	N	6.7	3.0	.2	1.20	0	3	1	DISCO	1		3-24	2-01	99	730	3.2	23		139	100%	2250	70	-121	-20		B 9-15
171	430681	L	7.9	2.4	.2	1.44	0	3	1	SUSAN	1		3-25	0-11	98	820	3.4	28		158	103%	2950	101	+606	+12		B 9-16
172	421961	A	4.8	2.4	.1	.84	0	2	1	BUTRFLY	1		3-13	1-00	110	530	3.8	20		98	105%	1720	65	-671	-25		B 9-04
173	458529	L	7.4	2.7	.2	1.36	0	2	1	TAFFY	1		5-07	0-10	55	380	3.7	14		75		2290	84	-276	-10		B 10-29
174	415479	A	9.5	2.3	.2	1.78	0	2	1	HPOPIN	1		3-16	1-00	107	940	3.5	33		178		3130	110	+793	+22		B 9-07
175	448288	A	6.0	2.3	.1	1.08	0	2	1	SISSY	1		5-22	0-11	40	210	2.9	6		37							B 11-13
176	320791	L	7.8	3.8	.3	1.39	0	3	1	KATE	3		2-09	2-11	32	250	3.6	9	X	44							B 8-03
177	428189	A	7.5	2.3	.2	1.36	0	3	1	MINT	1		6-02	1-01	29	180	2.8	5		39							B 11-24
HERD AVG.	% IN DOES MILK									% PROT NEEDED			B/WT														
	30 100 30 IN MILK		7.9 7.9	2.7 2.7	.2 .2	1.42	0	3		11		45	120	2-04	105						104%	2450	90	+148	+3		
				STATUS-DATE							* * *		L A C T A T I O N S * * *														
140	428193	A		SOLD		6-17			1	REVALRA	1		4-09	0-11	70	400	3.8	15	$	74		1980	74	-607	-20		

Appendix 1.

HERD SUMMARY

HERD CODE	BREED	COW MONTHS	LENGTH	FROM	THRU SAMPLE DATE	RECEIVED AT LAB	RECEIVED	MAILED	ASSC	SUPERVISOR	PROTEIN OR PLM	OPTIONS REQUESTED	RECORD PLAN
73-45-9063	RC	31.9	32	5-30	6-30-82	7-08	7-14	7-17	95	MS	NO		OFFICIAL DHIR

REPRODUCTIVE SUMMARY

GROUP	REPLACEMENT FEMALES	PRODUCING FEMALES NUMBER	AVG DAYS SINCE FRESH	NO. OF ANIMALS OPEN < 40 DAYS	40-130 DAYS	> 130 DAYS	AVG DAYS OPEN
PREGNANT							
POSSIBLY PREGNANT							
OPEN	13	30	108	5	11	13	108

SUMMARY OF ANIMALS TO BE MILKING, DRY, OR FRESH

	JUL	AUG	SEP	OCT	NOV	DEC	JAN	FEB	MAR	APR	MAY
REPLACEMENTS TO FRESHEN											
PRODUCING ANIMALS TO FRESHEN											
EXPECTED TO BE MILKING	30	30	30	30	30	30	30	30	30	30	30
EXPECTED TO BE DRY											

IDENTIFICATION SUMMARY

AGE GROUP	NUMBER OF FEMALES	AVERAGE AGE	NUMBER IDENTIFIED SIRE	DAM
0-6 MONTHS	11	0-04	7	11
7-12 MONTHS				
OVER 12 MONTHS	2	1-05		2
REPLACEMENTS	13	0-06	7	13
LACTATION 1	15	1-04	14	15
LACTATION 2	9	2-02	9	9
LACTATION 3	3	3-10	3	3
LACTATION 4+	3	6-04	3	3
PRODUCING FEMALES	30	2-04	29	30
% IDENTIFIED			84	100

HERD SIRE EVALUATION

DRY DAYS SUMMARY

NUMBER < 40	NUMBER 40-70	NUMBER > 70	TOTAL NUMBER	AVG DRY DAYS
5	7	2	14	45

LACTATION SUMMARY

	305 2X ME MILK	FAT	DIFFERENCE FROM HERDMATES MILK	FAT
LACTATION 1	2,445	86	+66	-4
LACTATION 2	2,660	101	+379	+15
LACTATION 3	2,210	86	+129	+5
LACTATION 4+	1,987	74	-207	-11
PRODUCING FEMALES	2,448	90	+148	+3

← SERVICE SIRE AVERAGES

BIRTH AND INVENTORY SUMMARY

STAGE OF LACTATION PROFILE

DAYS IN MILK: LESS THAN 100; 100-200; GREATER THAN 200; TOTAL

FEEDING SUMMARY

SAMPLE DAY FEED REPORTED FIRST STRING ONLY	FEED CODE		$ TON	DM			LBS PER COW ANNUALLY
HAY	52	4	80	90	57	16	1,459
CONCENTRATES	10		180	90	82	13	1,173

	NUMBER	AVG. SAMPLE DAY LBS MILK	AVG. BODY WEIGHT 120	SAMPLE DAY	ANNUAL AVERAGE
FIRST	13	7.6	PROTEIN NEEDED IN GRAIN	11	
OTHER	14	9.4	% FORAGE D.M. BODY WT.	3.0	3.0
ALL	27	8.5	LBS MILK PRODUCED / LB GRAIN		1.7

COST AND RETURN SUMMARY

	$ PER COW SAMPLE DAY	$ PER COW 365 DAYS	$ PER HERD SAMPLE DAY	$ PER HERD 365 DAYS
FORAGE COST	.15	58	5	1,949
GRAIN COST	.00	105	0	3,563
TOTAL FEED COST	.15	163	5	5,512
MILK VALUE	1.57	390	47	13,187
INCOME OVER FEED COST	1.42	227	42	7,675

LBS SHIPPED DAILY / LBS SAMPLE DAY	236	FEED COST CWT MILK	2.12	8.33
		RETURN / FEED COST	9.40	2.39
		MILK PRICE PER CWT	19.92	19.93

NEW OR CORRECTED IDENTIFICATIONS

COMPUTER NUMBER	BARN NAME	ANIMAL IDENTIFICATION	BR	BIRTH DATE	SIRE IDENTIFICATION	BR	DAM IDENTIFICATION	BR
130	PAMLRA	R 26	C	3-13-81			372868	C
173	TAFFY	458529	L	7-05-81	329278	L	320791	L
176	KATE	320791	L	2-11-79	4851	L	4479	L
177	MINT	428189	A	4-05-81	366840	A	193718	A

73-45-9063 95 MS

John A. Dairyman
Midtown, OK

PRODUCTION SUMMARY

SAMPLE DATE	DAYS IN TEST PERIOD	FEMALES ON FARM	% IN MILK	COW DAYS ON TEST	TEST INTERVAL DAILY AVG. MILK	FAT	FRESH INTERVAL	ROLLING 365 DAY AVERAGE MILK	FAT %	FAT	PROT	PROT
7-08-81	8	36	89	313	6.5	3.09		1,955	3.53	69		
8-15-81	38	37	93	1391	6.8	3.26		1,952	3.59	70		
9-15-81	31	37	97	1148	6.7	3.64		1,961	3.62	71		
10-15-81	30	37	95	1110	5.8	4.01		1,948	3.70	72		
11-27-81	43	36	84	1584	4.2	4.52		1,930	3.73	72		
12-29-81	32	34	59	1117	2.6	4.69		1,969	3.76	74		
2-10-82	43	30	50	1408	2.4	4.54		2,020	3.76	76		
3-20-82	38	26	65	1106	4.2	4.35		2,026	3.75	76		
4-23-82	34	30	79	1053	6.2	3.77		2,012	3.78	76		
5-29-82	36	28	96	1154	7.4	3.65		1,967	3.76	74		
6-30-82	32	30	100	971	7.9	3.24		1,955	3.79	74		
11	365			12355		33.85	HERD AVERAGE					

Appendix 2

HERD RANKING and SUMMARY DHIA-204

02-82 CUT OFF DATE

John A. Dairyman
Midtown, OK

96 DAIRY GOAT BREED

73-55-9035 HERDCODE

RANKED ON EPA FOR MILK

COMPUTER NO.	IDENTIFICATION NO.	SIRE NO.	RPT. %	PREDICTED DIFFERENCES MILK	PREDICTED DIFFERENCES FAT	DAM NO.	DAM REC'S	DAM MILK	DAM FAT	MATERNAL SISTERS NO.	MATERNAL SISTERS REC'S	MATERNAL SISTERS MILK	MATERNAL SISTERS FAT	DAUGHTERS NO.	DAUGHTERS REC'S	DAUGHTERS MILK	DAUGHTERS FAT	HERSELF REC'S	HERSELF MILK	HERSELF FAT	BARN NAME	ESTIMATED PRODUCING ABILITY MILK	ESTIMATED PRODUCING ABILITY FAT	ESTIMATED AV. TRANSMITTING ABILITY MILK	ESTIMATED AV. TRANSMITTING ABILITY FAT
DAIRY	GOAT																					RANK			
36	284754	223894	*23	-53	-1	181048	1	+788	+40	2	3.5	+236	+23					3	+1148	+61	COOKIE	+861	+46	+106	+11
20	223397	202586	* 8	+89	+4	173935				1	1.0	+1510	+85	2	1.0	+1103	+40	3	+721	+20	KRISTY	+541	+15	+233	+6
25	262946	240058	*14	+26		200324												3	+666	+27	FROSTY	+500	+20	+113	+4
43	297266	286257	*20		+1	246481	4	+111	+16									2	+545	+43	HEIRESS	+364	+29	+75	+7
30	284755	223894	*23	-53	-1	181048	1	+788	+40	2	3.5	+630	+39					3	+360	+29	CUPCAKE	+270	+22	+94	+7
50	322083	286257	*20		+1	266927	3	-758	-31									1	+280	+12	MOONCHI	+140	+6	-33	-1
35	262947	240058	*14	+26		215740				1	1.0	-209	-9					3	+164	+0	DOTTIE	+123	+0	+30	+4
61	373824	305106	* 5	+9		298870	2	-431	-20									1	+187	-5	CLEO	+94	-3	-9	-2
17	246481	223894	*23	-53	-1	181048	1	+788	+40	2	3.0	+754	+45	1	2.0	+545	+43	4	+111	+16	PRISSY	+89	+13	+82	+7
47	362567	227004	*10	+64	+3	262950	1	+442	+22									2	+88	+7	LACY	+59	+5	+56	+3
56	326673	286257	*20		+1	171043	2	-1008	-44	2	3.0	+13	+5					2	+64	+4	ALI	+43	+3	-40	-1
55	326671	286257	*20		+1	171043	2	-1008	-44	2	3.0	+23	+4	1	1.0	-1504	-65	2	+45	+6	CHARM	+30	+4	-95	-3
51	335354	298829	*10	-80	-3	220106												0	+0	+0	STARSON	+0	+0	+0	+0
1	171043	163592	* 5	-50	-2	152095												0	+0	+0	BELLE	+0	+0	+0	+0
9	220106	199024 *	*29	+8		171043	2	-1008	-44	2	2.0	+55	+5					4	-19	+3	ANGIE	-15	+2	-50	-2
29	262949	240058	*14	+26		200312				1	1.0	+442	+21	1	1.0	-562	-25	3	-121	-1	AMY	-91	-1	-18	-1
49	298873	266292	*13	-67	-2	266294				1	1.0	-1139	-44					1	-251	-12	ANNE	-126	-6	-80	-3
37	279764	223894	*23	-53	-1	182861												2	-236	-10	LADY	-157	-7	-51	-2
45	298870	241169	*16	+46	+1	266290				2	1.0	-277	+2	1	1.0	+187	-5	2	-430	-20	GINGER	-287	-13	-44	-2
57	342465	286257	*20		+1	207902	4	-736	-29	1	1.0	+60	+30					1	-698	-32	BEAUTY	-349	-16	-119	-4
38	310413	281214	*15	-6		210726	3	-9	+20	1	1.0	+405	+34					3	-604	-24	BONNY	-453	-16	-62	-2
19	266927	221089	*28	-12		221233	3	-231	-9	1	1.0	+293	+8	1	1.0	+280	+12	3	-758	-31	TESS	-569	-23	-109	-4
62	386007	298829	*10	-80	-3	326671	2	+45	+6									1	-1504	-65	FIESTA	-752	-33	-167	-7

NOT SAME HERD AS SAMPLE DAY AND LACTATION REPORT OR HERD SUMMARY.

HERD GENETIC EVALUATION

AVERAGES AND COMPARISONS	NO. OF ANIMALS	EST. PRODUCING ABILITY MILK	EST. PRODUCING ABILITY FAT	SIRES PRED. DIFF. MILK	SIRES PRED. DIFF. FAT	EST. AV. TRANS. ABILITY MILK	EST. AV. TRANS. ABILITY FAT
YOUR HERD AVERAGES	10	+250	+15	-2	+0	+65	+4
TOP 25% OF HERDS	10	+194	+6	+16	+0	+55	+2
ENTERED DURING YEAR	2	-329	-18	-36	-2	-83	-5
LEFT DURING YEAR	13	-168	-8	-15	+0	-52	-2
CULLING PROGRESS THIS YEAR		+236	+13	+7	+0	+66	+3.4
TOTAL PROGRESS THIS YEAR		+225	+12	+10	+0	+66	+3

SEE REVERSE SIDE FOR EXPLANATIONS

MANAGEMENT AVERAGES	305-2X-ME MILK	305-2X-ME FAT	% LEFT HERD	FRESHENING INTERVAL	LENGTH OF LACTATION	DRY PERIOD	DAYS OPEN	REPRODUCTIVE EFFICIENCY	MILK PER DAY
CURRENT HERD	1439	68	56	379	323	56	231	111	4.
LAST YEAR	1451	67	44	347	245	101	176	129	4.
BREED AVERAGE	1552	69	49	371	271	93	222	138	4.

HERD PROGRESS RESULTS FROM TWO SOURCES

1. THE QUALITY OF REPLACEMENT FEMALES THAT ENTERED THE MILKING HERD THIS YEAR
2. IMPROVEMENT DUE TO CULLING BELOW AVERAGE MILKING FEMALES

Appendix 3

55

DEVELOPMENTS IN GOAT REPRODUCTIVE BIOLOGY

Maurice Shelton,
Janet Lawson

In the U.S. and worldwide, goats are increasing in numbers of animals as are interested producers. Generally, goats are classified as meat, milk, or fiber types; however, meat is the only harvested product for most of the goat population. Good reproduction performance, at least to the extent of providing replacement stock or freshening of milk stock, is a requirement of all types. In the case of goats kept for meat production, reproductive level is a direct component of efficiency of production. Thus, a thorough knowledge of the reproductive process is needed in efforts to maximize fertility. The goat is most unusual in some aspects of the reproductive process, and recent research has provided new insights into our understanding of reproduction in this species.

SEASONALITY

The phenomenon of seasonality of reproduction is common to many plant and animal species. The length of the photoperiod is generally recognized as the primary controlling factor, but in some cases other factors such as food supply and temperature, can be shown to be part of the picture. Seasonality of mating remains to some degree a limitation to goat producers in 1) providing a year-round milk supply, 2) accelerated kidding (kidding more than one time per year), or 3) kidding at optimum times to meet special market opportunities.

Most established dairy breeds experience a period of anestrus during February to May in the Northern Hemisphere. Reports on dairy herds around the 1930s showed that the majority of kids were born from matings occurring September through December, with a few kids born each month of the year (Asdell, 1926; Turner, 1936). The extent to which the makeup of the goat population has changed since the early 1900s is not known. The Spanish goat in Texas is reported by ranchers to have an extended breeding season; some kidding has been observed in most months of the year. Additionally, ranchers claim that their does often kid twice a year. Because the meat-type goat in Texas is a composite of the non-Angora population, it was thought that observations on these animals might reflect trends for the rest of the United States goat population--exclusive of Angoras, which are known to be highly seasonal.

In an effort to delineate and quantify the seasonality of the Spanish goat in Texas, a group of 50 Spanish does were run continuously with males from May 1972 to May 1981 (Shelton and Lawson, 1982). In a correlated study, another 327 does were slaughtered or laparotomized in each month over a 5-year period to further identify seasonal cycling activity. In this study, as evidenced by the presence of corpora lutea, does appeared to be totally in anestrus during March and April. The majority of does cycled from August through January, with February, May, June, and July as transitional months.

The percentage of parturitions by month correlated with the ovulation data, with few kids being born in July, August, or September. Almost 63% of the kid crop were dropped in the winter season; a second smaller peak was observed in May. This second peak probably represents does that kidded early in the season (i.e., November and December) and rebred before going into anestrus. This is shown by the bimodal peaks (figure 1) that were obtained by graphing parturition intervals for does kidding in October, November, and December.

The seasonality of Spanish does in this study is not as extreme as that reported by Cortell (1977) for French Alpine does or by Turner (1936), Asdell (1926), or Yazman (1981) in U.S. and British dairy-goat herds. Data in figure 2 compares the kidding pattern of this study with that of Turner (1936) for dairy breeds in the years prior to 1936 and with data provided by Yazman (1981). The two dairy herds represented show little evidence of change in breeding season over the long time-lapse involved. By contrast, the data from the present study suggest that the breeding season of meat-type does in Texas is 60 to 90 days earlier than that shown in the other two reports. This may be due to geographic location or to genetic differences. The experimental site involved is between 30° and 31° N. latitude. In evaluating the kidding data, note that the does would have been bred at the initiation of estrual cycling; thus, this type of data alone does not give a true indication of the length of the breeding season.

STIMULATING REPRODUCTIVE EFFICIENCY

An early return to estrus postpartum and subsequent conception should promote greater reproductive efficiency in the doe. However, most does experience a prolonged postpartum interval (PPI). This may be due to postpartum anestrus, seasonal anestrus, or to concerted action of both these factors.

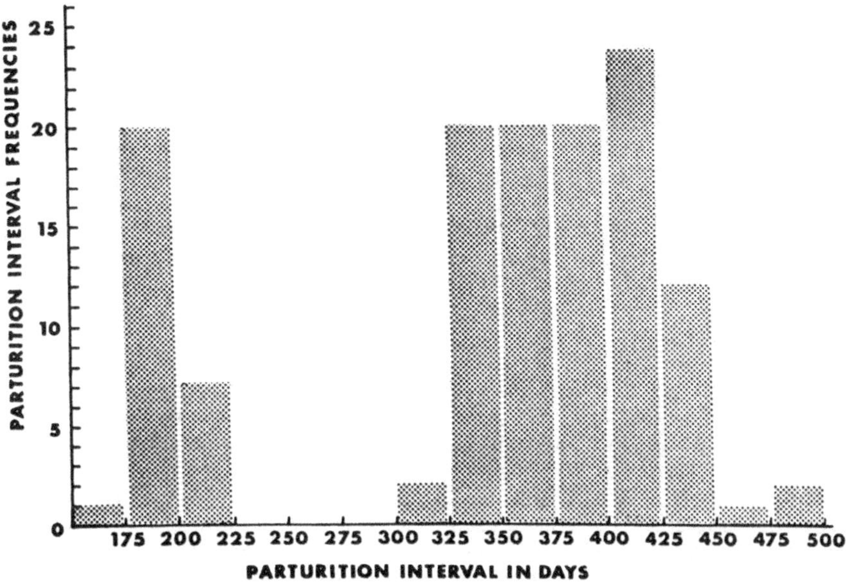

Figure 1. Subsequent Kidding Intervals for Does That Kidded in October, November, or December

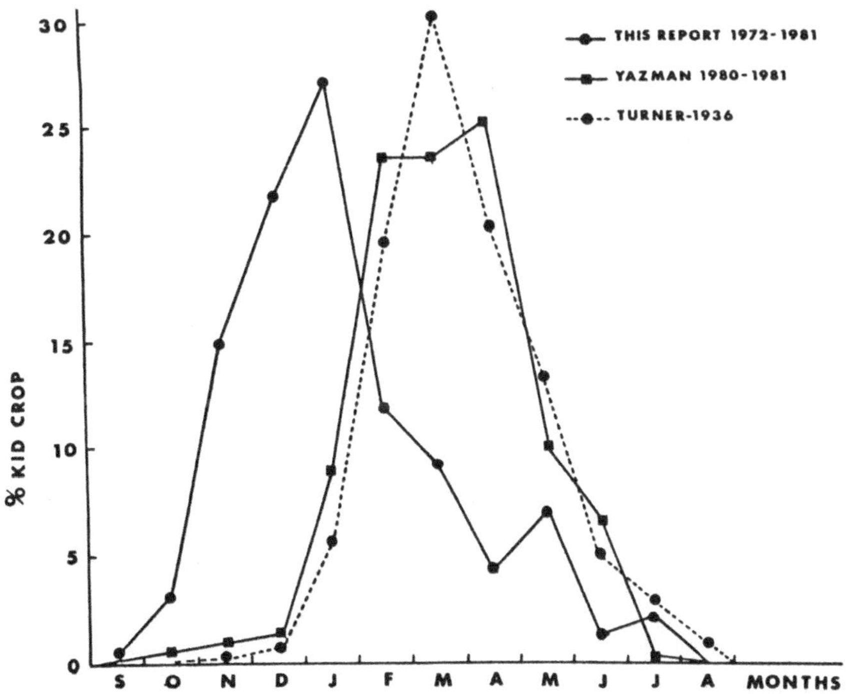

Figure 2. Birth patterns for three goat populations

Numerous experiments have studied use of hormonal therapy or light modification to trigger resumption of hypothalamic-pituitary-ovarian function and to promote out-of-season or early postpartum estrus. These treatments, while somewhat successful, may actually decrease overall kid crop--although they can contribute to production of out-of-season kids and a more even distribution of milk for commercial purposes (Nordfelt et al., 1982; Foote et al., 1982). There are practical limitations to the use of these treatments. Use of the male stimulating effect and/or allowing only limited nursing may provide alternatives to hormonal treatment.

Effect of Suckling

For some time it has been speculated that lactation or the neural stimulus of suckling inhibits ovulation and extends the postpartum interval. That this actually occurs has been documented in women (Tyson and Friesen, 1972), nonhuman primates (Plant et al., 1980), cows (Randel, 1981), and other species. The issue is more equivocal in ewes and even less researched in does. Recent work with beef cows suggests that manipulation of the suckling stimulus, such as early weaning, short-term calf removal, or allowing limited nursing will shorten the postpartum interval (Randel et al., 1981). In contrast, there has been a disparity of results in regard to early weaning treatments and resumption of postpartum cycling of ewes. Hunter (1968) faulted many of these earlier studies because comparisons were made among ewes that had lost their lambs at or soon after birth, rather than with animals randomly assigned to suckled vs nonsuckled groups. However, more recent reports on well-designed studies concerning suckling effect upon reproductive function in the ewe remain inconclusive (Moss et al., 1980).

Suckling manipulation with goats generally has been aimed at increasing milk production without loss of kids (Hadjipanayiotou and Louca, 1976). Current work with Spanish goats directed at increasing kid production suggests that the use of suckling manipulation may enhance early rebreeding postpartum during the normal breeding season. In the seasonality study previously mentioned, 20% of a group of does were run continuously with males for nearly a decade kidded in October and November; almost 45% of those rebred before the end of the estrual season. But of the total that kidded in October, November, and December (44%), only 21.7% kidded from matings before the start of anestrus. Thus, does that kid in December are less likely to rebreed before anestrus than those that kid in October or November. To increase kid production, practices are needed that stimulate does to kid early in the season and to rebreed while nursing a kid before going into anestrus.

To evaluate feasibility of such practices, a group of 200 does located at three locations in Texas were equally divided into four groups, dependent on their kidding dates, which ranged from October 15 to January 15. These four groups were subdivided into three groups when their kids reached 30 days of age. Groups were as follows: 1) early weaned (EW) to creep feed or oat pasture, 2) limit suckled (LS) 1 hour per day and 3) control or unlimited nursing (CO). Does that aborted or lost their kids at or soon after birth were considered as a fourth group (A/D). Of the CO, LS, EW, and A/D groups, 81%, 88%, 95%, and 62%, respectively, exhibited estrus within 40 to 50 days postpartum. The percentage that kidded within each group was CO = 35.5%, LS = 57.7%, EW = 79.5%, and A/D = 50%. When examined by season, there appeared to be a decreasing ability for the treatments to influence rebreeding as the anestrus season neared.

Effect of the Male

Shelton (1960) first documented the ability of the male goat to stimulate synchronization and early resumption of seasonal breeding activity in Angoras. A number of researchers since then have confirmed the so-called "male effect" in other goat breeds (Ott et al., 1980; Chemineau, 1982). In Shelton's work, does exhibited estrus approximately 10 days after introduction of the male. More recent studies have noted ovulation occurring in some does as early as 2.8 days (Chemineau, 1982) and 5.5 ± 1.3 days (Ott et al., 1980) after introduction of the male. In previous studies ovulation as early as 2 to 3 days after exposure to the male may have gone undetected because of silent estrus.

Sheep typically experience one to two silent ovulations before exhibiting estrus 19 to 23 days after ram introduction (Knight et al., 1980). In contrast, the doe more readily exhibits estrus after male introduction, but this first estrus is often anovulatory; 55% vs 11% for ewes (Camp et al., 1982; Knight et al., 1980).

In a photoperiod study involving yearling does, Bon Durant et al. (1981) noted a small rise of luteinizing hormone (LH) as late as 15 days after introduction of the male and 5 days before the ovulatory LH surge and mating. These researchers conceded that the does may respond hormonally as soon as 10 hours after male introduction--as reported to occur in ewes by Chesworth and Tait (1974). Sampling was, however, too infrequent to determine if such an effect took place. Corteel (1982) reported that the "male effect" reaction interval (RI) ranged from 1 to 30 days--and the RI as seen in the Bon Durant study may be due to the intensity of the doe's anestrus condition, rather than an inability to express estrus due to an earlier LH rise or ovulation.

Removing the buck from the herd at least several weeks before the end of the anestrus season and then reintroducing him has been proposed as a means for obtaining early rebreeding (Chemineau, 1982; Ott et al., 1980).

SHORT CYCLING

Within recent years, short cycling has become a controversial topic in both cows and does (Odde et al., 1980; Reeves and Gaskins, 1981; Bretzlaff et al., 1982). Short cycling in does has been defined as an estrus interval of less than 15 days where the norm is 20 to 21 days. In cases where short cycling has been well-documented, cycles from 2 to 7 days were characteristic. Though short cycling may be a natural phenomenon, there are situations where it becomes an aberration of a particular treatment. An increased incidence of short cycling has been noted in beef cows and ewes where hormonal or suckling treatments were used for induction of estrus (Knight et al., 1978; Reeves and Gaskins, 1981; Lofsted et al., 1981). Recently, Bretzlaff et al. (1982) reported that a series of short cycles followed prostaglandin-induced abortions in dairy does. Short cycles also occurred in animals of unknown reproductive status. Because prostaglandin ($PGF_2\alpha$) may be used as a tool for synchronization, and short cycles are thought to be anovulatory, this may have some interest to the dairy goat industry where timed insemination is practiced.

Short cycles have been noticed to occur with a frequency ranging from 10% to 20% in the doe (Simplicio et al., 1982; Gonzales-Stagnaro and Madrid-Burg, 1982). One-third of the Spanish does subjected to the limited suckling treatment previously mentioned were found to be short cycle. Camp et al. (1982) observed that as many as 44% of the cycles of a group of Nubian does were short, but 88% of these occurred during the first month of the breeding season. Smith (1981) cited reports indicating that the first estrus is anovulatory and that the second estrus, if it follows shortly, is also "not always ovulatory." However, Camp et al. (1982) found, by laparoscopy, that only half of these cycles were anovulatory. Corteel (1977) also reports that the short cycle may be more fertile in the doe than formerly presumed--because there are some species that have regular short cycles of approximately 8 days.

Thus, an insufficiency of LH seems to result in no ovulation half of the time and, where ovulations do occur, LH levels are inadequate to maintain corpora luteum function.

ABORTION

Abortion has long been recognized as a problem with Angora goats. Almost all Angora producers experience some loss, and individual cases have been reported of almost total losses. More recently, it has been recognized that non-Angora goats (meat or dairy types) also suffer losses from abortion. The conditions required to induce or initiate abortion in the non-Angora types appear to be more severe. Goats in general tend to abort more readily than other species, such as sheep. It might be of some interest to speculate why this is the case. No doubt one contributing factor is the fact that the goat is a corpus luteum dependent species, i.e., a functional corpus luteum must remain on the ovary to maintain pregnancy, and thus anything that interferes with this will generally result in abortion. In other species such as the sheep or the cow, the placenta is a major source of progesterone and thus pregnancy may be maintained in the absence of a corpus luteum on the ovary. Another most unusual and possibly unique feature of the goat is that they do not readily deposit a large amount of fat, with the result that even relative short periods of undernutrition may result in stress to the maternal or fetal organism. This is more true with Angora goats because a large portion of their nutrient intake goes into fiber production and may explain their relatively greater tendency to abort. A number of disease entities can be shown to cause abortion in the goat, but it is generally thought that most problems are noninfectious in nature.

It has long been recognized that abortions in Texas Angora flocks tend to occur between the 90th and 120th days of gestation and that they largely occur among underdeveloped and/or undernourished goats subjected to a period of stress (Shelton and Groff, 1974). When these specific conditions occur, heavy losses are often observed. These general conclusions had been obvious for some time, based on observations and rancher experience. However, the mechanisms inducing this abortion and its concentration in the midpart of the gestation period was not understood until recently. A series of studies in South Africa appear to have explained this mechanism.

Observations in the U.S. (Texas) had suggested that abortions occur in response to stress of small, undernourished does and that the fetus is delivered in a fresh and often live state. When workers in South Africa undertook to study this problem, they collected Angora does with an abortion history to concentrate their efforts on does most likely to abort. Their initial findings suggested some marked differences as compared to Texas experience--abortions were often in large does and the fetus was generally expelled after having been dead for some time in an edematous or autolized state. Thus, it became evident that there were, in fact, two kinds of abortion (Wentzel, 1973; Wentzel and Roelofse, 1975) that are not referred to as habitual and stress aborters. The relationship between these is not completely understood, but the method of dealing with each of them is reasonably clear.

A next step was to attempt to explain the mechanism of stress abortion and why it tends to be concentrated in the period between 90 and 120 days of gestation. The end result of a series of studies suggests that undernutrition and/or stress of small Angora does rapidly results in low-blood glucose levels (Wentzel et al., 1974, 1975, 1976, 1978; Wentzel and Viljoen, 1975). In the early stages of pregnancy, fetal demands are low; thus, these factors do not interfere with fetal development. Also, any losses that occur at this time would not be recognized. The period between 90 to 120 days corresponds with the period of rapid fetal development; thus, low-blood glucose value results in stress to the fetus. At this time, the fetal adrenal is not fully developed and, in response to stress, it produces estrogenic precursors that have estrogenic properties and are potent abortifacients. The result is often "abortion storms" with live or fresh fetae being observed. In the latter stages of gestation, the fetal adrenal and other endocrine organs are more mature and elaborate the more normal adrenal steroids. These are less potent or slow-acting abortifacients that result in fewer abortions in this period and the absence of synchronized abortion storms, and may well present a different picture in the abortus. The habitual aborting Angora does are thought to abort as a result of hyperadrenalism on the part of the doe herself. The result is the death of the fetus in utero and expulsion at some later date. It is interesting to speculate on the relationship of this condition to a high level of good (fine) quality mohair production. It is not known if some stress aborters eventually develop into habitual aborters; or if they do, if it is a result of a genetic predisposition or to an altered adrenal function resulting from the initial abortion experience. Brief and preliminary studies in the U.S. (Shelton et al., 1981) tend to support the results of the South African research.

Reasonably good recommendations can be made for dealing with the two types of abortion. The simpliest recommendation is for the habitual aborter, which should be culled unless mohair is sufficiently valuable that it may be desirable to keep them only for hair production. Any offspring that they may have produced should not be used for breeding purposes. Sress abortion can be prevented largely by 1) having well-developed does, 2) maintaining an adequate level of (energy) nutrition, and 3) preventing stress that might interrupt feeding and normal activities.

REFERENCES

Asdell, S. A. 1926. Variation in the onset of the breeding year in the goat. J. Agric. Sci. 16:632-639.

Bon Durant, R. H., B. J. Darien, C. J. Munro, G. H. Stabenfeldt, and P. Wange. 1981. Photoperiod induction of fertile oestrus and changes in LH and progesterone concentrations in yearling goats. J. Repro. Fert. 63:1-9.

Bretzlaff, K. N., K. McEntee, J. Hixon, and R. S. Ott. 1982. Short estrous cycles after PGF-induced abortions in goats. Third Int. Conf. Goat Production and Disease.

Camp, J. C., D. E. Wildt, P. K. Howard, L. Stuart, and P. K. Chakraborty. 1982. Behavioral, endocrine, and ovarian relationships during the estrous cycle of the Nubian goat. Proc. Third Int. Conf. Goat Production and Disease. p 310.

Chemineau, P. 1982. Control of ovarian activity induced by teasing in the creole meat goat. In press.

Chesworth, J. M. and A. Tait. 1974. A note on the effect of the presence of rams upon the amount of luteinizing hormone in the blood of ewes. Anim. Prod. 19:107.

Corteel, J. M. 1977. Management of artificial insemination of dairy seasonal goats through oestrus synchronization and early pregnancy diagnosis. Management of Repro. in Sheep and Goat Symposium. Wisc.

Foote, W. D., W. C. Foote and E. A. Nelson. 1982. Induced out-of-season activity in dairy goats. Proc. Third Int. Conf. Goat Prod. and Disease. p 340.

Gonzalez-Stagnaro, C. and N. Madrid-Burg. 1982. Sexual season and estrous cycle of native goats in a tropical zone of Venezuela. Proc. Third Int. Conf. Goat Prod. Disease. p 311.

Hadjipanayiotou, M. and A. Louca. 1976. The effects of partial suckling on the lactation performance of Chios sheep and Damascus goats and the growth rate of the lambs and kids. J. Agric. Sci. 87:15-20.

Moss, G. E., T. E. Adams, G. D. Niswender, and T. M. Net. Effects of parturition and suckling on concentrations of pituitary gonadotropins, hypothalamic GnRH and pituitary responsiveness to GnRH in ewes. J. Anim. Sci. 50:497.

Nordfelt, W., C. R. Nordfelt, W. Foote, E. Nelson, and D. Foote. 1982. Induced breeding in dairy goats by increased photoperiod. Proc. Third Int. Conf. Goat Prod. and Disease. p 570.

Odde, K. G., H. S. Ward, G. H. Kiracofe, R. M. McKee, and R. J. Kittok. 1981. Short estrous cycles and associated serum progesterone levels in beef cows. Therio. 14:105.

Ott, S. S., D. R. Nelson, and J. E. Hizon. 1980. Effect of presence of male on initiation of oestrus cycle activity of goats. Theriogenology 13:183.

Plant, T. M., E. Shallenburger, D. L. Hess, J. T. McCormack, L. Dufy-Bark, and E. Knobil. 1980. Influence of suckling on gonadotropin secretion in the female rhesus monkey (Macaca Mulatta). Biol. Repro. 23:760.

Randel, R. D. 1981. Effect of once-daily suckling on postpartum interval and cow-calf performance of first-calf Brahman x Hereford heifers. J. Anim. Sci. 53:755.

Reeves, J. J. and C. T. Gashius. 1981. Effect of once-a-day nursing on rebreeding efficiency of beef cows. J. Anim. Sci. 53:889.

Shelton, M. 1960. The influence of the presence of male goat on the initiation of estrous cycling and ovulation of Angora does. J. Anim. Sci. 19:368.

Shelton, M. and J. L. Groff. 1974. Reproductive efficiency in Angora goats. Texas Agri. Expt. Sta. BU 1136.

Shelton, M., G. Snowder, M. Amoss, and J. E. Huston. 1981. The relation of certain blood parameters to abortion in Angora does. Texas Agri. Expt. Sta. PR 3896.

Shelton, M. and J. Lawson. 1982. The effect of season on reproductive activity of meat-type goats in Texas, U.S. Proc. Third Inter. Conf. Goat Production and Disease. p. 341.

Simplicio, A. A., G. S. Riera, and J. R. Nunes. 1982. Estrous cycle and period evaluation in an undefined breed type (SRD) for goats in Northeast Brazil. Proc. Third Int. Conf. Goat Prod. and Disease. p 310.

Smith, M. C. 1981. Caprine. In: D. A. Morrow (Ed.) Theriogenology. W. B. Saunders Co., Philadelphia.

Turner, C. W. 1936. Seasonal variation in the birth rate of the milking goat in the United States. J. Dairy Sci. 19:619.

Tyson, J. E. and H. G. Friesen. 1972. Human lactational and ovarian response to endogenous prolactin release. Science 177:897.

Wentzel, D. 1973. The habitually aborting Angora doe: Recognition and endocrinology of two distinct types of abortion. Ph.D. Thesis., University of Stellenbosch.

Wentzel, D., J. C. Morgenthal, C. H. Van Niekerk, and C. S. Roelofse. 1974. The habitually aborting Angora doe. II. The effect of an energy deficiency on the incidence of abortion. Agroanimala 6:129.

Wentzel, D., K. S. Kiljoen. 1975. The habitually aborting Angora doe. VI. Induction of abortion by administration of exogenous oestrogens. Agroanimalia 7:41.

Wentzel, D., J. C. Morgenthal, and C. H. Van Niekerk. 1975. The habitually aborting Angora doe. V. Plasma oestrogen concentration in normal and aborter does with special reference to the effect of an energy deficiency. Agroanimalia 7:35.

Wentzel, D. and C. S. Roelofse. 1975. The habitually aborting Angora doe: Induction of abortion by administration of cortisone acetate. Agroanimalia 7:45.

Wentzel, D., M. M. Leroux, and L. J. J. Botha. 1976. Effect of the level of nutrition on blood glucose concentration and reproductive performance of pregnant Angora goats. Agroanimalia 8:59.

Wentzel, D., J. M. Van Der Westhuysen, J. H. P. Van Der Merwe and K. S. Viljoen. 1976. Effect of introfoetal administration of 4-Androstene-3, 17-Dione on circulating maternal oestrogen levels and subsequent reproductive outcome of pregnant Angora does. Agroanimalia 8:39.

Wentzel, D., J. J. E. Celliers, and L. J. J. Botha. 1978. Time-course of decreasing progesterone levels in prostaglandin-treated Angora goat does. Agroanimalia 10:55.

Yazman, J. 1981. Personal communication. Winrock International Livestock Research and Training Center.

56

REPRODUCTION MANAGEMENT FOR DAIRY GOATS

Samuel B. Guss

Goats are the most prolific ruminant species. In temperate areas of the world, goats produce an average of 1.5 kids per doe per year. However, in tropical areas of the world, where nutrition is adequate, they produce 3 litters of 1.5 each 2-year period.

Moving farther away from the tropics, the breeding season is confined to the period of shortening day length (autumn), and that period is shorter for goats that are farther from the equator. Thus, in southern California, Florida and Mexico, does begin to cycle and show estrus in late July and August.

In northern United States and Canada, the breeding season extends from mid-September through the middle of January.

VARIATIONS IN BREEDING

In temperate regions, Nubians appear to have a breeding season beginning at least 1 month earlier than that of the Swiss breeds. In the latitude of Pennsylvania, the Nubian breeding season begins in mid-August and ends in late January. The Swiss breeds show estrus from mid-September through December.

Drying off dairy goat does that are not pregnant seems to stimulate onset of cycling through the normal breeding season and 1 month beyond.

Dairy goat buck kids may be fertile by the time they are able to thrust their penis beyond the prepuce; this can be as early as 3 1/2 months of age. Therefore, they must be maintained separately from doelings after they reach 3 months of age.

Well fed doelings may start cycling by the time they are 4 1/2 months of age. To prevent dystocia problems, bucks should not be allowed access to doelings until they reach 7 months of age and weigh at least 70 pounds.

Lifetime production and breeding performance are better when doelings are well grown and healthy and are bred to freshen the first time between 12 and 15 months of age.

HEAT CYCLES

Normal heat cycles during the breeding season are similar to those of cattle--18 to 23 days. The length of heat may be normally more erratic than it is in cattle, but the usual heat period extends from 8 hours to 1 1/2 days.

NATURAL BREEDING

Young well-fed healthy males become sexually active and fertile after 4 months of age. They can be bred to as many as 30 females in 1 month if they are hand bred to females in heat.

Contrary to a notion among dairy goat owners, young males are not harmed by breeding activity. I am aware of a Toggenburg male that began breeding does when he was 135 days old. As a result, 23 does conceived, and all later produced kids within the same 10-day period. I believe that it is safe to say that a young dairy goat male older than 4 months of age can be successfully used as long as his libido remains active and he can copulate successfully. This, of course, is a great advantage in obtaining proof of the buck's genetic worth. Within 20 months beyond the time the youngster is allowed to breed the first does, the milk production of his daughters and their conformation (type) will reveal valuable information regarding the young sire's worth.

MALE SEXUAL BEHAVIOR

Goat bucks are literally breeding machines during the normal breeding season. The size of their testicles increases dramatically as much as 35% at the beginning of the breeding season when females show evidence of cycling.

Normally, mature sires collect females in small harem groups and fight off any rival males that try to associate with their females. Mature males may be kept together, apart from the female herd, if they are away from sight or sound of the females. However, to avoid fighting, it is best to confine them separately.

Young males should never be confined with mature males where there is any breeding activity going on. It is not uncommon for a mature male to attack and kill a young male when they are confined together and females are showing estrus nearby.

PREPARATION FOR THE BREEDING SEASON

It is well-known that goats respond similarly to the way sheep respond when they are healthy and on a rising plane of nutrition by the onset of the breeding season (flushing). This results in production of more ova and more offspring. The size and birth weight of kids born as a result of good preparation of their parents for the breeding season is a very important management consideration. For example young, undersized, underfed females may produce few ova resulting in large fetuses and a tragic dystocia situation at kidding time.

Because males engage in frenzied activity throughout the breeding season, they usually lose weight, feed erratically, and even become emaciated by the end of the breeding season.

Both males and females in the breeding herd should be examined and treated for internal parasites about 3 weeks before breeding season begins. In herd situations where severe infection is found, repeat treatment again in 14 to 17 days should be given, and the beginning of breeding postponed for an additional 2 weeks.

Both males and females should be fed more and better feed to ensure that they will be gaining weight when they are to be mated. Females that are excellent milk producers may not only show increase in milk production from the additional food supply, but they should be clearly improving in body condition by the time they are bred.

In the absence of a male, it is often very difficult to detect signs of heat. Fortunately, there are some simple ways to bring cycling and heat about.

The Presence of the Buck

During the breeding season, the sight, odor, and presence of a buck--even when on the other side of a good tight fence or partition--will start females cycling and showing heat. The shortest time possible for this to occur is within 40 to 50 hours.

Synchronizing Heat - Use of a Buck Jar

A group of females isolated from a male can often be brought into nearly synchronous heat by this simple procedure: Take a piece of toweling or carpet and rub it well onto the head of an odoriferous, sexually active buck. Place the material in a wide mouthed gallon jar and seal it. Open the jar when you want the females to start cycling and allow them to smell it. They normally show intense interest in the odor emanating from the jar and the majority of them will be in standing heat within 48 to 60 hours.

Use of Prostaglandins ($PGF_2{}^a$) to Induce Heat

Drugs can be used to induce does to come into heat and to synchronize heat (estrus) in females. This can be most helpful when using artificial breeding (AI) and to ensure that the great majority of kids are born at a time most suitable for the owner.

For example, a dose of 1.25 mg PGF_2a administered to does and repeated 11 days later, brought 97% of one group of does into standing heat (36 of 37 does) within 50 hours after the second dose.

Signs of Heat

Anxiety, bleating, tail wagging, and mounting by other does are good signs of heat. Duration of standing heat is variable from about 8 through 24 hours.

When to Breed

Ovulation occurs as it does in cattle within 12 hours of the end of standing heat. Whether does are bred artificially, or by a buck, it is usually best to breed them in late standing heat. Breeders have many systems for controlling the number of services per estrus, but it is very simple and convenient to allow the buck to copulate with the doe when she has been in good standing heat for about 8 hours and again 12 hours later if she will still accept him.

MANAGEMENT OF DOES IN EARLY PREGNANCY

Stress

Avoid the undue stress of feed changes, medication, etc. during the first 30 days of pregnancy. Using natural service, the bred doe can be brought into or allowed "through the fence" contact with the buck between 18 and 25 days post-service to determine whether or not she has returned to heat.

The Milk Progesterone Assay Test

This laboratory test of milk from animals bred from 22 to 26 days is highly successful in detecting pregnancy. It is performed by a number of laboratories throughout the country that conduct the test on thousands of dairy cows.

The laboratories supply test kits that provide for milk samples from the does to be shipped with proper preservative to the laboratory.

The Hulet Method for Pregnancy Detection

This method developed by Dr. Hulet of the University of Utah is a simple, safe, rapid method for detecting pregnancy in sheep and goats bred from 70 to 110 days. The doe is placed on her back with a person on each side to hold a foreleg and rearleg together. This relieves pressure on the abdomen.

A 3/4 inch smooth rod made with a blunt rounded end is lubricated and passed rectally along the dorsal (backbone) side of the abdomen to a depth about 4 to 6 inches anteriorly (in front of) the udder. Then the end of the rod is directed toward the belly wall. The operator then attempts to feel the rod, using his free hand, slowly moving the rod from side to side. The end of the rod can readily be detected just beneath the belly wall in nonpregnant females. It is not necessary nor advisable to spend much time pressing and searching for the rod. It is easily found if the doe is not pregnant.

This procedure should take only a few minutes time and it should be done gently and patiently. However, it is safe and simple and highly efficient for pregnancy detection.

MANAGEMENT THROUGH PREGNANCY

Most dairy goat milk production decreases rapidly by the middle of pregnancy (60-80 days). The ratio of pounds of grain fed to pounds of milk produced can be widened to prevent the doe from becoming excessively fat. The condition of the doe is important during pregnancy. She should be in good lean condition with adequate exercise and forage to keep her alert and active. Substantial savings in grain-fed and real health benefits can be attained in late pregnancy from careful management of does through the first 4 months.

In the last month of pregnancy, does again should be fed on a slowly rising plane of nutrition so that they will gain in condition without becoming fat. It should be the goal of every dairy goat herdsman to maintain producing females in lean active condition through all phases of the milking and reproduction cycle. The notion that more milk will be produced by animals fattened in late pregnancy has long been refuted.

So-called "dry feeds," which are excessively rich in energy, are not recommended for does in late pregnancy. Milking grain mixtures should be fed and their protein content determined by the quality of the forage fed.

There is no advantage in feeding supplementary mineral mixtures, or in adding more minerals to grain rations fed, in late pregnancy. Milk fever is a real hazard in freshening mature does that have been fed excessive calcium in mineral mixes or calcium rich alfalfa (lucerne) hay in late pregnancy.

Synchronization of Kidding

Synchronizing of kidding of does in the last 10 days of pregnancy offers an opportunity for handling the kidding chore and arranging kidding of several, or all, of the does that conceive within a few days to go through parturition (kidding) within a relatively short time.

Administration of prostaglandins (PF_2a) at 145-148 days gestation makes possible synchronization of kidding, especially for herds where breeding is synchronized. Injection of 10 to 20 mgm of PF_2a after 145 days will induce normal parturition in 32 hours.

Kids that are born as much as 5 to 10 days ahead of time, apparently are born healthy and vigorous. There are great advantages to herd owners to get the "kidding ordeal" finished. Kids born within a few days are much easier to manage in groups than are kids with as much as 2 to 3 weeks difference in age.

Normal Parturition

At the onset of parturition, most does show anxiety and restlessness. At this time, they should be placed in maternity pens with about 20 square feet of floor space; the pens should be completely free from manure and the sides washed down preferably with nonodorous disinfectant.

Maternity pens and kid pens should have three solid sides and one side open to minimize drafts and allow plenty of light to enter. Just before each doe is placed into her maternity pen, it should be bedded with clean, bright, dust-free chopped or shredded straw.

THE THREE STAGES OF PARTURITION

First Stage

During the first stage or preparatory stage for kidding, the birth canal dilates including the cervix. Even before this begins, close observation will reveal a loosening or relaxation of the tail head and the ligaments around it. A thick clear stream of mucous is usually observed just before the onset of active labor. The first stage of parturition may last as long as 3 or 4 hours.

Second Stage of Parturition

Regular, strong labor contractions occur in the second stage and the doe may lie down and rise again several times. This stage of labor more closely resembles that of the mare than it does of the cow.

In the goat, the second stage of active contractions and straining rarely requires more than 30 minutes for the first kid to be born, and does may have as many as four kids within an hour. As soon as the water bag (a light pink or whitish balloon filled with fluid) bursts, the first evidence of the fetus is seen. Normally, kids are presented forefeet first with the nose and head lying betwen their forelegs. With a little assistance, does can easily expel a normal live kid, even if it has one foreleg retained.

Kids also can be born normally presenting their hind feet first with extended rear legs.

If the doe is observed to be straining unsuccessfully for more than a few minutes, an examination should be made. Wash your hands and the vulva of the doe with clean warm detergent solution. Carefully pass your hand into the birth canal and try to put the kid into normal birth position gently. If anything appears to be abnormal and the kid is not presented very soon after the beginning of active labor, call your veterinarian for immediate assistance. While active assistance for dystocia (abnormal birth) can often be delayed for several hours with a cow, it is imperative that the goat receive immediate assistance, if possible.

Breech presentations (with the hind legs flexed and directed forward), or presentations with the head deflected downward or to one side, do not allow passage of the fetus. If birth is delayed in an abnormal situation for more than a half hour, the cervix may close with tragic results. (Normal and abnormal birth presentations are illustrated in Sheep and Goat Handbook Vol. 2, p 161.)

The Last Stage of Parturition

Multiple kids are usually born within 30 minutes to an hour. If straining continues, the doe should be examined to be certain that all fetuses are born. The fetal membrane (placenta or afterbirth) usually makes its appearance within an hour or 2 and is normally expelled within 24 hours. Usually, gentle, light traction with a twisting motion will aid expulsion, but if that does not occur readily, the doe should have the placenta removed manually by a veterinarian.

Following kidding, evidence of uterine discharge after 3 or 4 days justifies examination or uterine antibiotic infusion. Because the "rest period" between kidding and the next service is normally much longer than that of a dairy cow's, uterine infections or birth canal trauma can be expected to have sufficient time for correction with proper treatment.

PROLAPSED OR EVERTED VAGINA OR UTERUS

The prolapsed or everted vagina may occur in does before kidding and the prolapsed uterus following kidding. It is much less common in goats than in cows.

The best means of prevention seems to be maintaining does in late pregnancy in active vigorous condition with opportunity for daily exercise. Pessaries used for ewes can be inserted for everted vaginas; everted uteri can be replaced and corrected manually by a veterinarian.

TWO KIDDINGS PER YEAR

Some dairy goat herd owners have been producing more kids for replacements or sale and stabilizing their herd's milk production by producing two crops of kids each year. To accomplish this does are bred early in the breeding season so that they will kid in late December or January. When the does have been milking about 2 months, the barn is darkened and the does receive 14 hours of artificial light a day for at least 2 weeks. Within a 2-week period the light exposure is reduced to 9 hours a day.

This same procedure will bring buck goats into breeding condition. Good breeding results can be expected (70% to 80% of the does pregnant). Late summer or fall kidding will help provide a more uniform milk supply. It is possible and practical to have herd does kid 3 times within a 2 year period, but it is very difficult for dairy goat does to be brought into more than three lactations within any 2 year period.

ARTIFICIAL INSEMINATION

Artificial insemination has been performed routinely and successfully in cattle for the past 40 years; not until recently has it been used in dairy goats. When superior sires can be identified, using production records, herd mate companions, and classification of daughters, there will be real incentive for using AI in dairy goats to improve milk production.

The processing, storage, and handling of frozen goat semen is generally considered to be the same as for cattle. French workers insist that spermatoazoa "washing" is necessary, but the rest of the world does not share that view.

Does are inseminated in late standing heat or within 4 to 6 hours after standing heat. An ordinary sterile plastic AI catheter is used. The operator, with the aid of a head light, or ordinary flashlight, and a 3/4" diameter plastic speculum, passes the speculum into the cervix and discharges the thawed semen from the straw directly into the uterus.

Conception results have been about as successful as those achieved with dairy catte. Improved technology will undoubtedly achieve the excellent results now confidently expected by dairy cow herdsmen.

SUMMARY

Reproduction management for dairy goat herds is not substantially different from that of cattle herds. Cattle technology in this area has improved dramatically in the last 40 years. Dairy goat owners can follow in the cattle owners' footsteps and achieve goals that have eluded all but a few highly successful breeders.

Controlled heat induction, synchronized estrus, and synchronized kidding have improved the reproduction efficiency, production, and type of dairy goat breeding herd. Savings in labor are great incentives for using the new technology.

REFERENCES

Bretzleff, Katherine. 1981. Induction of Parturition in the doe. Dairy Goat Journal.

Guss, S.B. "Management and Diseases of Dairy goats". Published by Dairy Goat Journal.

Ott, R. 1980. Use of prostaglandins for control of reproduction in dairy goats. Dairy Goat Journal.

57

FACTORS AFFECTING KID PRODUCTION OF ANGORA GOATS

Maurice Shelton

Angora goat production has been an important industry in Texas since shortly after the goat was first intoduced in 1849. Texas has long supplied approximately one-third of the total world production of mohair. Other important producing countries are South Africa and Turkey, with Lesotho, U.S.S.R., Australia, and Argentina also producing significant amounts. With the favorable prices for mohair in recent years, a number of other areas are becoming interested in Angora goats. There is no reason why Angora goats cannot be sucessfully produced in many areas, but where there is no browse to be utilized and/or where long winter feeding periods may be required, there is less reason to predict a long--term success for the industry.

At present, most of the Angora goats in Texas are in the Edwards Plateau where the industry began. Before losses to predatory animals became so severe and before the mohair market break in 1964, large numbers were also found in other areas such as the Grand Prairie and Cross Timbers (Central and North Central Texas) where they were used to advantage in land clearing and bush control or utilization. Large areas of the South Texas Plains now have excellent forage resources for goats, but they are not being used for the purpose.

Poor reproductive efficiency has long been a problem with the Angora. However, this problem may have become more serious over the years due to a reduced labor supply to care for the breeding flocks and to a continuously increasing genetic potential for fiber production. Assuming a constant feed supply or forage quality, an increased level of mohair production occurs at the expense of body development and reproductive efficiency. Thus, to maintain a high level of reproduction, producers must evolve practices that circumvent or correct for the above factors. No reliable statistics on the kid crop weaned are available, but values in the range of 50% are often suggested as an overall average. Individual flocks may vary from 0% to well above 100%. The genetic potential for reproduction in the Angora is not the limiting factor. Environmental or nutritional constraints are the primary limitations. Producers need to know what practices can be used to circumvent these constraints and if their implementation will provide an economic response. The latter is almost certainly dependent on mohair prices and its resultant influence on demand for replacement kids.

A high reproductive rate is important because (1) of the sale value of the kids produced, (2) it permits the industry to respond more rapidly to changes in demand for mohair, (3) it facilitiates genetic improvement through providing a greater selection differential and (4) it permits improved mohair production (both quantity and quality) through lowering the average age of the flock. As goats become older, both the quantity and quality (fiber diameter) of mohair deteriorates; thus, if the total goat population remains stable, the average age is a direct function of reproductive rate.

BASIC REPRODUCTIVE PHENOMENA

An understanding of the basic reproductive phenomena is necessary to discuss approaches to improving reproductive efficiency.

Age at puberty is the initial or minimal age at which the animal becomes reproductively active, i.e., does start ovulating and are capable of becoming pregnant; males are capable of siring offspring. Angora goats are highly seasonal, and they reach puberty either during their first season at 6 to 8 months or 1 year later at approximately 18 months. Individual well-developed kids will reach sexual maturity their first season. However, to prevent the occasional breeding of doe kids, the two sexes should be separated and isolated from mature animals during the breeding season. Since kids require or deserve special treatment, they should be managed separately throughout much of the year. Many animals will not breed satisfactorily as yearling to kid at 2 years, but this failure is more a result of lack of condition and development than of age.

In respect to breeding season, Angora goats are seasonally polyestrous; that is, they are seasonal breeders and the females have reccurring estrual periods during the season if they are not bred. Angora goats have not been widely studied in this respect, but many other species, including sheep, have been, and it seems safe to interpolate across species in this respect. The phenomenon of seasonal breeding is known as photoperiodism, or respose to the length of the daylight period, and is found in many plants and animals. The Angora goat is somewhat unique for domestic animals in that both the males and the females are seasonal. The mating season of the male is easily detected by the characteristic odor and rutting behavior. The Angora is also unique in that the females do not normally start cycling until they are stimulated by the presence of the male. Later in the season other stimuli can serve this same purpose. The breeding season for Angora goats is generally from September to February. October is the most typical month for mating.

Length of the estrus cycle for the Angora doe is considered to be 20 days with a range of 19 to 21 days. Values outside this may be observed but are more often a result of erroneous determination of estrus.

Length of the estrus period for goats in general is over 30 hours. However, reports dealing specifically with the Angora report values of 20 to 24 hours.

Gestation length has been reported in two studies (Shelton, 1961; Van Rensburg, 1970). Both reports suggest 149 days with a range of 143 to 153. Since goats are known to abort readily, it is difficult to distinguish a late abortion from an early normal parturition. If the short periods (i.e., 143 to 145 days) are removed, the average gestation length would be more on the order of 150 days.

The ovulation rate of well-developed Angoras may be high with a large percentage of the does showing two ovulations. In one study (Shelton and Stewart, 1973) of goats run under commercial conditions, 10.2% did not ovulate, 69.7% had single ovulations, and 20.1% had twin ovulations. Although it may be advantageous for an individual producer to obtain twin kids, it may well be questioned if this is a desirable goal for the industry as a whole.

The birth weight of Angora kids averages about 6 or 6.5 lbs for single female or males, respectively. Twin kids are likely to be approximately 1/2 lb lighter. Some kids will be born with very light weights, most of which will not survive under natural conditions. Unusually large kids with resultant dystocia or difficult birth are not frequent with goats.

Successful reproduction, at least in terms of kid-crop weaned, consists of a number of discrete and somewhat independent functions that are:

a. Occurrence of estrus.
b. Ovulation and associated variation in ovulation rate.
c. Conception that may include both fertilization and implantation of the embryo.
d. Embryo survival to and through parturition.
e. Survival of the kid from parturition to weaning.

Losses or failure of reproduction may occur at each of these stages. In a given flock it is possible by appropriate research procedures to partition these losses into segments. If one assumes a constant ovulation rate of one or two, it is possible to express the losses in percentage with the remainder being the net kid crop. However, since ovulation rate is a somewhat continuous variable, the partitioning is not a simple mathematical process. One study (Shelton and Stewart, 1973) suggested that major losses occurred in items (b) and (e) in the above listing, but this will no doubt vary with different flocks or populations. Except for item (c) in the above listing, the causes of variation or causes of failure are generally well understood and can, in most cases, be prevented.

Failure to show estrus is almost totally explained by lack of sexual maturity (age), lack of physiological maturity, or to a debilitated state due to poor nutrition. Age is not itself a major direct limitation since few producers attempt to breed them earlier than yearlings to kid at two years of age. However, age is related to poor reproductive efficiency in that two-year-old does often produce a poor kid crop. This is almost totally explained by lack of development. Ideally, yearling does would weigh 60 lbs at breeding, but few actually do so under commercial conditions in Texas. Many of the does weighing 50 to 60 lbs will breed, but few of those below 50 lbs will reproduce successfully.

Variation in ovulation rate is also largely a function of size. It may be safely assumed that the occurrence of at least a single ovulation (the release of one ovum) is synonymous with the occurrence of estrus. At increasing weights, the frequency of twin ovulations increases in a direct and almost linear manner (figure 1). These data were taken from

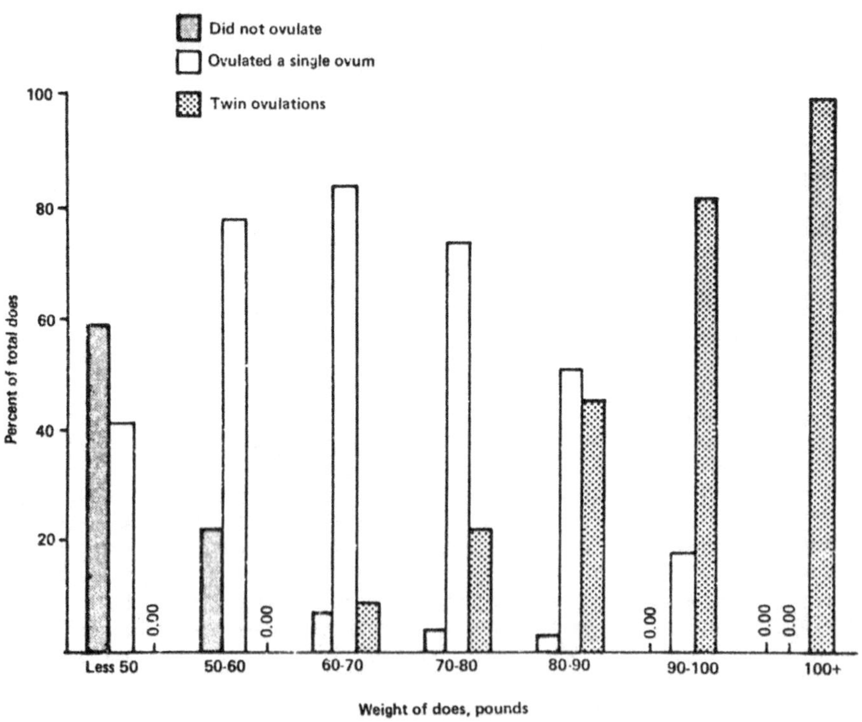

Figure 1. The relation of body weight to ovulation rate of Angora does.

Shelton and Stewart (1973). The higher weights reported in this figure are somewhat unrealistic for Angora goats under commercial conditions. Unfortunately, a large number of goats fall into the groupings on the lower range of this scale.

A failure of conception (fertilization and/or implantation) is more difficult to explain. Among flocks with a reasonably high ovulation rate, the disparity between the ovulation rate and kidding can be quite great. Biological or physiological impediments to conception in the female is normally low (1-2%). Also infertility of male goats is low if it is assumed that they are "rutting" and if they are sufficiently physically developed and strong to carry on a normal mating load. These situations can normally be ascertained or evaluated by appearance. The testes should also be examined as some young males have small, under-developed testes and should not be used. The writer also has observed males that were subjected to such heavy breeding loads, such as synchronized matings, that they became inactive for a period of time thereafter. In no case should reliance be placed on a single male. In case of single-sire matings, a cleanup or followup sire should be used. If the above factors are adequately taken care of and the male is not the limiting factor, other factors should be explored. Does that mate but fail to conceive are often very small, undeveloped, or emaciated. This can be corrected by appropriate management factors. Another major source of disparity between the ovulation rate and the number of embryos is among does with multiple ovulations, but in which only one embryo is present. If a doe ovulates only one ovum that is not fertilized, the doe will recycle. On the other hand, if she ovulates 2 or 3 ova and only one is fertilized and implanted, the doe will not recycle. This is a major source of loss of potential. Unfortunately, there are few clues as to how to overcome this problem. It appears to be largely chance that one vs two will be fertilized. Also, there is a reduced implantation rate for multiple ovulations. This may be explained by crowding or the necessity for intrauterine migration.

Embryo survival, or the reciprocal of this--abortion, or death and resorption in utero, is recognized as a problem with Angoras. A number of infectious diseases can be shown to cause abortion, but most of the problems with Angora are considered to be noninfectious. This type of abortion appears to be a reasonably understood phenomena (Shelton and Groff, 1974; Shelton, et al. 1981; and Wentzel, 1982). Much of the basic work on this problem has been done in South Africa (to be discussed in more detail in another paper in this series). Almost all flocks suffer a few losses from abortion. Normally, this would be in the range of 1% to 5%. Losses in the range of 15% to 30% have been observed by the author and much higher losses have been reported. Low-level losses are usually explained by habitual aborters and the percentage of these in the flock increases with age.

Losses due to habitual aborters can be terminated only by their being culled from the flock. The higher loss values reported above are due largely to sporadic abortion storms or stress abortion. These are largely explained by a high genetic priority for mohair production, a generally low level of nutrition, or short-term undernutrition due to environmental or management factors that interrupt feeding. A combination of more than one of these factors is more likely to lead to heavy losses from abortion. Corrective measures consist of removing one or more of the predisposing factors.

Death loss of the kids dropped is probably the single greatest cause of loss on an industry-wide basis. These losses are usually explained by:

- Birth of weak kids.
- Cold and/or wet weather during kidding.
- Poor mothering, no milk, nonfunctional teats, or abandonment by doe.
- Predation losses.

Each of these may be important in a given circumstance. Weak kids tend to result from long-term underdevelopment or undernutrition of the doe, and can be corrected. Cold or wet weather during kidding is outside the control of the producer. This source of loss may be minimized by delayed kidding or by kidding in confinement. Poor mothering or the absence of milk may be associated with poor nutrition, but another common problem is does whose teats have been cut off during shearing. Abandonment by the doe appears to be a major source of loss under range conditions. Goats tend to plant or bed down the young and feed away from them. If either the doe or kid is disturbed, or perhaps for other reasons, the doe may never recover the kid. Measures used to prevent this consist of kidding in confinement or kidding in small traps or pastures where the doe is unable to move very far away from the kid.

Losses to predation is also a major problem with goats. Angora kids are particularly susceptible to predation due partially to the habit of does grazing away from the kids. Other factors are their color and small size. Losses may occur not only to traditional predators such as foxes, coyotes, bobcats, and eagles, but they may also be taken by smaller mammals as well. Corrective measures consist largely of intensive predator removal prior to the kidding season.

SUMMARY AND RECOMMENDATIONS

The producer must decide for himself the relative priority of importance to be placed on reproductive efficiency. Maximizing kid crop almost certainly brings about some increase in production costs or a diminution in the relative priority placed on fiber production. At times of favorable mohair prices, it is almost certainly economic to maximize reproduction.

REFERENCES

Shelton, M. 1961. Factors affecting kid production of Angora does. Tex. Agri. Expt. Sta. MP 496.

Shelton, M., G. Snowder, M. Amos and J.E. Huston. 1981. The relation of certain blood parameters to abortion in Angora goats. Tex. Agri. Expt. Sta. PR 3896.

Shelton, M., and J. Groff. 1974. Reproductive efficiency in Angora goats. Tex. Agri. Expt. Sta. BU 1136.

Shelton, M., and J. Stewart. 1973. Partitioning losses in reproductive efficiency in Angora does. Tex. Agri. Expt. Sta. PR 3187.

Van Rensburg, S.J. 1970. Reproductive physiology and endocrinology of normal and habitually aborting Angora goats. Thesis. Faculty of Vet. Science, Univ. of Pretoria, Pretoria, Union of South Africa.

Wentzel, D. 1982. Noninfectious abortion in Angora goats. Proceedings Third World Conference on Goat Production. p 155.

Part 12

NUTRITION AND MANAGEMENT OF GOATS

58

COMMERCIAL PRODUCTION OF DAIRY GOAT MILK IN THE U.S.

James A. Yazman

The rapidly expanding population of dairy goats and dairy goat owners in the U.S. reflects several sources of new interest--the increase in numbers parallels the growth of vegetable gardening and seems to be part of a movement toward food self-sufficiency and improved nutrition. Dairy goats also are increasing in popularity as pets, 4-H projects, and among people who enjoy the competition of showing and producing high milk records. Although there are organized milk markets (local, statewide, and regional) in various locations, probably fewer than 20% of the dairy goats in the U.S. are milked on commercial dairy farms. Thus, the increased number of dairy goats does not stem from market demand for goat milk, rather, as with recent increases in horse numbers, it seems to stem from an interest in dairy goat ownership.

CATEGORIES OF DAIRY GOAT OWNERSHIP

Dairy goat owners can be classified into four broad categories based primarily on size of herd and goal of ownership: 1) the homesteader, 2) the hobbyist, 3) the semicommercial owner, and 4) the commercial dairyman.

The homesteader milks one to five or more does and uses most of the milk on the family table and for raising goat kids for replacements, for meat, and to sell as breeding stock. Some milk may be sold or given to friends and neighbors, and some surplus may be used to raise pigs and calves. The homesteader owns goats because they integrate well into his overall plan of food self-sufficiency and improved nutrition. He may or may not compete in dairy goat shows. Usually he is concerned more with the milk yield per doe that can be achieved with on-farm resources, rather than with pushing for high milk records by feeding larger amounts of purchased feeds.

The hobbyist has dairy goats primarily for the pleasure of competition and ownership. The herd size varies, as in all categories, but is usually no more than 15 to 20 does, or the number that can be managed with little problem on a part-time basis and moved between shows. Sale of breeding stock is the major source of income from the herd and much of the milk goes into raising kids. High milk records and type classification are essential to achievement of goals. Surplus milk may be fed to calves and pigs as a source of extra income and some may be sold on-farm to friends and neighbors. Generally, both the hobbyist and the homesteader have little interest in commercial-level herd ownership. However, their contribution to the development of the species through selective breeding for high production and good type should not be overlooked.

When the national supply of dairy goat milk is evaluated, a third category, the semicommercial owner, is important. He operates a herd of 15 to 60 does with the goal of utilizing family labor and surplus farm resources to supplement an income derived from sources other than his goat herd, usually an off-farm job. Much of his milk is marketed through pigs, calves, and breeding stock. However, with the development of organized markets for dairy goat milk and adequate prices, many semicommercial producers would convert their herds to commercial operations. As such, this category represents an untapped reservoir of dairy goat milk.

The commercial goat dairyman operates his dairy herd to make a profit from the sale of milk, meat, cull adults, surplus males, and breeding stock. The number of herd owners in this category is probably 5% or less of the total. Interestingly, many commercial operators began as homesteaders and hobbyists and originally purchased a goat to provide milk for a child allergic to cow's milk.

LIMITATIONS OF THE DAIRY GOAT FOR COMMERCIAL MILK PRODUCTION

Seasonality. Dairy goats in the northern latitudes, like deer and most breeds of sheep, are seasonal breeders. Dairy goats tend to be anestrus between the months of March and August with the peak of estrous activity occurring October through December. Seasonality of reproduction is the principal limitation to a consistent monthly supply of milk and therefore to stabilized milk marketing. The effect of reproductive seasonality is seen in table 1 derived from the study of 51,531 DHIA records by Grossman and Wiggans (1980). Their analysis indicates that 80.3% of the total kiddings took place in the months of February through May. Some extension of the kidding season may be attained by breeding does in December and January for kidding in May and June. However, because milk production by the doe tends to decline in the fall as day length decreases, late kidding reduces lactation yield per doe (table 1). There are breed differences; compared to other breeds, Nubians having a greater tendency to show estrus and conceive during the months of July and August. Selection for early breeding may have potential to reduce seasonality in reproductive behavior. Controlled environmental lighting and hormonal intervention also offer possibilities and have shown potential under commercial conditions in the U.S. and France (Ashbrook, 1982; Corteel et al., 1982; Nordfelt et al., 1982).

TABLE 1. DISTRIBUTION OF KIDDINGS BY MONTH, AND EFFECT OF MONTH OF KIDDING ON TOTAL LACTATION YIELD

	Jan	Feb	Mar	Apr	May	June
% of kiddings	6.9	19.9	26.7	21.1	12.6	6.0
Avg milk yield (kg)	876	907	893	862	829	783
	July	**Aug**	**Sept**	**Oct**	**Nov**	**Dec**
% of kiddings	2.4	0.6	0.3	0.6	0.9	2.0
Avg milk yield (kg)	787	763	785	689	717	795

Source: Grossman and Wiggans, 1980.

Volume of milk. Even with high herd lactation averages (1,000 kg to 1,500 kg milk), costs of production and marketing per unit of milk are quite high relative to cow milk. Principal among these costs are labor and milk transportation. To achieve profitability, close attention must be paid to measures of efficiency such as milk produced per hour of labor and milk yield per doe.

Body size. Body size can be both an advantage (see below) and a disadvantage. Required fencing may be more costly because dairy goats can pass through fences and barriers that restrain dairy cattle. Dairy goats are more easily led than driven. Their small size also makes them difficult to drive down lanes because they turn quickly.

Size also affects nutritional efficiency. Table 2 compares low, medium, and high milk yields in a 70 kg dairy goat and 650 kg dairy cow. While the milk yield per unit of body weight may be higher in the dairy goat, considering maintenance requirements that are a function of metabolic bodyweight (BW 0.75), the dairy goat is at a disadvantage. The 70 kg dairy goat must produce 5.7 kg milk to be equivalent to the 650 kg dairy cow producing 40 kg in terms of Mcal of ME for maintenance per kg of milk (table 2). Body size and level of production will determine the relative efficiencies of the two species. Feed costs and the ability of dairy goats and cows to consume protein and energy above maintenance level will determine the relative economic efficiency of the two species in terms of conversion of feed to milk.

TABLE 2. COMPARISON OF ENERGY REQUIREMENTS FOR MAINTENANCE (MCAL ME) PER KG BODY WEIGHT AND PER KG MILK PRODUCED DAILY IN DAIRY GOATS AND DAIRY COWS AT THREE DIFFERENT LEVELS OF PRODUCTION

	70 kg dairy goat[1]			650 kg cow[2]		
Level of production (kg):	2.5 Low	5.0 Med	7.5 High	20 Low	40 Med	60 High
Mcal ME/kg BW for maintenance	2.45	2.45	2.45	17.12	17.12	17.12
Mcal ME/kg milk for maintenance	0.98	0.49	0.33	0.86	0.43	0.29
Kg milk yield/kg BW	0.04	0.07	0.11	0.03	0.06	0.09

[1] Source: NRC (1981).
[2] Source: NRC (1978).

Ability to select among diet components. Balancing dairy goat rations is more difficult than that of dairy cows because of a goat's keen ability to select among diet components. Dairy goats can pick alfalfa leaves off the stems--and corn out of coarse rations--where as cows tend to consume all the material without selection. Because alfalfa leaves are very high in protein, feeding a 16% concentrate to dairy goats that are feeding on alfalfa hay as a forage is probably a waste of protein. Conversely, a coarse ration containing rolled or whole corn may not provide enough protein, if the dairy goat selects the corn and leaves behind the rest of the ration. Feed strategies must consider selection among dietary components of differing palatability if adequate rations for dairy goats are to be formulated.

Health. The scope of this report does not allow an extensive discussion of the health problems of dairy goats under commercial management conditions. Dairy goats experience many of the metabolic, bacterial, and viral diseases suffered by dairy cattle, and add a few of their own. These include abscesses, both external and internal, induced by Corynebacterium pseudotuberculosis and D. pyogenes; parasitic infections; and caprine arthritis encephalitis (CAE). The dairy goat is highly susceptible to mastitis caused by Staph. aureus. Mastitis is particularly a problem for herd owners who expand their herds to commercial level and install milking machinery.

Brucellosis has not been a problem in dairy goats in the U.S., but leptospirosis has been reported. Abortions are a problem but often are not associated with a specific disease complex.

The young dairy goat suffers from many of the same health problems experienced by lambs, including enterotoxemia and coccidiosis. Nutritional diarrhea, especially when the young goats are fed milk replacers high in lactose, can result in high mortality. Close attention should be paid to sanitation and nutrition if mortality and morbidity are to be minimized in kid rearing programs.

Genetic selection. Lactation yields in the range of 2,000 kg to 2,500 kg of milk have been recorded under official test. However, because artificial insemination has not been widely accepted among dairy goat producers, testing of sires to identify superior gene pools has been difficult. Few sires have daughters on test in more than one or two herds, which lessens the reliability of comparisons among sires. Until sire tests can be developed, choosing bucks to use in breeding programs is, at best, a high risk procedure based upon advertising, performance in the show ring, and recognition of herd name.

Marketing of cull stock and surplus kids. Marketing of stock for meat purposes is an important source of income to dairy cow milk producers. Meat from cull dairy cows is usually graded low (utility, cutter, canner) but is in demand for use in processed foods. Young bull calves bring prices of $50 to $150 and are used for production of veal and lean beef. This level of market demand does not exist for dairy goat meat unless the operation is located near an ethnic population with a tradition for consuming goat meat. Selling cull adults at $15 to $20 per head and surplus kids at $5 to $10 does not make a significant impact on farm income.

ADVANTAGES OF THE DAIRY GOAT FOR COMMERCIAL MILK PRODUCTION

Prolificacy. Twin and triplet births are very common in adult dairy goats. First-parturition does will often have singles, but twins are not uncommon. Nutrition and breeding management may play a role in rate of reproduction. Whatever the cause, the tendency toward twins, coupled with a usual overall balance between male and female offspring, indicates that rapid genetic progress can be made by identifying superior-producing dams and using their offspring as replacements. Prolificacy can only be an advantage if management programs minimize kid mortality and problems of pregnancy toxemia (ketosis) that dairy goats share with sheep.

Early sexual maturity. The kid born in April is usually ready to breed in November and will enter the milking herd the next April. Coupled with prolificacy, genetic progress can be very rapid in dairy goats because the selection differential can be high and the generation interval low. If heritability for productive characteristics (i.e., milk yield) is similar to dairy cattle and sire proofs can be developed to isolate superior gene pools, it would be possible to obtain much more rapid increases in milk yield with dairy goats than with dairy cows.

Size and ease of handling. Dairy goats are smaller than dairy cattle and, therefore, equipment can be less expensive. More dairy goats can be housed on a unit of land compared to cattle, and where land is very expensive, milk yield per acre may be a measure of efficiency favoring dairy goats. Dairy goats are more easily handled by women and children than the dairy cows, which may be an important factor in using surplus family labor.

Ability to browse. Dairy goats can convert forages such as shrubs, vines, and forbs to milk and meat. These often grow on marginal lands and may not be consumed by dairy cows. This has been the basis of the growth of the dairy goat industry in regions such as the Ozark Mountains of Arkansas where the land is rocky and forested. It also opens the possibility of dual-species grazing systems with goats consuming "brush" and sheep or cattle consuming grass.

RESOURCES AVAILABLE TO THE COMMERCIAL GOAT DAIRYMAN

The principal components of the dairy goat milk production system are the dairy goat, facilities and equipment, and feeds and forages.

The dairy goat. The five major breeds available to the commercial goat dairyman in the U.S. are the Alpine, La Mancha, Nubian, Saanen, and Toggenburg. Minor breeds such as the Oberhasli also are found, but less frequently in commercial herds. Each breed has its own history and production characteristics. The Alpine, Saanen, and Toggenburgs are the higher yielding milk-producing breeds. Grossman and Wiggans (1980) analyzed DHI records for milk and fat yield, fat percentage, and percentage of records that reached 305 days (table 3). Saanens and Alpines produce higher levels of milk but a lower percentage of butterfat, as compared to Nubians and La Manchas. Toggenburgs often have problems maintaining adequate fat levels during lactation. As in the dairy cow, butterfat levels of all breeds react to such environmental stimuli as disease, fiber level in the diet, and stage of lactation. Although the Nubian produces a milk relatively high in fat, the other breeds of dairy goats do not produce fat levels greatly different from the higher-producing breeds of dairy cattle. This has important implications in the nutritional energetics of producing dairy goat milk. Also important to commercial milk production is the percentage of records that reach 305 days--ranging from 44.2% in the Toggenburg down to 25.9% in the Nubian.

The production results in table 3 have many implications for commercial goat milk production. Where no premium is placed upon fat content, the Nubian will be a less-desired breed due to its low milk yield and tendency toward short lactations. The Toggenburg produces an adequate level of milk but fat content may fall below 3% more often than with other breeds. For commercial production, where high per-doe yield of milk with normal (i.e., 3.5%) fat content is the goal, the Alpine and Saanen are the breeds of choice. The La Mancha, with a higher milk yield than the Nubian, higher fat content in the milk than the Alpine and Saanen, and greater tendency toward lactations of 305 days, shows potential to perform adequately under commercial conditions.

TABLE 3. AVERAGES AND STANDARD DEVIATION FOR MILK AND FAT YIELD, FAT PERCENTAGE AND PERCENTAGE OF RECORDS 305 DAYS IN LENGTH[1]

Breed	Milk (kg)[2]	Fat (kg)[2]	% Fat	% Records 305 days
Alpine	952.3 (304.8)	33.2 (11.0)	3.5	37.9
La Mancha	816.6 (262.9)	31.3 (9.4)	3.8	39.0
Nubian	805.7 (254.4)	36.9 (12.4)	4.6	25.9
Saanen	962.2 (312.8)	33.8 11.0	3.5	41.2
Toggenburg	921.1 (280.2)	30.4 (8.4)	3.3	44.2

[1] Source: Grossman and Wiggans (1980).
[2] Numbers in parentheses are standard deviations.

Equipment and facilities. Very little research has been done in the U.S. on equipment and facilities for commercial goat dairies. In Europe, especially France, milking equipment designed specifically for dairy goats is available. Only one U.S. equipment company manufactures a milking machine claw specifically for dairy goats. With the exception of the claw, which has two stems for milking inflations instead of the four needed for the dairy cow, the remaining equipment necessary for milking dairy goats is the same as found in cow dairies. Many commercial goat dairymen have successfully adapted used equipment from cow dairies for milking goats. Little is known about vacuum levels and pulsation ratios necessary for milking of dairy goats. Where installed correctly and maintained according to manufacturer's specifications, use of milking machinery designed for cows has proven satisfactory when operated at vacuum levels of 10 in. to 12 in. of mercury and 60:40 or 50:50 milk:rest pulsation ratios.

In designing milking and housing facilities for dairy goats, several factors must be kept in mind. Foremost is to minimize the amount of labor required in all phases of the operation. Specific to the milking operation is the need to recognize that dairy goats produce less milk than cows and therefore finish milking more quickly. With average cows yielding 10 kg to 15 kg per milking in a typical herringbone milking parlor, one man may be able to handle three to four machines, but with goats producing an average 2 kg to 3 kg per milking, only two to three machines can be managed. Parlor configuration is critical to sanitary, safe, and efficient milking of both goats and cows. Milking facilities must be designed keeping in mind the movement of does between the milking site and the housing site. Even with proper training and inducement, dairy goats tend to be more difficult to move than dairy cows.

Feeds and forages. Digestible nutrient intake is a function of nutrient density, voluntary intake, and digestibility. Goats have long been used under laboratory conditions to evaluate feedstuffs and to act as a model for the cow, but relatively little is known about intake and nutritional response in dairy goats under commercial conditions. Several observations can be made from practical experience.

Contrary to the popular misconception about goats consuming garbage, they are highly selective in their feeding behavior. They have a highly developed sense of taste and smell that, in many cases, limits the range of feedstuffs available to the commercial producer. Under intensive management using mixed, pelleted concentrates, intake is often a problem--even on feeds known to be highly palatable to dairy cattle. A very important characteristic of dairy goats is that, even with hay as the roughage source, they retain a keen ability to select among diet constituents. With alfalfa hay, leaves are often eaten and stems pushed aside in contrast to cattle that consume the whole plant. This may be an adaptation to small reticulorumen size and low rumen retention time (Huston, 1978). The consequence of this behavior is that goats will show a greater response to energy supplements, such as corn and molasses, than to protein supplements such as a 16% mixed concentrate (Morand-Fehr and Sauvant, 1978). This is especially true when fed forage that allows them to select among plant parts that have leaves high in protein such as alfalfa. In evaluating forages for dairy goats, percentage leaf composition and chemical composition of separated leaves and stems is essential to an accurate prediction of nutritive value. Diets of alfalfa hay supplemented with high-energy concentrates such as corn, oats, and barley are common among commercial producers. Complete feeds using ground or chopped forage as a base often fail to elicit a predicted response due to selective intake of diet components by goats. Even coarsely ground, mixed concentrates can fail to provide a balanced diet if components show differences in palatability. The feeding of dairy goats under commercial conditions to support levels of production in the range of 900 kg to 1,300 kg milk per doe per year requires a high degree of nutritional management.

Very little research attention has been given to the feeding of silage and grazing of dairy goats on improved pasture under commercial conditions. The goat evolved in the hot, dry, semiarid regions of the world where brush and trees were the principal components of the diet. Forages high in moisture, such as silages, and those which must be grazed, such as grass, are less preferred to drier shrubs and vines, which are consumed at eye-level or above. Feeding systems for dairy goats involving silages and pasture grass can probably be developed but need careful evaluation before adoption on a large scale.

RESEARCH TO SUPPORT COMMERCIAL DAIRY GOAT MILK PRODUCTION

If the production of a national supply of dairy goat milk on private, commercial farms is to be realized, the goat dairyman needs support in the areas of research and development. A "shopping list" of needs includes the following:

National sire proof. Purchasing frozen semen for AI or buying bucks for herd sires under the current system of identifying superior gene pools is not conducive to rapid genetic improvement. As with dairy cows, only a small fraction of the annual crop of male goats should be considered for testing based on pedigree. An even smaller fraction should be made available as frozen semen to goat dairy men based upon preliminary proofs from daughters milking in a number of different herds. With the prolificacy and early sexual maturity of the dairy goat, a sire-proving system adopted by goat dairy men could make rapid genetic improvement in the potential for milk production of the nation's commercial dairy goat herd.

Management programs to reduce seasonality of reproduction. Research needs include methods of manipulating the environment of the dairy goat (or to change the animal itself through selection) to reduce seasonality in reproduction and to promote a smooth annual flow of milk from the herd. Controlled lighting systems may provide the answer but other systems, including hormonal, nutritional, and genetic, should be evaluated.

Equipment and facilities. Much of the common equipment and facilities designed for sheep and cattle can be adapted for dairy goats, but some cannot. Given the high labor cost per unit of milk, labor reduction must always be considered along with safety, health, and sanitation. Milking equipment designed for milking cows appears to be adaptable for dairy goats, although research is needed to address such issues as pulsation rate and ratio and vacuum levels. Durable facilities for feeding need to be designed to reduce wastage and spoilage and to provide for labor-efficient rearing of replacement stock.

Nutrition. The dairy goat has long been used as a model for the dairy cow in research on nutrition and physiology of lactation. Less work in the U.S. has focused on requirements of the dairy goat for high milk production. Adapting recommendations for dairy cows to dairy goats will fail because of the particular nutritional demands created by the doe carrying a litter, by small size and rumen capacity, by the susceptibility to internal parasites, and by the ability to select among feed components. Nutritional strategies, or "how" to feed dairy goats, may represent more of a research problem in managing dairy goats than nutritional requirements, or "how much" to feed. The recently published NRC Requirements for Goats (NRC, 1981) provides a foundation for research but should not be regarded as an end point in itself. Attention needs to be paid to research completed in other countries such as France, Germany, and Japan to determine their applicability to U.S. conditions. Because feed, and the labor involved in feeding, are very high-cost items in the commercial dairy goat operation, nutritional management must be optimal if profit is the goal.

Health. The dairy goat suffers from several diseases of economic importance including abscesses and caprine arthritis encephalitis. Mortality and morbidity in a commercial operation are also a result of pregnancy toxemia, milk fever, and mastitis. In many cases, conventional therapy and preventive management will relieve the problem. However, with many conditions, such as mastitis, dairy goats do not respond as well as do dairy cattle. In other situations, drugs available to treat dairy cattle are not approved for treating dairy goats. The nationwide production of a consistent, safe supply of sanitary dairy goat milk will require research that considers the dairy goat as a unique species with unique health problems.

Milk-product development. Research needs in the area of product-formulation and marketing techniques are beyond the scope of this paper. The commercial goat dairyman must realize that market development is a corollary to progress in the production of dairy goat milk.

SUMMARY COMMENTS AND OBSERVATIONS

The genetic and nutritional resources for high levels of production of milk from dairy goats exist but because of the high labor cost per unit of milk produced, the price received for goat's milk must be from 1.5 to 2.5 times that received by the cow dairyman. Levels of production per lactation of 1,000 kg to 1,500 kg per doe will be essential in herds averaging in size from 60 to 120 does. The lack of milking equipment and therapeutic drugs specific for dairy goats may put an upward ceiling on herd size. This limit would be defined as the number of goats a herd owner can manage with a minimum input of hired labor and maximum control over mastitis, infectious disease, internal parasites, and feed quality.

Whether or not the research and development effort necessary to the dairy goat industry is forthcoming depends largely upon the future demand for dairy goat products. Drug companies, universities, and equipment manufacturers will recognize the need for research to solve the problems of the dairy goat producer only if production of dairy goat milk for commercial markets becomes a widespread agricultural enterprise involving a great amount of land, labor, and capital resources. Increases in efficiency of milk production from dairy goats will continue to be made as costs rise and the market for dairy goat products expands, but a consistent volume of milk from the national dairy goat herd will only result when market demand defines prices that lead to profitable goat dairying. Therefore, market development must go hand-in-hand with research if a commercial dairy goat industry is to be a reality in the U.S.

REFERENCES

Ainslie, H. R. 1979. New York DHI data - herd test year ending April 30, 1979. Farm Flashes 10:2.

American Dairy Goat Association. 1980. The 1980 ADGA Yearbook. Spindale, North Carolina. pp 3-11.

Ashbrook, Paul F. 1982. Year-round breeding for uniform milk production. Proc. Third Intl. Conf. on Goat Prod. and Disease. Tucson, Arizona. p 153.

Corteel, J. M., C. Gonzalez and J. F. Nunes. 1982. Research and development in the control of reproduction. Proc. Third Intl. Conf. on Goat Prod. and Disease. Tucson, Arizona. p 584.

Goth, J. and Frank D. Murrill. 1979. Dairy goats in California DHI program. Univ. Calif. Coop. Ext. Serv. Berkeley, California.

Grossman, M. and G. R. Wiggans. 1980. Dairy goat lactation records and potential for buck evaluation. J. Dairy Sci. 63(11):1925.

Huston, J. E. 1978. Forage requirements and nutrient requirements of the goat. J. Dairy Sci. 61:988.

Morand-Fehr, P. and D. Sauvant. 1978. Nutrition and optimum performance of dairy goats. Livestock Prod. Sci. 5:203.

NRC. 1978. Nutrient Requirements of Domestic Animals. No. 3. Nutrient Requirements of Dairy Cattle. National Academy of Science. National Research Council, Washington, D. C.

NRC. 1981. Nutrient Requirements of Domestic Animals. No. 15. Nutrient Requirements of Goats: Angora, Dairy, and Meat Goats in Temperate and Tropical Countries. National Academy of Science. National Research Council, Washington, D. C.

Nordfelt, Wesley, Carol Ruppel Nordfelt, Warren Foote, Edward Nelson and Darrel Foote. 1982. Induced breeding in dairy goats by increased photoperiod. Proc. Third Intl. Conf. on Goat Prod. and Disease. Tucson, Arizona. p 570.

59

NUTRITIONAL MANAGEMENT OF THE DAIRY GOAT

James A. Yazman

INTRODUCTION

Commercial production of dairy goat milk is a relatively new agricultural enterprise in the U.S. Though official statistics are not available, the number of dairy goats milked to supply commercial markets is not significant in comparison to the U.S. dairy cow population. Because of the limited production sector, research efforts designed to solve the problems of the commercial-goat dairyman have begun only recently in the U.S. Extensive research programs are found, however, in other countries, especially France and India, where goat milk and meat are important food commodities. When designing feeding programs under commercial conditions, the goat dairyman in he U.S. has had to rely on practices recommended from research studies with dairy cattle and on experiences of the cow dairyman. Unfortunately, feeding guidelines designed for dairy cattle may not produce the same results when applied to dairy goats, particularly in economic terms.

Our knowledge about the nutritional physiology, nutrient requirements, and feeding behavior of the goat (both dairy and meat types) has been greatly expanded by the recent publication of symposia research reports and comprehensive tables of nutritional requirements (NRC, 1981). Especially important to the design of commercial feed management programs is an understanding of the feeding strategy of the goat in consumption and utilization of forage. Except where specialized markets exist of goat milk and milk products, the advantage of milking goats, as opposed to cows, has been in the ability of the goat to utilize low-quality forage sources such as browse. Balancing rations for dairy goats with the objective of minimizing feed costs requires an understanding of the characteristics of forages important to the goat and of how the goat selects its diet. The traditional belief has been that the goat has an ability to digest fiber to a greater degree than does the cow. However, selection of more digestible species and plant parts probably plays a more important role in the goat's survival in environments of low forage quality than does digestion.

Nutritional Physiology of the Goat

Huston (1978) and McCammon-Feldman, et al. (1981) have reviewed and summarized research results on the nutritional physiology of the goat. Points important to the goat dairyman in comparing goats to sheep and cattle include:

- Digestive tract volume in ruminants is proportional to body size, with the result that cattle have a greater digestive capacity.
- Goats tend to have a faster rate of passage of food particles through the digestive tract and a reduced retention time in the reticulo-rumen.
- As a result of the faster passage rate in goats, there is a tendency to pass larger food particles from the reticulo-rumen to the abomasum and intestines.
- On equal diets, cattle tend to digest cellulose to a greater degree. Cellulose is a major component of plant cell wall.
- Goats tend to have a higher rate of dry-matter intake (weight of dry matter consumed per unit of body weight).

Demment and Van Soest (1982) proposed that rapid passage rate in small ruminants is a constraint imposed by a higher ratio of metabolic rate (Kcal/kg live weight/time) to digestive tract capacity than in large ruminants. Uden (1978, summarized by Van Soest, 1980) found retention times of mordanted particles to be approximately 70 hours for sheep and cattle compared to 52 hours for goats. As plant cell wall is principally digested by the microbial population of the reticulo-rumen, faster passage out of that organ will result in reduced cell-wall digestibility. The other component of forage dry matter, cell contents, is almost totally utilized (Van Soest, 1982). Therefore, where cell-wall digestibility is reduced, dry-matter digestibility is also reduced. The faster passage rate and concomitant reduction in call-wall digestibility is disputed by several workers (Devendra, 1978, 1982; Gihad, et al., 1980; Sharma, et al., 1982) who claim a superior ability to digest fiber (cell wall) for the goat. Owen and Ndosa (1982) recently compared digestion of temperate grass hay and soybean meal-supplemental straw in sheep and goats and found no significant difference between the two ruminant species. Differences in results of feeding trials with goats often result from a failure to consider selection among components of the diet offered and level of voluntary intake. Cell-wall digestibility values tend to higher where feed intake is restricted and selection opportunity is high (Van Soest, 1982).

Ruminant species are adapted to the consumption and utilization of feedstuffs high in cell wall, such as grasses, forbs, and browse. Cattle are adapted to consumption of large quantities of forage material that they retain in their reticulo-rumen for long fermentation periods greater than 48 hours). The reduced volume of the digestive tract creates a special problem for small ruminants, such as sheep and goats that consume the same forage diets as cattle. Fermentative digestion takes place in the reticulo-rumen and is a curvilinear process. The more easily fermentable constituents of dry matter (sugars, amino acids, vitamins, and soluble minerals of the cell contents) are fermented rapidly, while fermentation of cellulose, hemicellulose, and other cell-wall constituents requires a longer residence time in the reticulo-rumen.

The schematic of Huston (1978, figure 1) illustrates the effect of the smaller reticulo-ruminal volume of the goat as compared to that of the cow. With a smaller retention time (RT_2), the passage rate of particles out of the goat's reticulo-rumen (R_3) is greater than that of the cow with retention time RT_1. As a consequence, R_2 in the goat (the rate of reticulo-ruminal digestion) is lower compared to the cow's; therefore, the extent of digestion of cell-wall (and by extension, dry matter) is lower. Intestinal digestion rate (R_4) and fecal excretion rate (R_5) are closely related to R_3--the rate of passage of undigested particles out of the reticulo-rumen. Huston cites several references in support of his model that proposes the cow's relatively greater efficiency of dry-matter digestion because of a longer retention time in the reticulo-rumen.

Maintenance requirements for dietary energy are a function of body weight ($wt^{0.75}$), thus the smaller size of the goat suggests a higher requirement for energy per unit of live weight or compared to that required by the cow. For example, the ME (metabolizable energy) requirement for maintenance for a 550 kg cow is 15.11 Mcal, or 0.027 Mcal per kg of live weight (NRC, 1978). For an average 60 kg goat, the maintenance requirement for ME is 0.037 Mcal per kg live weight (NRC, 1981). Thus, the goat must achieve a relatively higher intake of energy per unit of body weight, to compensate for a reduced ability to extract energy from normal ruminant diets high in cell wall. Gestation and milk production in dairy goats creates a demand for higher levels of energy, protein, and other metabolic nutrients, which must be satisfied from the diet if live weight is to remain constant. To achieve a relatively greater intake of metabolic nutrients, the goat must either consume more dry matter per unit of body weight or consume a diet more concentrated in digestible nutrients than that selected by the cow. McCammon-Feldman, et al. (1981) summarize extensive literature that compares dry-matter intake in goats, sheep, and cattle on grass and legume diets. The summary shows an average 27.2% greater rate of intake for goats compared to cattle (2.29% of body weight compared to 1.80% for cattle). Rates of voluntary intake as high as 6.0% (Le Houerou, 1980) and 8.0% (French, 1970) of body weight have been reported for goats on browse. Often rates of voluntary intake are calculated from a combination of body weight gain, nutrient requirements, and estimated nutritive value of the forage available. Espęcially where differential quality among species and plant parts within species exists, reported values for voluntary intake in goats are often overestimated, if intake is not directly observed. Even if the goat consumes dry matter at levels of 5% of body weight or higher, increases in voluntary intake show a "diminishing returns" effect as digestibility of cell wall is depressed at higher levels of intake (Van Soest, 1982).

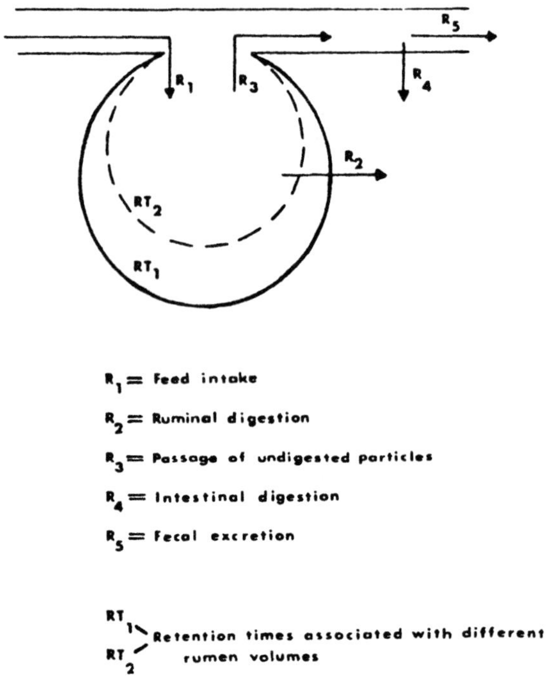

Figure 1. Schematics of the various rates associated with feed digestion in the ruminant and the influence of rumen volume (Huston, 1978)

Because of the greater nutritional requirements of goats and their reduced ability to digest cell wall (compared to cattle on the same diet), an increase in the goat's voluntary intake of dry matter is not enough to assure an adequate supply of metabolizable nutrients in most production situation. Especially in dairy goats producing high levels of milk (4.0 kg per day or greater), the diet supplied must be of high enough quality to compensate for reduced digestive efficiency. For goats on unimproved pasture, brushy range, or forested browse, selective intake of dry matter high in digestible nutrients is a necessity for survival.

Feeding Strategy of the Goat

The goat has evolved unusual physical and behavioral characteristics that allow it to survive in ecosystems where the average nutritional quality of the available forage is low, but where there is wide variation in quality among species and plant parts within species. On diets of uniformly low quality, such as chopped wheat straw, the advantage is to the animal able to ingest large quantities that are held in the reticulo-rumen and fermented for long periods (greater than 48 hours). Where diet quality varies, the advantage is to the animal able to "select the best and reject the rest." Several of the morphological and behavioral traits that are important to the feeding "strategy" of the goat are summarized by McCammon-Feldman, et al. (1981):

1. The goat possess a narrow muzzle, mobil upper lip, and a prehensile tongue that allows for the browsing of leaves protected by thorns or attached to petioles and branches of low nutritive value.
2. The goat exhibits a high degree of dexterity and is able to stretch upward on its hind legs to browse leaves of trees and vines.
3. The goat has a tendency to range over greater distances than sheep and cattle in their feeding activities. In so doing, it exposes itself to a greater range of forage species and a greater volume of feedstuffs.
4. The goat will vary its diet according to the seasonal availability of grasses, forbs, and browse.
5. The goat exhibits an ability to select plant species and plant parts within species that have a high nutritive value.

Huston (1978) and McCammon-Feldman, et al. (1981) interpret the goat's feeding strategy as an adaptation to the limitations imposed by a reduced digestive capacity. In addition to consuming dry matter at a greater rate than cattle, the goat compensates for a reduced ability to digest cell wall by selecting a diet low in cell wall, such as the leaves of shrubs and vines. Because goats often consume forage that sheep and cattle refuse, their feeding strategy serves to reduce competition in their ecological niche. In a temperate range ecozone grazed by goats, sheep, cattle, and deer, McMahon (1964) found the deer were the only species to significantly overlap the goat in terms of dietary preference.

Browse as a Nutritional Resource

In many commercial goat-milk operations in the temperate zones and national dairy-goat-development programs in the tropical zones, browse is expected to contribute significantly to the feeding program. Browse is defined by McCammon-Feldman, et al. (1981) as the leaves and twigs of woody species. A characteristic of browse important to ruminants is the differentiation in nutritional quality between leaves and stems. Browse leaves are typically higher in nutritive value than are stems, especially in content of crude protein and soluble carbohydrates (Wilson, 1969). In temperate regions, this difference is reduced during the growing season (Short, 1973); where as it is consistently high across seasons in the tropical zones (Van Soest, et al., 1978).

Relatively few studies have examined browse as a nutritional resource for ruminant livestock production. Under normal circumstances of forage availability in the temperate zones, it has little or no role in the diet of cattle (which are mainly grazers), and sheep (which prefer grass but will consume browse when necessary). In semiarid zones where availability of grass and forbs is low, cattle will consume browse such as woody shrubs and trees.

Wilson (1969) reviewed and summarized results of studies on the nutritive value of browse for ruminants. In general, the crude protein content of browse is m;ore consistent than that of grasses, which have a high content at the beginning of the growing season but decline rapidly as they mature. An important characteristic of Australian browse species is that they maintain their organic matter digestibility throughout the dry season, whereas grasses decline in digestibility with maturity. The leaves of browse plants are high in chemical entities, especially lignin and cutin, that are refractory to fermentative digestion. However, these entities are concentrated in the cell wall, which makes up a lower proportion of total dry matter in browse as compared to that of grasses--especially in the tropics. In Nicaragua, McCammon-Feldman (1980) compared the chemical composition and nutritive value of Hyparrhenia rufa, a tropical grass to that of two tropical browse species, Cordia dentata and Pithecolobium dulce. In this study of stall-fed goats, the grass was lower in lignin content of the dry matter and as a percentage of the cell wall, but voluntary intake was higher for the two browse species (table 1). Although the lignin content of the browse species showed reduced in vivo apparent cell-wall digestibility (38.4% and 42.2% compared to 53.1% for the grass), the in vivo apparent organic-matter digestibility was higher for the two browse species.

The leaves of browse species are high in polyphenolic compounds that are related to lignin; however, these compounds are soluble and include tannins, flavones, and alkaloids (Van Soest, 1982). Such compounds are thought to reduce digestibility of plant leaves through an inhibition of cellulose activity in the rumen or through formation of chemical bonds with plant constituents, which make them refractory to rumen fermentation. As an example, tannin forms an insoluble complex with protein, a chemical process important in "tanning" leather. Frenton (1982) found the the high content of phenolic compounds could explain the low digestibility of alder leaves (Alnus rubra bong) by goats and sheep. He suggests that the goat's ability to digest the dry matter of alder leaves to a greater degree than do sheep (63.9% vs 58.0% apparent dry-matter digestibility, $P<.05$) is a result of evolved metabolic strategies. McCammon-Feldman, et al. (1981), however, warn of possible errors in interpretation of results from digestibility trials involving forages high in compounds such as tannins, where longer than normal adaptation periods were not allowed for adjustment of rumen microbial populations.

TABLE 1. CHEMICAL COMPOSITION, IN VIVO DIGESTIBILITY OF ORGANIC MATTER AND CELL WALL AND AVERAGE DAILY DRY-MATTER INTAKE OF ONE TROPICAL GRASS AND TWO BROWSE SPECIES CONSUMED BY GOATS IN NICARAGUA

Species	CP	CW	L	L/CW	OM	CW	ADDMI
	----(% of DM)----			-------(%)-------			(g/kg BW)
Hyparrhenia rufa	4.5	60.4	4.2	7.0	47.2	53.1	13.9
Cordia dentata	11.4	53.5	15.8	29.5	52.9	33.4	27.6
Pithecolobium dulce	20.4	43.9	12.3	28.0	58.1	42.2	25.5

CP = crude protein, CW = cell wall, L = lignin, OM = organic matter
BW = body weight, ADDMI = average daily dry matter intake.
Source: McCammon-Feldman (1980).

Despite the often-stated belief that the conversion of browse to goat meat and milk is an ideal method of utilizing a forage resource of little alternative value, there are limitations to the value of browse in practical feeding programs for goats, especially dairy goats. To convert the browse plant, which is fairly low in nutritive value in its total above-ground growth, to a feedstuff of dietary importance, the goat selects plant parts high in nutritive value. Selectivity implies availability of a sufficiently large forage mass to allow for much of the dry matter to be refused. On range with mixed populations of grass, forb, and browse species, predicting carrying capacity for goats is not simply a matter of converting the smaller body weight to a cow equivalent. As cattle are grazers and will consume forage of higher cell-wall content, forage utilization (forage dry matter harvested as a percentage of dry matter available) will tend to be higher than with goats. Because of their capacity to retain and digest forage dry matter of high cell-wall content, the consequences of forcing cattle to graze to a lower canopy level in a pasture will be less than those suffered by goats when they are restricted in range and forced to consume a greater than normal percentage of the forage than is typical of a free-ranging situation.

For goats on browse, forcing the consumption of low-value twigs and stems will probably lead to reduced productivity. Thus, in predicting land area carrying capacity for goats, it is more critical to have an understanding of what goats will consume and its nutritive value than is the case when assessing total dry matter availability for grazing cattle.

Survival of browse stands is another consideration when grazed by goats. Goats, like deer, prefer leaves and young twigs. Removal of leaves can seriously damage browse plants, causing reduced annual growth or death of the plants. Lay (1965), over a 10-year period, clipped common plants browsed by deer in a southeast texas pine forest understory and found an optimum removal rate of 25% of annual growth to maintain stands. Some plants, such as vines, withstood 100% clipping with no loss of stand. Although the procedure used by Lay was to obtain clippings once each year (in October), goats, as well as deer, browse year-round.

Utilizing browse in the nutrition of goats requires skillful management. Except in circumstances of where rainfall and temperature contribute to continuous browse growth, carrying capacity of the land area required to support browsing goats will probably be quite low (two head or less of mature does per acre). For commercial milking operations requiring a large volume of milk, browse will play a limited role, unless land and labor are relatively inexpensive. Where browse is judged to have potential, the design of feeding programs for goats must consider the forage that is actually consumed by the goat in selective browsing--not simply available forage.

The Design of Feeding Programs for Dairy Goats

The ration-balancing procedure in design of diets for ruminants starts with measured or assumed levels of intake of forage. Then nutritional requirements are balanced through supplementation of deficient nutrients from concentrated feedstuffs. In determining intake of nutrients from forage, a crucial assumption is that laboratory evaluations of nutrient content are indicative of the forage actually consumed. Thus, because of the above mentioned differences in nutritive value between species and between plant parts within species, forage samples must be harvested with the same degree of selectivity as used by the animal when feeding--otherwise intake of specific nutrients cannot be accurately estimated. Such inaccuracy results in varying degrees of over- or under-feeding of essential nutrients, such as total protein.

Accuracy of estimation is especially difficult in balancing rations for dairy goats that selectivity consume forage species having wide differences in nutritive value. Except where the forage is harvested and finely ground, dairy goats would probably be more selectivity than would be cattle fed the same forage. Design of economical feeding programs requires knowledge of what is being consumed, as well as how much. Table 2 shows the intake of total protein by a 65 kg dairy goat producing 4 kg per day of 3.5% fat milk and consuming three different theoretical forages at three different ratios of leaf to stem consumption. Forage A, with 12% total protein content in the leaf compared to 7% in the stem would be typical of a mature, temperate pasture grass. Forage B could be an immature woody shrub in the temperate zone during the growing season. Forage C has a total protein distribution that might be characteristic of a mature, leguminous shrub where much of the plant's nutritional value is concentrated in the leaves.

TABLE 2. ESTIMATED INTAKE OF TOTAL PROTEIN (G/DAY) BY A 65 KG DAIRY GOAT CONSUMING THREE DIFFERENT THEORETICAL FORAGES[1]

Forage	TP content of: Leaf	Stem	Leaf:stem ratio (% of forage consumed) 50:50	75:25	100:0
	--(% of DM)--				
A	14	7	205	239	273
B	20	5	244	217	390
C	30	3	322	454	585

[1] TP = total protein, DM = dry matter.
Assumption: 65 kg doe producing 4 kg of 3.5% butterfat milk per day. Forage consumption at 3.0% of live weight per day (1.95 kg).

If the rate of voluntary intake is assumed to be 3% of the 65 kg body weight, then 2.28 kg of dry matter would be consumed daily. How much total protein is consumed depends upon the ratio of leaf to stem. At a rate of 50% leaf to 50% stem, daily intake of total protein from A, B, and C would be 205 g, 244 g, and 322 g, respectively. If, however, the degree of selection was such that only leaf was consumed, daily intake of total protein would be much higher for A, B, and C at 273 g, 390 g, and 585 g, or an increase over the 50:50 selection ratio of 33.2%, 59.8%, and 81.7%, respectively.

The program illustrated by the above theoretical situation is how to determine how much total protein must be provided in the form of a concentrate supplement to meet the nutritional requirements of the 65 kg dairy goat. Table 3 shows that the amount of total protein required from supplement varies from 201 per day when forage A is consumed at the 50:50 selection ratio, down to no required supplementation when forage C is consumed at the 75:25 and 100:0 ratios.

TABLE 3. REQUIRED SUPPLEMENTATION OF TOTAL PROTEIN (G/DAY) FOR 65 KG DAIRY GOAT CONSUMING THREE DIFFERENT FORAGES[1]/

Forage	TP content of: Leaf	Stem	Leaf:stem ratio (% of forage consumed) 50:50	75:25	100:0
	--(% of DM)--				
A	12	7	201	167	133
B	20	10	162	189	16
C	25	5	84	*	*

[1]/ TP = total protein, DM = dry matter.

Assumption: 65 kg doe producing 4 kg of 3.5% butterfat milk per day. Forage consumption at 3.0% of live weight per day (1.95 kg).

Daily requirement for crude protein is 134 for maintenance plus medium activity plus 272 g for milk production = 406/day (NRC, 1981).

* Consumption of total protein from forage exceeds requirement.

Generally concentrate supplements high in total protein content are expensive and/or scarce. Use of such feedstuffs without regard to the actual consumption of total protein from the forage results in inefficient use of an expensive resource. Proper ration balancing for dairy goats in cases where foraged B and C are available in such abundance as to allow leaf:stem ratios of selectivity of 75:25 to 100:0 would involve feedstuffs low in total protein and high in energy, such as corn or molasses. Morand-Fehr and Sauvant (1980) have shown that dairy goats producing average daily levels of milk (4 kg or less) and fed high quality, leafy forages do respond with increased milk production to energy, rather than protein, supplementation.

SUMMARY CONCLUSIONS

The ability to predict the response in terms of growth rate and milk yield to intake of feedstuffs of known nutritive value is very important to successful management of dairy goats for commercial milk production. With many years of research and experience behind the, cow dairymen have been able to design systems of feeding that result in herd lactation-milk-yield averages of 1,000 kg or more. High producing dairy goats may, or may not, perform well under similar systems of feeding. Practical nutritional research that considers the unusual physiological characteristics and feeding behavior of the dairy goat needs to be undertaken. From the research results that have been reported, the following conclusions can be reached:

- Higher rates of voluntary dry-matter intake compared to dairy cattle are essential to high levels of milk production by dairy goats. Feedstuffs which are high in cell wall will be limiting to dry-matter intake and (Van Soest, 1981) may therefore reduce milk production.

- Chopping or grinding and pelleting of feedstuffs of uniformly low quality (high in cell wall which is poorly digestible) can be expected to result in a greater response in terms of milk yield than if the same feedstuff is fed in whole, loose form. The increased productivity would probably be largely attributable to increased voluntary dry-matter intake as digestibility would most likely be depressed as a result of a higher rate of particle passage from the rumen.
- Selectivity by dairy goats will result in consumption of an unbalanced diet if differential palatability exists among dietary components. Grinding with or without pelleting of nutritionally complete diets may be of significant importance in managing dairy goats for high milk production.
- Where browse, improved pasture, or unchopped leafy hay is the principal forage source and grazing pressure or feeding rate allows for a high degree of selectivity, energy supplementation will be of greater importance to high milk yield with dairy goats than will protein supplementation. Especially on browse, estimation of intake of crude protein and energy by dairy goats requires observation as to plant species and parts within plant species actually consumed. Balancing actual intake with the proper supplement will be critical to increased profitability where protein supplements are expensive relative to energy supplements.

REFERENCES

Demment, M. W. and P. J. Van Soest. 1982. Body size, digestive capacity and feeding strategies of herbivores. Winrock Int., Morrilton, Arkansas.

Devendra, C. 1979. The digestive efficiency of goats. World Rev. Anim. Prod. 14:9.

Devendra, C. 1982. The utilization of fiber by goats. Proc. Third Int. Conf. on Goat Prodn. and Disease. Tucson, Arizona. p 364.

French, M. H. 1970. Observations on the goat. FAO Agricultural Studies No. 80, Rome, Italy.

Frinton, Peter A. J. 1982. Digestion of alder and associated systemic effects in goats and sheep. Proc. Third Int. Conf. on Goat Prodn. and Disease. Tucson, Arizona. p 79.

Gihad, E. A., T. M. El-Bedawy and A. A. Mehrez. 1980. Fiber digestibility by goats and sheep. J. Dairy Sci. 63:1701.

Huston, J. E. 1978. Forage utilization and nutrient requirements of the goat. J. Dairy Sci. 61:988.

Lay, D. W. 1965. Effects of periodic clipping on yields of some common browse species. J. Range Mgt. 18:181.

Larson, Bruce L. 1978. The dairy goat as a model in lactation studies. J. Dairy Sci. 61:1023.

Le Hourtou, H. 1980. The role of browse in the management of natural grazing lands. proc. of an Intl. Symp. on Browse in Africa. International Livestock Center for Africa. Addis Ababa, Ethiopia, (in press).

McCammon-Feldman, B., U.S. Garrigus and P. J. Van Soest. 1980. Differences in digestive response to grass and browse species by goats. J. Anim. Sci. 51:242 (Abstr.).

McCammon-Feldman, B., P. J. Van Soest, P. Horvath and R. E. McDowell. 1981. Feeding strategy of the goat. Cornell International Agriculture Mimeograph 88. Cornell Univ., Ithaca, New York.

Morand-Fehr, P. and D. Sauvant. 1980. Composition and yield of goat milk as affected by nutritional manipulation. J. Dairy Sci. 63:1671.

National Research Council. 1978. Nutrient requirements of domestic animals. No. 3. Nutrient requirements of dairy cattle. National Academy Press. Washington, D.C.

National Research Council. 1981. Nutrient requirements of domestic animals. No. 15. Nutrient requirements of goats: Angora, dairy, and meat goats in temperate and tropical countries. National Academy Press. Washington, D.c.

Owen, E. and J. E. M. Ndosa. 1982. Goats vs sheep: roughage utilization capacity. Proc. Third Int. Conf. on Goat Prodn. and Disease. Tucson, Arizona. p 362.

Sharma, V. V., P. C. Murdia and N. K. Rajaura. 1982. Comparative utilization of roughage by ruminant species. Proc. Third Int. Conf. on Goat Prodn. and Disease. Tucson, Arizona. p 363.

Short, H. L., R. M. Blair and E. A. Epps, Jr. 1973. Estimated digestibility of some southern browse tissues. J. Anim. Sci. 36:792.

Uden, P. 1978. Comparable studies on rate of passage, particle size, and rate of digestion in ruminants, equines, rabbits, and man. Ph. D. Thesis. Cornell Univ., Ithaca, New York.

Van Soest, P. J. 1980. Impact of feeding behavior and digestive capacity on nutritional response. Paper presented at the Technical Consultation on Animal Genetic Resources, Conservation, and Management. Rome, Italy, June 2-6, 1980.

Van Soest, P. J. 1982. Nutritional ecology of the ruminant. O and B Books, Inc. Corvallis, Oregon.

Van Soest, P. J., D. R. Martens and B. Deinum. 1978. Preharvest factors influencing quality of conserved forage. J. Anim. Sci. 47:712.

Wilson, A. D. 1969. A review of browse in the nutrition of grazing animals. J. Range Mgt. 22:23.

Wilson, A. D. 1977. The digestibility and voluntary intake of the leaves of trees and shrubs by sheep and goats. Australian J. Agr. Res. 28:501.

60

DAIRY GOAT REARING MANAGEMENT WITH MILK AND MILK SUBSTITUTES

Miguel A. Galina,
J. Ruiz

INTRODUCTION

The feeding of young goats has been extensively reviewed by French authors who concluded that feed represented 75% of over-all production costs (Morand-Fehr et al., 1982). In evaluating the growth characteristics of kids, Morand-Fehr et al. (1982) analyzed the growth of 170 alpine kids that were weaned at 5 to 6 weeks, then received a milk replacer from week 2 to weeks 5 and 6, and fed ad libitum variable amounts of concentrate feed (100 to 600 g per day). The feed supply always satisfied the requirements for growth, with a rather rectilinear growth curve (figure 1). The mean live weight was in the range of 31 kg at the age of 30 weeks. Individual variability of live weight decreased during the growth period parallel to the regression of some factors favoring this variability, i.e., birth weight, litter size, and weaning conditions.

The Department of Animal Sciences in the National Agronomy Institute obtained experimental results with a dairy goat herd under intensive production in a temperate area in the middle of France (Morand-Fehr and Sauvant, 1978). Average daily weight gain and average live weight in midperiod are shown in figure 2.

Feeding methods for young goats have been extensively discussed and comparable results have been found comparing goat milk, cow milk, and milk replacer (Morand-Fehr et al., 1982; Mowlem, 1979, 1981, 1982; Owen and DePaiva, 1982; Arora et al., 1982; Havrevoll, 1982; Opstvedt, 1969; and Skjevdal, 1974). Due to the high cost of goat milk, artificial rearing has been successfully implemented by several researchers (Morand-Fehr et al., 1979).

Several experiments obtained some weight gain with goat milk, cow milk (with or without antibiotics), and milk replacer but such results were obtained when the amount of milk replacer exceeded the amount of goat milk by 50% (Opstvedt, 1969). This was probably because of the lower fat content of milk replacer and the slightly lower digestive utilization (as discussed by Morand-Fehr et al., 1982). Fehr (1971) compared results obtained with goat milk

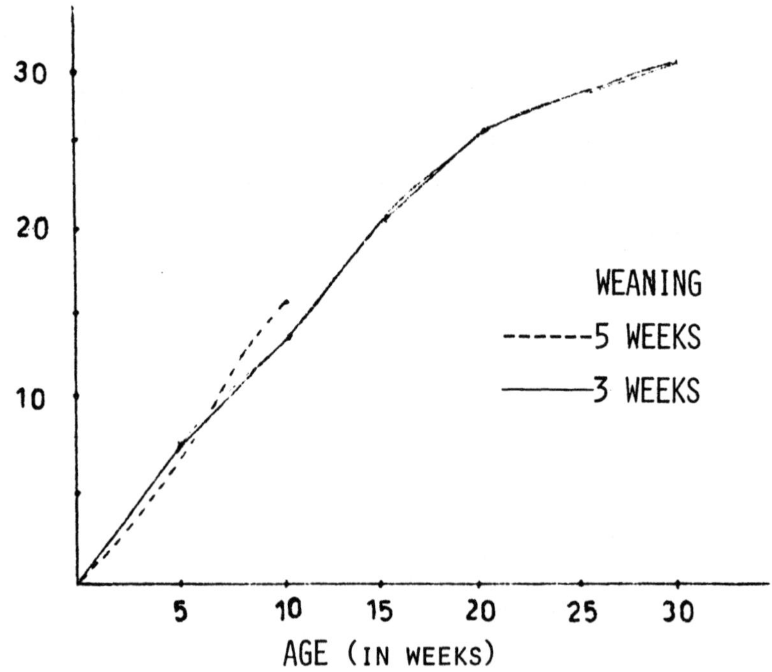

Source: Morand-Fehr et al. (1982).
Figure 1.

FIGURE 2. RECOMMENDED WEIGHT AND WEIGHT GAINS FOR YOUNG FEMALE GOATS

Age	Average live weight in midperiod	Average daily gain in grams
Birth-30 days	6.5	165
30-60 days	11.5	165
60-90 days	16.3	155
90-120 days	20.7	140
120-150 days	24.5	115
150-180 days	27.7	90
180-210 days	30.0	70

Source: Morand-Fehr et al. (1982).

and milk replacer fed to kids weaned very early (3 weeks) with results obtained with cow milk and milk replacer fed to animals weaned at 5 weeks. The growth rates were similar and not significantly different. However, with very early weaning, goat milk reduced the weaning shock (Morand-Fehr et al., 1982).

Also, in animals weaned at 5 weeks, cow milk did not seem to provide better results than those obtained with milk replacer; similar comparative results using cow milk and milk replacer were obtained in Norway (Skjevdal, 1974). On the other hand, Mowlem (1981, 1982) has shown better growth rates in kids fed milk replacers than with kids fed goat milk. These results can be explained by the very high concentration of the milk replacer used (18% to 20%).

Milk temperature, amount of feed offered, milk concentration, and number of meals have a direct influence on the performance of weight gain, as reported fully by Morand-Fehr (1978).

Feed needs in terms of dry matter, energy, protein, calcium, and phosphorus in young kids have been reported by Peraza (1981) and the importance of an adequate supplementation of phosphorous and energy for the female kid has been demonstrated (Hernandez, 1978). Dairy goat milk has been extensively used for the production of a caramel candy (cajeta) or cheese; commercial goat milk shows profits of \$0.70 to \$0.80 (U.S.) per liter compared with \$0.10 to \$0.13 (U.S.) per liter cost of milk replacer. A series of experiments have been developed to observe the productive performance of young goats managed under three feeding programs (Ruiz et al., 1982; Freixanet, 1982).

MATERIAL AND METHODS

A series of observations were developed on 111 kids under three feeding diets. All animals in the experiment were separated from their mothers at birth and colostrum or goat milk fed by bottle for 1 week.

Diet 1 consisted of 100% goat milk fed by bottle for the first 3 weeks and in a pail for the rest of the experiment. The amounts of milk increased (from 500 ml in week 1) by increments of 100 ml per week to complete the experiment at 10 weeks with 1500 ml of milk.

Diet 2 consisted of a 50/50 mixture of goat milk with 50% powdered skimmed cow milk (96 g per liter) with 40 g per liter. This group was fed 600 ml of the mixture in week 1 and the mixture was progressively increased by 100 ml per week finishing at 10 weeks with 1600 ml.

Diet 3 consisted of a mixture that was 25% goat milk and 75% milk replacer with 50 g of vegetable oil per liter. The diet started with 650 ml in week 1 and increased week by week until 1650 ml were fed in week 10 before weaning.

Details of the method were presented by Ferixanet (1982).

RESULTS

Daily weight gains of 127 g were shown in diet 1, and 126 g for diet 2 (no significance in ANOVA $P < 0.001$). In diet 3, the gain was 113 g, which was significantly different.

Severe weather was reported during 2 weeks of the experiment, and the daily gains were almost zero during this period. Statistically, when variances from those 2 weeks are controlled, the results are 150 g for diet 1, 144 g for diet 2, and 134 g for diet 3.

DISCUSSION

With animals housed in individual cages, previous investigations (Owen, 1982) have found daily weight gains of 233 g in animals fed with milk replacer ad libitum, or 197 g in kids fed 3 times a day. Similar results have been obtained consistently by Mowlem (1979, 1981, 1982). However, large concentrations of powdered milk (from 124 g to 300 g per liter) can explain those results. The economical feasibility of such schedules under Mexican farm conditions should be carefully examined before they are implemented.

The observations made in the study are in better accord with those made by French investigators under their farm conditions (Morand-Fehr and Sauvant, 1978; Morand-Fehr et al., 1981).

The objective of the experiment was to develop technology that easily could be reproduced by local farmers. High cost of labor discourages feeding several times a day and the problems of technological implementation limit the automatic feeders. Thus, this experiment was oriented to a twice-a-day milking schedule.

REFERENCES

Arora, S., R. Chapra and P. Atreda. 1982. Relative performances of kids fed milk and milk replacers on growth rate. Proc. III Int. Conf. on Goat Production and Disease. Dairy Goat J.:492.

Fehr, P. 1971. Methodes d'alimentation des chevrettes destinnes a la production laitiere. 10th Int. Congress of Animal Sciences, Versailles, France.

Freixanet, M. E. 1982. Evaluacion de la eficiencia productiva de un rebano caprino (varias razas) en Jilotepec Estado de Mexico. Thesis. Facultad de Estudios Superiores, Cuautitlan, National Autonomous University of Mexico (UNAM), Mexico.

Hernandez, P. 1978. Efectos de la nutricion sobre la presentacion de la pubertad en las cabras. Thesis. Facultad de Medicina Veterinaria y Zootecnia, National Autonomous University of Mexico (UNAM), Mexico.

Havrevoll, O. 1982. Milk and milk replacers as feed for rearing goats. Proc. III Int. Conf. on Goat Production and Disease. Dairy Goat J.:492.

Morand-Fehr, P. 1981. Growth. In: C. Gall Goat Production. pp 252-283. Academic Press.

Morand-Fehr, P., J. Hervieux, P. Bas and D. Sauvant. 1982. Feeding of young goats. Proc. III Int. Conf. on Goat Production and Disease. Dairy Goat J.:90.

Morand-Fehr, P. and D. Sauvant. Alimentation des jounes en l'alimentation des ruminantes. INRA-ITOVIC. Versailles, France.

Morand-Fehr, P., D. Sauvant, J. Hervieux, R. Disset, M. DeSimiane, G. Toussaint and M. Babin. 1979. L'alimentation des jeounes. INRA-ITOVIX. Bulletin Technique. pp 52-77.

Mowlem, A. 1982. Rearing dairy goats using milk replacers. Proc. III Int. Conf. for Goat Production and Disease. Dairy Goat J.:491.

Mowlem, A. 1981. Recent advances in kid rearing. British Goat Soc. J. March:41.

Mowlem, A. 1979. Milk replacers for kid rearing. British Goat Soc. Yearbook:54.

Owen, E. and P. DePaiva. 1982. Artificial rearing of goat kids. III Int. Conf. on Goat Production and Disease. Dairy Goat J.:491.

Opstvedt, J. 1969. Norwegian experiments on nutrition. In: Grassland Sheep and Goat Production. Oslo, Norway.

Peraza, C. 1981. Algunas consideraciones actuales sobre la alimentacion de la cabra lechera. Memorias capinas. I. Encuentro Nacional sobre la produccion de Ovinos y Caprinos. Facultad de Estudios Superiores, Cuautitlan, National Autonomous University of Mexico (UNAM), Mexico. pp 161-201.

Ruiz, J., M. Galina and A. Gutierrez. 1982. Cria artificial de cabrito bajo tres regimenes alimenticios. Memorias de Buriatris. (In press.)

Skjvadl, T. 1974. Milk feeding of kids. Report 173, Vol. 1. University of Norway, Oslo, Norway.

61

COST BENEFIT OF INTENSIVE MANAGEMENT OF A DAIRY GOAT HERD UNDER ZERO GRAZING

Miguel A. Galina, M. Guerrero,
M. Gutierrez, N. Celis

INTRODUCTION

Estimates used here were made from official statistics of the Mexican government and sample data obtained by the researchers for the years 1970 to 1980. These data suggest that the number of goats in Mexico has been stable at 93 million to 96 million (±4%). However, the slaughter of goats has increased by 15%, with a 384% rise in meat prices. At the same time, the national milk production increased from 17 million liters to 27 million liters, an increase of 60% (Galina et al., 1982). Commercialization of goat milk production has resulted in prices of $.60 to $.80 (U.S.) per liter, which has encouraged new goat producers in dairy farming (Galina, 1982).

Sound financial cost-benefit programming should be one of the factors that has to be carefully evaluated before establishing profitable goat farming. For example, in a technical and financial evaluation of a cooperative project in northern Mexico, it was concluded that major shortcomings were the high expectations that were not achieved and credit with a fixed limit that was quickly expended (Galina and Juarez, 1982). Other considerations such as land reform and social status of the goat producer must be included before programs can be launched in any particular region, (Galina et al., 1982). Land reform in Mexico has developed small agricultural units that could be utilized in intensive livestock enterprises with a sound profitable return.

French technology has developed such units successfully (LeJaouen and DeSimiane, 1981). Such enterprises have two main objectives:

- To provide dairy goats with feed of sufficient nourishment value to permit minimum use of concentrates that are usually of higher cost.
- To feed large numbers of goats per unit of cultivated area because of the limited land area in France.

Thanks to intensification of forage use and intensive management, profitable returns are common for dairy goat farmers in France. A detailed feeding program for such enterprises has been presented previously (LeJaouen and DeSimiane, 1981; Huguet et al., 1974; DeSimiane et al., 1980; Morand-Fehr and Hervieau, 1981; DeSimiane et al., 1981; and Juarez and Peraza, 1981).

Juarez (1981) has shown that feed constitutes from 70% to 80% of the yearly cost of a dairy goat enterprise.

Unfortunately, detailed studies of the economical feasibility of such farms have not been developed under local conditions. Thus, in 1981, a program was developed to study the cost-benefit of an intensive dairy goat farm managed on zero grazing of two hectares of private property on which 80% of the feed had to be purchased from elsewhere. The first evaluations of this program are discussed next.

MATERIAL AND METHODS

The farm was located in Jilotepec Estado de Mexico (19°20' and 20°20' latitude and 100°15' and 99°20' longitude; 300 m above sea level; 700 mm of rainfall; average temperature 18.3°C, with a maximum of 29.3°C and a minimum of 5.7°C).

A herd of creollo dairy goats was purchased from local producers and 7 pure breed sires were introduced to the females. Cost of production was carefully recorded and first lactation profit were evaluated. The detailed methodology of the farm management has been previously described (Freixanet, 1982).

RESULTS

Reproduction results are summarized in figure 1. A 7% abortion rate was observed in the herd.

FIGURE 1. REPRODUCTIVE PERFORMANCES RATES

Conception	92%
Fertility	86%
Prolificacy	115%
Survival at 15 days	96%
Survival at weaning	93%

The feed program was compared against that suggested by French workers (Morand-Fehr, 1981) as summarized in figure 2.

Cost of production is summarized in figure 3.

FIGURE 2. FEED PROGRAM FOR THE DAIRY GOAT HERD

	Dry matter (g)	Digestible protein (g)	energy (Mcal)
Maintenance 40 kg goat			
French	135	32	2.10
Jilotepec	106	76	2.54
Gestation 50 kg goat			
French	130	103	3.80
Jilotepec	162	113	3.60
Lactation	-	-	-
Goats producing 1.5 kg of milk daily and 40 kg body weight			
French	170	114	3.67
Jilotepec	150	107	3.45
Goats producing 2 kg of milk daily and 45 kg of body weight			
French	200	142	4.22
Jilotepec	164	116	3.67

Source: Freixanet (1982).

FIGURE 3. COST OF PRODUCTION IN THE FIRST YEAR

	From September 1981 to September 1982		
	Cost	% of yearly cost	% of total cost
Wages & services	219,532	40.1	17.6
Feed	326,928	59.8	26.2
Constructions	312,719		25.1
Initial herd	382,000		30.7
	1,241,179	100	100
Benefits first year			
Milk	150,000		
Sale of females	150,000		
Sale of kids	17,900		
	317,900		
Capital accumulated			
70 goats	490,000		
49 female kids	245,000		
	735,000		
Accumulated total	1,052,900		

*Values calculated in Mexican pesos 1 U.S. dollar = 50 Mex. pesos.

DISCUSSION

Production performance of dairy goat herds has been discussed elsewhere (Galina et al., 1982; Ruiz et al., 1982). In this paper, the author has compared the production performance data of an experimental herd against data obtained by others (Corteel, 1968; Ricordeau and Bouillion, 1975; Corteel et al., 1982; Riera, 1982; Montaldo, 1981; Juarez and Peraza, 1981a, 1981b).

Results obtained in Brazil (Simplicio et al., 1982) also were similar to those discussed in the present investigation. Expected milk yields were achieved and a quota set for profitable production was obtained by 70% of the animals purchased at the beginning of the experiment.

Milk production in the remainder of the herd was satisfactory, taking into account that during the first lactation only 65% to 70% of the potential milk yield could be achieved in one milking per day.

Wages were calculated by the accumulation of the minimum wage of the area for the 2 workers who handled the whole operation.

Rent was considered as part of the services due for the use and wear of the facilities, including medicines and various needs on the farm.

All feed was considered "purchased", even when produced on the farm, and paid for by the enterprise at current crop value.

Results indicated that almost 60% of the yearly cost was directly due to feed. Purchase of the initial herd and construction cost amount to 55% of the initial investment. Those calculations are similar to those presented by Juarez (1981) elsewhere in Mexico. Benefits of the operation included the milk production, excess livestock sales (females that did not reach profitable returns in the first 10 weeks of lactation), and kids for slaughter.

Capital accumulated included 70 dairy goats ready for a second lactation and 49 female kids that would enter into productivity in 1983.

Economical returns of such an enterprise are sufficient to cover initial investment returns by the end of the second year, permitting a profitable agricultural enterprise by the third year of operation.

REFERENCES

Corteel, J. M. 1968. La reproduction de l'espece caprine. La Chevre Numero special 24.

Corteel, J. M., C. Gonzalez and J. Nunes. 1982. Research and development in the control of reproduction. Proc. III Int. Conf. on Goat Production and Disease. Dairy Goat J.:162.

DeSimiane, M. and H. Miossec. 1975. Utilisation des fourrages deshydrates par la chevre laitiere. LaChevre 89:29.

DeSimiane, M., L. Drilleau and M. Lacombe. 1980. Utilisation des ensilages (herbe, mais) par la chevre. La Chevre 117:29.

DeSimiane, M., S. Giger, G. Blanchart and L. Huguet. 1981. Valeaur nutritionale et utilisation des fourrages cultives intensivement. Nutrition et systemes d'alimentation de la chevre INRA-ITOVIC Tours France 275-299.

Freixanet, M. E. 1982. Evluacion de la eficiencia productiva de un rebano caprino (varias razas) en Jilotepec Estado de Mexico. Thesis. Facultad de Estudios Superiores, Cuautitlan, National Autonomous University of Mexico (UNAM), Mexico.

Galina, M., M. Murguia and J. Hummel. 1982. Current status of the goat industry in Mexico. Proc. III Int. Conf. on Goat Production and Disease. Dairy Goat J.:505.

Galina, M. 1982. Unpublished data.

Galina, M. and A. Juarez. 1982. Social consequences of the transfer of technology in the improvement of the peasant agriculture. Proc. III Int. Conf. on Goat Production and Disease. Dairy Goat J.:331.

Galina, M., M. Guerrero, V. Rojas, M. Ruiz and V. Vasquez. 1982. Social status of the goat industry in Mexico. Proc. III Int. Conf. on Goat Production and Disease. Dairy Goat J.:420.

Galina, M., M. Freixanet and M. Guerero. 1982. Comportamiento Reproductivo de un rebano caprino en establacion total cero pastoreo. Memorias, Buriatria (In press).

Huguet, L., B. Broqua, R. Disset and M. DeSimiane. 1974. Problemes generaux pases par l'utilisation des fourrages verts. ITOVIC Paris, France.

Juarez, L. A. and C. Peraza. 1981a. Systemes d'alimentation en elevage caprin semi-intensif et intensif au Mexique. Nutrition et systemes d'alimentation de la chevre. INRA-ITOVIC. Tours France.

Juarez, L. A. and C. Peraza. 1981b. Systemes d'alimentation en elevage caprin extensif au Mexique. Nutrition et systemes d'alimentation de la chevre. INRA-ITOVIC. Tours France.

Juarez, L. A. 1981. Resultados economicos en tres sistemas de explotacion caprina lechera. Memorias caprinas. I. Encuentro nacional sobre produccion de ovinos y caprinos. Facultad de Estudios Superiores, Cuautitlan, National Autonomous University of Mexico (UNAM), Mexico.

LeJaouen, J. C. and M. DeSimiane. 1981. Cero pastoreo, manejo de un rebano caprino en sistemas intensivos. Memorias caprinas. I. Encuentro nacional sobre la produccion de ovinos y caprinos. Facultad de Estudios Superiores, Cuautitlan, National Autonomous University of Mexico (UNAM), Mexico.

Montaldo, H. 1981. Produccion lechera de algunas razas caprinas utilisadas en Mexico y metodos de seleccion. Memorias caprinas. I. Encuentro nacional sobre la produccion de ovinos y caprinos. Facultad de Estudios Superiores, Cuautitlan, National Autonomous University of Mexico (UNAM), Mexico.

Morand-Fehr, P., and J. Hervieau. 1981. Strategie de la distribution des aliments concentres chez la chevre en lactation. Nutrition et systemes de alimentation de la chevre. INRA-ITOVIC. Tours France.

Ricordeau, G, and J. Bouillion. 1975. Observation sur la duree du cycle sexuel et le taux de reussite en debut de la seison chez les caprines. In: Reproduction et Selection. Journeed de la Recherche Ovine et Caprine INRA.

Riera, S. 1982. Reproductive efficiency and managements in goats. Proc. III Int. Conf. on Goat Production and Disease. Dairy Goat J.:162.

Ruiz, J., M. Galina and A. Gutierrez. 1982. Cria artificial del cabrito bajo tres regimenes alimenticios. Memorias de Buriatris (In press).

Simplicio, A., E. de Figuereido, G. Riera and F. Lima. 1982. Reproductive and productive performances of the undefined (SRD) genotype of goats under the traditional management system of northeast Brazil. Proc. III Int. Conf. on Goat Production and Disease. Dairy Goat J.:349.

NAMES AND ADDRESSES OF THE LECTURERS AND STAFF

GEORGE AHLSCHWEDE
Sheep and Goat Specialist
Texas Agricultural Extension Service
Route 2, Box 950
San Angelo, TX 76901
--Sheep Specialist-Tour Coordinator

JOE B. ARMSTRONG
Associate Professor and
Extension Horse Specialist
Animal Science Department
New Mexico State University
Las Cruces, NM 88003
--Horse Geneticist

ROY. L. AX
Assistant Professor
Department of Dairy Science
University of Wisconsin
Madison, WI 53706
--Dairy Physiologist

HAROLD KENNETH BAKER
Meat and Livestock Commission
Box 44, Bletchley
Milton Keynes
MK2-2EF ENGLAND
--Beef Cattle Specialist & Geneticist

R. L. BAKER
Visiting Professor of Animal
Breeding & Genetics
Animal Science Department
University of Illinois
Urbana, IL 61801
--New Zealand Geneticist

R. A. BELLOWS
Location Research Leader
USDA-ARS
Route 1, Box 2021
Miles City, MT 59301
--Beef Cattle Physiologist

W. T. BERRY, JR.
Executive Vice-President
National Cattlemen's Association
P. O. Box 3469
Englewood, CO 80155
--Beef Cattle Organization Leader

HENRY C. BESUDEN
Vinewood Farm
Route 2
Winchester, KY 40391
--All-Time Great (Sheepman)

RONALD BLACKWELL
Executive Secretary-General Manager
American Quarter Horse Association
Amarillo, TX 79168
--Horse Organization Leader

BILL BORROR
Tehama Angus Ranch
Route 1, Box 359
Gerber, CA 96035
--Cattle Breeder

MELVIN BRADLEY
Professor of Animal Science and
State Extension Specialist
University of Missouri
Columbia, MO 65211
--Horse Specialist

B. C. BREIDENSTEIN
Director
Research and Nutrition Information
National Livestock and Meatboard
444 North Michigan Avenue
Chicago, IL 60611
--Meat Scientist

JENKS SWANN BRITT, D.V.M.
Veterinarian/Dairyman
Logan County Animal Clinic/J&W Dairy
Route 1
Russellville, KY 42276
--Veterinarian

HERB BROWN
Research Associate
Lilly Research Laboratories
P. O. Box 708
Greenfield, IN 46140
--Animal Scientist

O. D. BUTLER
Associate Deputy Chancellor
for Agriculture
Texas A&M University
College Station, TX 77843
--Animal Scientist

EVERT K. BYINGTON
Range Scientist
Winrock International
Route 3
Morrilton, AR 72110
--Range Scientist

B. P. CARDON
Dean
College of Agriculture
University of Arizona
Tucson, AZ 85721
--Animal Scientist

ARTHUR CHRISTENSEN
Manager
Christensen Ranch
P. O. Box 186
Dillon, MT 59725
--Sheep Producer

ROBERT L. COOK
Wildlife Biologist
Shelton Land and Ranch Company
P. O. Box 1107
Kerrville, TX 78028
--Wildlife Management Specialist

CARL E. COPPOCK
Professor
Animal Science Department
Texas A&M University
College Station, TX 77843
--Dairy Nutritionist

DICK CROW
Publisher
Western Livestock Journal
Crow Publications Inc.
P. O. Drawer 17F
Denver, CO 80217
--Journalist

STANLEY E. CURTIS
Professor of Animal Science
University of Illinois
Urbana, IL 61801
--Animal Scientist

A. JOHN DE BOER
Agricultural Economist
Winrock International
Route 3
Morrilton, AR 72110
--Agricultural Economist

WAYNE L. DECKER
Professor
Department of Atmospheric Science
University of Missouri
Columbia, MO 65211
--Meteorologist and Agri Weather Specialist

R. O. DRUMMOND
Laboratory Director
U.S. Livestock Insects Laboratory
P. O. Box 232
Kerrville, TX 78028
--Entomologist

ED DUREN
Extension Livestock Specialist
P. O. Box 29
Soda Springs, ID 83276
--Animal Scientist

WILLIAM EATON
Clear Dawn Angus Farm
R. R. 1
Huntsville, IL 62344
--Registered Cattle Breeder

WILLIAM D. FARR
Farr Farms Company
Box 878
Greeley, CO 80632
--All-Time Great (Cattleman)

H. A. FITZHUGH
Animal Scientist
Winrock International
Route 3
Morrilton, AR 72110
--Animal Scientist

MIGUEL A. GALINA
Professor
Universidad Nacional Autonoma de Mexico
A.P. 25, Cuautitlan Izcalli
Edo de Mexico, MEXICO
--Animal Scientist

HENRY GARDINER
Route
Ashland, KS 67831
--Cattleman

DONALD R. GILL
Extension Animal Nutritionist
Oklahoma State University
005 Animal Science
Stillwater, OK 74078
--Nutritionist

HUDSON A. GLIMP
Blue Meadows Farm, Inc.
Route 2, Box 407
Danville, KY 40422
--Sheep Producer

MARTIN H. GONZALEZ
President
ECO TERRA SA de CV
Fernando de Borja 208
Chihuahua, Chihuahua
MEXICO
--Range Scientist

TEMPLE GRANDIN
Livestock Handling Consultant
Department of Animal Science
University of Illinois
Urbana, IL 61801
--Livestock Facilities Specialist

SAMUEL B. GUSS, D.V.M.
Professor Emeritus
Veterinary Science Extension
Pennsylvania State University
2410 Shingletown Road
State College, PA 16801
--Veterinarian

JAMES C. HEIRD
Assistant Professor
Department of Animal Science
Texas Tech University
P. O. Box 4169
Lubbock, TX 79409
--Horse Specialist

A. L. HOERMAN
Extension Livestock Specialist
Texas A&M University Extension Center
P. O. Drawer 1849
Uvalde, TX 78801
--Beef Tour Coordinator

DOUGLAS HOUSEHOLDER
Extension Horse Specialist
Texas A&M University
College Station, TX 77843
--Tour and Clinic Moderator

HARLAN G. HUGHES
Agricultural Economist
University of Wyoming
Laramie, WY 82071
--Agricultural Economist

CLARENCE V. HULET
Research Leader
U.S. Sheep Experiment Station
USDA, ARS
Dubois, ID 83423
--Sheep Physiologist

HENRYK A. JASIOROWSKI
Director, Cattle Breeding Research
Warsaw Agricultural University
AGGW-AR 02-528 Warszawa
Rakowiecka Str. 26/30
POLAND
--Animal Scientist

DONALD M. KINSMAN
Professor, Animal Industries Department
University of Connecticut
Storrs, CT 06268
--Meat Scientist

JACK L. KREIDER
Associate Professor
Horse Program
Texas A&M University
College Station, TX 77843
--Physiologist

JAMES W. LAUDERDALE
Senior Scientist
The Upjohn Company
Performance Enhancement Research
Kalamazoo, MI 49001
--Physiologist

ROBERT A. LONG
Professor
Department of Animal Science
Texas Tech University
P. O. Box 4169
Lubbock, TX 79409
--Animal Scientist

CRAIG LUDWIG
Director of TPR
American Hereford Association
715 Hereford Drive
P. O. Box 4059
Kansas City, MO 64101
--Cattle Organization Leader

JAMES P. MC CALL
Director of Horse Program
Stallion Station
Louisiana Tech University
P. O. Box 10198
Ruston, LA 71272
--Horse Specialist

WILLIAM C. MC MULLEN, D.V.M.
Large Animal Medicine & Surgery
Texas A&M University
College Station, TX 77843
--Veterinarian

JOHN W. MC NEILL
Beef Cattle Specialist
Texas Agricultural Extension Service
6500 Amarillo Blvd., West
Amarillo, TX 79106
--Beef Cattle Specialist

DOYLE G. MEADOWS
Manager
Robinwood Farm
2822 East 2nd Street
Edmond, OK 73034
--Horse Specialist

JOHN L. MERRILL
XXX Ranch
Route 1, Box 54
Crowley, TX 76036
--Range Scientist and Rancher

BRET K. MIDDLETON
Animal Science Department
Iowa State University
Ames, Iowa 50011
--Computer Cow Game Coordinator

J. D. MORROW
Executive Vice-President
International Brangus Breeders Association
9500 Tioga Drive
San Antonio, TX 78230
--Cattle Organization Leader

HARRY C. MUSSMAN, D.V.M.
Administrator
APHIS/USDA
Washington, D.C. 20250
--Veterinarian

CHARLES W. NICHOLS
Manager
Davidson & Sons Cattle Company
Route 2, Box 15-0
Arnett, OK 73832
--Cattleman

J. DAVID NICHOLS
Anita, IA 50020
--Registered Cattle Breeder

MICHAEL J. NOLAN
Executive Secretary
Health and Regulatory Committee
American Horse Council, Inc.
1700 K Street, N.W., Suite 300
Washington, D.C. 20006
--Horse Specialist

JAY O'BRIEN
P. O. Box 9598
Amarillo, TX 79105
--Cattleman

JERRY O'SHEA
Principal Research Officer
The Agricultural Institute
Moorepark, Fermoy Company
Cork, IRELAND
--Dairy Equipment and Mastitis Expert

RICHARD O. PARKER
Assistant to the President
Agriservices Foundation
648 West Sierra Avenue
P. O. Box 429
Clovis, CA 93612
--Physiologist

BRENT PERRY, D.V.M.
President
Rio Vista International, Inc.
Route 9, Box 242
San Antonio, TX 78227
--Veterinarian

GUSTAV PERSON, JR.
Extension Agent
Guadalupe County
Ag Building
Seguin, TX 78155
--Horse Tour Coordinator

L. S. POPE
Dean
College of Agriculture
New Mexico State University
Las Cruces, NM 88003
--Animal Scientist

DOUGLAS PRESLEY
Extension Agent
Bexar County
Room 310
203 W. Nueva
San Antonio, TX 78207
--Horse Clinic Coordinator

NED S. RAUN
Vice-President, Programs
Winrock International
Route 3
Morrilton, AR 72110
--Animal Scientist

PATRICK O. REARDON
Assistant General Manager
Chaparrosa Ranch
P. O. Box 489
La Pryor, TX 78872
--Range Scientist

RUBY RINGSDORF
President
American Agri Women
28781 Bodenhamer Road
Eugene, OR 97402
--Farm Organization Leader

DON G. ROLLINS, D.V.M.
Technical Veterinary Advisor
Mid-America Dairymen, Inc.
800 West Tampa
Springfield, MO 65805
--Veterinarian

BUDDY ROULSTON
Professional Horse Trainer
Brenham, TX 77833
--Horse Trainer

MANUEL E. RUIZ
Head
Programa de Produccion Animal
CATIE
Turrialba, COSTA RICA
--Nutritionist

CHARLES G. SCRUGGS
Vice-President and Editor
Progressive Farmer
P. O. Box 2581
Birmingham, AL 35202
--Journalist

RICHARD S. SECHRIST
Executive Secretary
National Dairy Herd Improvement Association
3021 East Dublin-Granville Road
Columbus, OH 43229
--Dairy Organization Leader

MAURICE SHELTON
Professor
Texas Agricultural Experiment Station
Route 2, Box 950
San Angelo, TX 76901
--Sheep Specialist

PAT SHEPHERD
Manager
South Plains Feed Yard
Drawer C
Hale Center, TX 79041
--Feedlot Manager

JOHN STEWART-SMITH
President
Beefbooster Cattle Ltd.
P. O. Box 396
Cochrane, Alberta
CANADA TOL OWO
--Cattle Breeder

GEORGE STONE
President
National Farmers Union
12025 East 45th Avenue
Denver, CO 80251
--Farm Organization Leader

JACK D. STOUT
Associate Professor
Extension Dairy Specialist
Oklahoma State University
Stillwater, OK 74074
--Dairy Specialist

MRS. BAZY TANKERSLEY
Al-Marah Arabians
4101 North Bear Canyon Road
Tucson, AZ 85715
--All-Time Great (Horsewoman)

MAURICE TELLEEN
Editor and Publisher
The Draft Horse Journal
Box 670
Waverly, IA 50677
--Horseman and Journalist

THOMAS R. THEDFORD, D.V.M.
Extension Veterinarian
Oklahoma State University
Stillwater, OK 74078
--Veterinarian

TOPPER THORPE
General Manager
Cattle-Fax
5420 South Quebec Street
Englewood, CO 80155
--Agricultural Economist

ALLEN D. TILLMAN
Rockefeller Foundation, Emeritus
523 West Harned Place
Stillwater, OK 74074
--Animal Scientist

JAMES N. TRAPP
Associate Professor
Agricultural Economics Department
Oklahoma State University
Stillwater, OK 74078
--Agricultural Economist

ROBERT WALTON
President
American Breeders Service
P. O. Box 459
DeForest, WI 53532
--All-Time Great (Dairyman)

RODGER L. WASSON
Executive Director
American Sheep Producers Council, Inc.
200 Clayton Street
Denver, CO 80206
--Sheep Association Leader

DOYLE WAYBRIGHT
Mason Dixon Farms
RD 2
Gettsburg, PA 17325
--Dairyman

GARY W. WEBB
Stallion Manager
Winmunn Quarter Horses, Inc.
Route 1, Box 460
Brenham, TX 77833
--Horse Specialist

RICHARD O. WHEELER
President
Winrock International
Route 3
Morrilton, AR 72110
--Agricultural Economist

DICK WHETSELL
Vice-Chairman
Board of Directors
Oklahoma Land & Cattle Company
P. O. Box 1389
Pawhuska, OK 74056
--Range Scientist

R. GENE WHITE, D.V.M.
Coordinator of the Regional College
of Veterinary Medicine
University of Nebraska
Lincoln, NE 68583
--Veterinarian

RICHARD L. WILLHAM
Professor of Animal Science
Iowa State University
Kildee Hall
Ames, IA 50011
--Animal Breeding Specialist

DON WILLIAMS, D.V.M.
Henry C. Hitch Feedlot, Inc.
Box 1442
Guymon, OK 73942
--Feedlot Manager and Veterinarian

JAMES N. WILTBANK
Professor
Department of Animal Husbandry
Brigham Young University
Provo, UT 84602
--Beef Cattle Physiologist

CHRIS G. WOELFEL
Dairy Specialist
Texas Agricultural Extension Service
218 Kleberg Center
College Station, TX 77843
--Dairy Tour Coordinator

JAMES A. YAZMAN
Animal Scientist
Winrock International
Route 3
Morrilton, AR 72110
--Dairy Goat Specialist and Nutritionist

Other Winrock International Studies Published by Westview Press

<u>Hair Sheep of West Africa and the Americas</u>, edited by H. A. Fitzhugh and G. Eric Bradford

<u>Future Dimensions of World Food and Population</u>, edited by Richard G. Woods

Other Books of Interest from Westview Press

<u>Carcase Evaluation in Livestock Breeding, Production and Marketing</u>, A. J. Kempster, A. Cuthbertson, and G. Harrington

<u>Energy Impacts Upon Future Livestock Production</u>, Gerald M. Ward

<u>Science, Agriculture, and the Politics of Research</u>, Lawrence Busch and William B. Lacy

<u>Developing Strategies for Rangeland Management</u>, National Research Council

<u>Proceedings of the XIV International Grassland Conference</u>, edited by J. Allan Smith and Virgil W. Hays

<u>Animal Agriculture: Research to Meet Human Needs in the 21st Century</u>, edited by Wilson G. Pond, Robert A. Merkel, Lon D. McGilliard, and V. James Rhodes

Other Books of Interest from Winrock International*

Ruminant Products: More Than Meat and Milk, R. E. McDowell

The Role of Ruminants in Support of Man, H. A. Fitzhugh, H. J. Hodgson, O. J. Scoville, Thanh D. Nguyen, and T. C. Byerly

Potential of the World's Forages for Ruminant Animal Production, Second Edition, edited by R. Dennis Child and Evert K. Byington

Research on Crop-Animal Systems, edited by H. A. Fitzhugh, R. D. Hart, R. A. Moreno, P. O. Osuji, M. E. Ruiz, and L. Singh

Bibliography on Crop-Animal Systems, H. A. Fitzhugh and R. Hart

*Available directly from Winrock International, Petit Jean Mountain, Morrilton, Arkansas 72110

For Product Safety Concerns and Information please contact our EU representative GPSR@taylorandfrancis.com Taylor & Francis Verlag GmbH, Kaufingerstraße 24, 80331 München, Germany

Batch number: 10725820

Printed by Printforce, the Netherlands